Evolution and the Fossil Record

Evolution and the Fossil Record

edited by
Keith Allen
and
Derek Briggs

Smithsonian Institution Press
Washington, D.C.

First published in the United States in 1990 by
Smithsonian Institution Press

Library of Congress Catalog number 89-64435

ISBN 0-87474-269-2 (cased)
0-87474-273-0 (paper)

Printed and bound in Great Britain

CONTENTS

LIST OF FIGURES

LIST OF TABLES

LIST OF CONTRIBUTORS

David W. Hughes, Department of Physics, University of Sheffield, Sheffield, S3 7RH, UK.

George R. McGhee, Jr, Department of Geological Sciences, Wright Geological Observatory, Rutgers University, New Brunswick, New Jersey 08903, USA.

Paul K. Strother, Department of Geology, 725 Commonwealth Avenue, Boston, Massachusetts 02215, USA.

Mark A. McMenamin, Department of Geology and Geography, Mount Holyoke College, South Hadley, Massachusetts 01075-1484, USA.

Christopher R.C. Paul, Department of Earth Sciences, University of Liverpool, Brownlow Street, P.O. Box, 147, Liverpool L69 3BX, UK.

Paul Selden, Department of Extra-Mural Studies, University of Manchester, Manchester M13 9PL, UK.

Dianne Edwards, Department of Geology, University of Wales College of Cardiff, PO Box 914, Cardiff CF1 3YE, UK.

Peter R. Crane, Department of Geology, Field Museum of Natural History, Roosevelt Road at Lake Shore Drive, Chicago, Illinois 60605. USA.

Jeremy M.V. Rayner, Department of Zoology, University of Bristol, Woodland Road, Bristol, BS8 1UG, UK.

Michael J. Benton, Department of Geology, University of Bristol, Wills Memorial Building, Queens Road, Bristol BS8 1RJ, UK.

Rhondda E. Jones, Department of Zoology, James Cook University, Townsville, Queensland 4811, Australia.

Keith C. Allen, Department of Botany, University of Bristol, Woodland Road, Bristol BS8 1UG, UK.

Derek E.G. Briggs, Department of Geology, University of Bristol, Wills Memorial Building, Queen's Road, Bristol BS8 1RJ, UK.

PREFACE

Many major aspects of evolution can only be explored through the fossil record and the evidence which it provides for the origination, diversification and extinction of groups of organisms through geological time. Such research has involved a shift of emphasis in palaeontology from documentation of the sequence of events in the history of life, to analysis of pattern and process. Some of the more significant results of such analyses have concerned the nature and causes of mass extinctions and the possibility that they were in some way periodic. New areas now commanding attention include the appearance rather than the demise of taxa–radiations as opposed to extinctions.

Although evolutionary studies have emerged as a major area of expansion in palaeontology, more fundamental aspects, including documentation of the occurrence, taxonomy and ranges of particular fossil taxa, remain critical to the health of the subject. Without an expanding data base the analysis of evolutionary patterns would rapidly stagnate. Thus the new evolutionary palaeobiology continues to stimulate field collecting and descriptive taxonomy, not primarily for biostratigraphic purposes, but as a means of testing evolutionary hypotheses. The detailed investigation of sections across the Cretaceous–Tertiary boundary, for example, is addressing the question of the duration of the extinction event, and providing constraints on possible causes.

Evolutionary palaeobiology is increasingly taught in undergraduate programmes for geologists and biologists (and in graduate courses), as a follow-up to an introductory course based on a systematic and functional treatment of major fossil groups (to provide the vocabulary and basic concepts in palaeontology). But there was hitherto no obvious textbook to cover evolution and the fossil record—at present both teacher and student have to consult an extensive primary literature in a range of journals and conference proceedings. This lack of a one-volume synthesis probably reflects the rapid rate at which the field of evolutionary palaeobiology is moving. We feel, however, that research on the major aspects of this field have now reached a stage where it is possible to fill this gap in the current textbook literature.

The range of expertise required to synthesize such a wide and rapidly moving field is too great to be achieved in a reasonable time by only one or two authors. Therefore, having planned the book in outline, we invited appropriate experts to write the individual chapters. This has the drawback of introducing a certain variability in style and treatment from chapter to chapter which naturally remains even after our intervention as editors! We are convinced, however, that any disadvantage is far outweighed by the gain in accuracy and authority that results. In addition, without the multiauthor approach, our plan for the book could not have been realised.

We commissioned an introductory chapter on the evolution of the Universe, to put the story of the Earth and its biosphere into a broader context for biologists and geologists. Chapters 3, 4, 6 and 8 cover major topics in the evolution of life—from Precambrian organisms including the early metazoans, through the colonisation of land, to the origin of flight. Chapter 2 discusses the major extinction events and their likely causes. Chapters 5, 7 and 9 review evolutionary patterns in invertebrates, higher plants and vertebrates respectively.

A number of general considerations, while treated in particular chapters, are relevant to the broader theme. For example, Chapter 2 enumerates various measures of turnover or diversity change through geological time, and reviews different causes of extinction. Chapter 5 includes a discussion of the completeness of the fossil record and the implications for the reliability of patterns deduced from it. Chapter 9 explains ways of expressing relationships, particularly the important distinction between monophyletic (holophyletic), polyphyletic and paraphyletic groups. The book concludes with a consideration of science and creation—not, however, with an account of how misleading and mischievous is creation 'science' (a topic extensively treated elsewhere)—but with a cautionary tale of how fundamentalist teaching leads to contradictory and confusing concepts of evolution in college and university students.

Each chapter is followed by an extensive reference list which will provide a lead into the broader literature of the subject. These lists have been prepared as citations (except in the case of Chapter 1 which is a more general introduction to an enormous subject) but the number for each chapter has been limited. This inevitably means that not every original source has been cited, but it has ensured a more readable text.

We are grateful to Iain Stevenson, Editorial Director of Belhaven Press, with whom this book was originally conceived and planned. He and his colleagues have achieved very rapid publication following completion of the project, which will ensure that the result is as timely and up-to-date as possible.

We hope that this book will encourage others, like ourselves, to develop courses to cover this exciting area of palaeontological and evolutionary research. The book is aimed, however, at a much wider audience than college and university students of the earth and life sciences. It provides a review of a major area of evolutionary study for scientists in other disciplines.

K.C. Allen
D.E.G. Briggs
University of Bristol, February 1989

Chapter 1

EVOLUTION OF THE UNIVERSE, STARS AND PLANETS

David W. Hughes

THE ORIGIN OF THE UNIVERSE

In the beginning there was the Big Bang. This was not like the explosions that take place today on Earth, where the resultant fireball progressively enshrouds more and more material. In the Big Bang everything exploded at once; all space exploded and all matter immediately started to separate and rush apart. The temperature was proportional to the inverse of the radius. After about one-tenth of a second this temperature had dropped to 3×10^{10} K. After three minutes the Universe was about as dense as water, had a temperature of 10^9 K and was cool enough for protons and neutrons to start combining. This combination ended, due to ever decreasing temperatures and densities, with about 23% of the mass being in the form of helium, 77% in the form of hydrogen, and traces of heavy hydrogen (deuterium), light helium and lithium.

After a few hundred thousand years the material had cooled down to a temperature of around 3000 K and the electrons then started to associate with the ions to form atoms. The Universe became transparent and electrically neutral, and the turbulent whirling gas started to congregate into clumps due to the influence of the force of gravity. At a time 100 Myr after the Big Bang, these clumps were congregating in the form of individual galaxies of condensing stars. These early stars were made of just those ingredients that were manufactured in the first three minutes after the Big Bang.

Galaxies are the basic building blocks of the Universe. They tend to congregate in groups and clusters, and these clusters appear to be arranged in filamentary strings that surround giant voids.

The form of individual galaxies depends on their total mass and angular momentum. The Milky Way, the galaxy that contains the Sun, is a typical spiral galaxy. It has considerable angular momentum per unit mass, and a total mass of around 3×10^{44} g. Over 90% of this mass is in the form of stars, there being about 10^{12} in all. The remainder is made up of gas (9%) and dust (1%).

The Big Bang took place around 15 000 Myr ago. This figure comes from an analysis of the rate of evolution of stars and specifically from the cooling rate of white dwarf stars (these are the stellar equivalent of old-age pensioners). By measuring the temperature and luminosity of the coldest white dwarf star and working backwards to calculate how long it took to get to that state, a value for the age of the Universe (as stated above) can be obtained.

The Universe is permeated by microwave radiation, which has a 2.7 K black-body spectrum. This is the cooling remnant of the initial Big Bang fireball. The discovery of this radiation, in 1965, led to the demise of the Continuous Creation Theory. According to this theory the Universe not only looked the same from all places in it, but also was thought to look the same at all times, expansion being compensated for by the creation of new material.

The Sun is around 4600 Myr old and is a typical second-generation star. It has been formed not only out of hydrogen and helium but also from the remnants of older, more massive, faster-evolving stars that have become unstable and blown up, littering the cosmic gas and dust clouds with nuclear ash.

The first stars to form in the newly-born galaxy contained only hydrogen and helium. Stars obtain energy from two sources, gravitational collapse and nuclear fusion. In the first, the star simply contracts and thus converts potential energy into kinetic energy. This is exhibited as an increase in the temperature of the stellar material which, in its turn, affects the amount of heat that the star radiates into space. In nuclear fusion simple elements, like hydrogen and helium, are converted into more complicated elements which have more protons and neutrons in their nuclei. Slight mass losses occur and these are exhibited as stellar energy, according to the Einstein formula $E = mc^2$. Here E is the energy released if a mass m is lost, and c is the velocity of light. Eventually the stellar mass will be made of such elements as carbon, sulphur, silicon, iron and oxygen.

The rate of energy production by a star is proportional to its mass raised to the power 3.5. A reasonable estimate for the energetic lifetime, t, of a star of mass M times that of the Sun is given by

$$t = 10^{10}/M^{2.5} \text{ years.} \tag{1}$$

The end product of stellar evolution is strongly mass-dependent. For $M < 1.4$ a white dwarf star is formed. This object is about Earth-sized, is stable and slowly cools with time. For $1.4 < M < 3$ the process produces a neutron star and for $M > 3$ a black hole. Evolution towards these last two end products is accompanied by considerable mass loss, all in the form of heavy elements. The normal distribution of stellar masses, at the present time, is such that only about Y per cent is in the form of stars with masses less than M, where Y is given by

$$Y = 100\ (1 - \exp\ (-1.4683\ (M - 0.055)^{1.458})) \qquad (2)$$

Assuming that this equation was also applicable when stars were first formed, and as the Universe is only 10 000 Myr old, equation (1) indicates that stars less massive than the Sun ($M < 1$) have not yet reached the end of their evolution. The original stars with $M > 1.4$ have evolved fully and during the latter phases of this evolution have ejected considerable amounts of heavy elements. Equation (2) indicates that about 10% of the total star mass is in this form. These ejected elements will become mixed in with existing clouds of hydrogen and helium and new stars can be born from these clouds. Equation (1) shows that stars of $M = 2$ (which make up about 2% of the total stellar mass) have an energetic lifetime of 1800 Myr. Ejecta from stars such as these have been recycled several times. Clearly the material in the cosmic Universe is becoming progressively richer in heavy elements.

A careful analysis of the material in the Sun and in primitive bodies in the solar system allows the cosmic composition at the time of the formation of these objects some 4600 Myr ago (and about 5000 Myr after the Big Bang) to be estimated. This composition is given in Table 1.1.

Table 1.1. *Cosmic abundance of the elements at the time of the formation of the Sun and the solar system*

Atomic Number	*Element*	*Number of atoms per million*	*Mass (g) per tonne*
1	H	920466	735000
2	He	78340	249000
8	O	608	7710
10	Ne	76	1220
7	N	84	950
6	C	304	2920
26	Fe	37	1645
12	Mg	24	474
14	Si	30	685
16	S	15	377
28	Ni	2	86
13	Al	2	44
11	Na	2	30
20	Ca	2	58
18	Ar	6	184

The mean age of the solar system, based on the radioactive decay of the element rubidium into strontium, is 4570 ± 30 Myr. Within the limits of uncertainty, this is the same age as the Sun. Therefore dating does not help in deciding whether the Sun and the planets were produced together or whether the Sun picked up the planets at a slightly later date. It does, however, seem highly likely that the Sun and the planets *were* formed together and that they condensed out of the same cloud of cosmic material. This supposition has not yet been proved, but its validity is assumed for the rest of this chapter. The hypothesis is that planet formation is a common, but not essential, adjunct to star formation.

THE FORMATION OF STARS

Stars are formed when galactic clouds of gas and dust condense. The flattened disc of our galaxy contains many such clouds and their stability will be considered first. Why did they not all condense to form stars long ago? An explanation is provided by the Jeans criterion. This considers the relative amounts of gravitational energy (which tends to promote increases in density) and thermal energy (which encourages expansion) in the cloud. Condensation only takes place if the total energy is negative, that is, if the thermal energy is less than the gravitational, i.e.

$$\frac{3}{2}\,\frac{RT_c M_c}{m} < \frac{C_1 G M_c^2}{R_c} \tag{3}$$

where the cloud has a mass M_c, an absolute temperature T_c, a radius R_c, and is made up of material with a mean molecular weight m; C_1 is a dimensionless constant that is a function of the density distribution in the cloud (and can normally be taken to be just less than unity); G is the universal constant of gravitation, and R is the ideal gas constant. Considering cosmic composition, this can be rewritten such that the mass of a cloud that will just collapse is given by

$$M_c = \frac{20 T_c^{3/2} M_\odot}{n_{H_2}} \tag{4}$$

where n_{H_2} is the number of H_2 molecules per cubic centimetre and $M_\odot$ is the mass of the Sun. A typical dense interstellar molecular gas cloud has a temperature of 10 K and about 1000 molecules of hydrogen per cubic centimetre. This leads to a critical mass of $20M_\odot$. This calculation, however, neglects the galactic magnetic field (approximately 3 μG) which generates an internal pressure that opposes contraction, and the fact that the cloud is spinning, which introduces a centrifugal force that also resists the tendency to collapse. Taken together these considerations increase the critical mass by a factor of about five.

A typical, easily visible, small cloud in our galaxy is the 'Coal Sack'. This can be seen as a dark absorbing region in the galactic plane in the constellation Crux. The cloud has a diameter of 8 pc (1 pc = 3×10^{18} cm) and a mass of $15M_{\odot}$. Like most clouds, it seems to be stable. Collapse would have to be triggered by some external influence, which would lead to an increase in the density. Two influences have been suggested: the first being the shock resulting from being hit by the expanding material from a

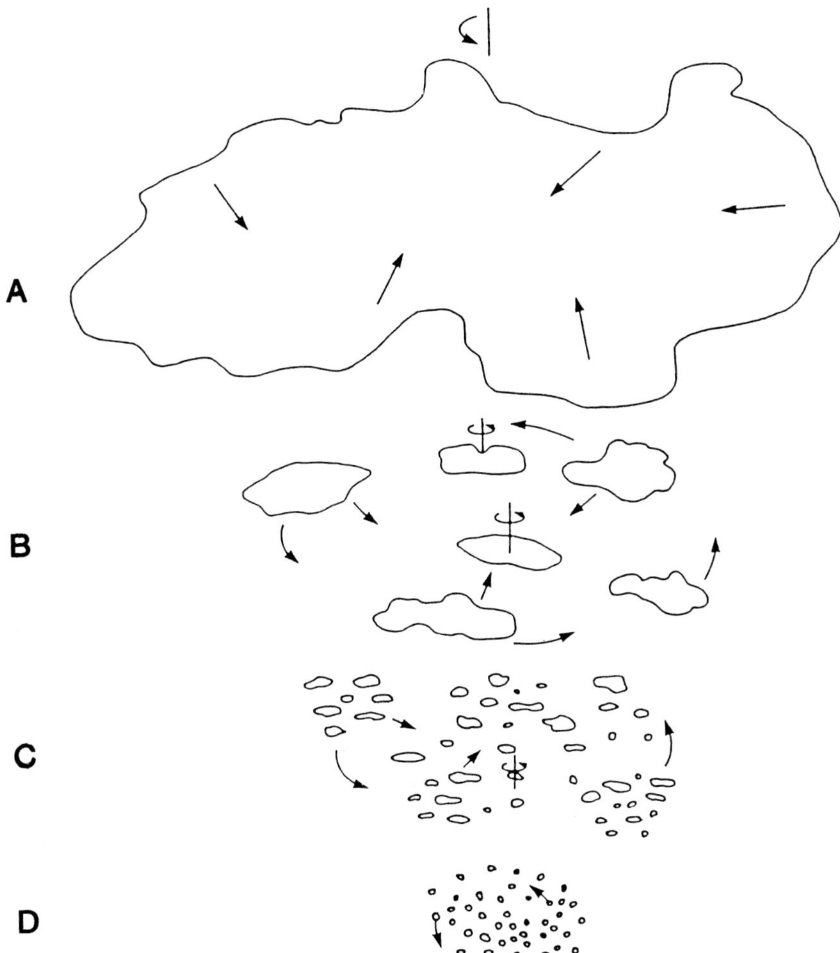

Figure 1.1. As an interstellar cloud of gas and dust collapses it spins faster, the period of rotation being proportional to the square of the diameter. Mass at the 'equator' becomes unstable when its velocity has the same order of magnitude as the escape velocity. The cloud fragments and then these fragments themselves break up. By this process most of the initial spin angular momentum is converted into the orbital angular momentum of the fragments about the common centre of mass. When the system has reached stage D it is similar to an open cluster.

supernova explosion; the second being the sharp density increase caused by entering one of the arms of the galaxy and being retarded by the density wave. A cloud at the Sun's distance from the galactic centre would enter an arm every 100 Myr.

A collapsing cloud becomes gravitationally isolated from other objects and thus conserves angular momentum. Due to this the period of rotation is proportional to the square of the diameter. So as the cloud gets smaller it spins faster, and soon becomes unstable. At this point the cloud starts to break up. Further condensation again leads to rotational instablity and this results in the fragments themselves fragmenting. By this process the total angular momentum of the initial cloud is shared out between the spin angular momentum of the fragments and their orbital angular momentum about the common centre of mass of the system (see Figure 1.1). The end product is an open cluster of stars. The Pleiades in the constellation of Taurus is a typical example. A single condensing cloud will break up to form a cluster of some 500 or so stars, occupying a volume of space some 5 pc in diameter. These clusters are loosely bound and after a few hundred million years, their stars have wandered off individually into the galactic disc. The Sun is just one of these wanderers and has left all its brother and sister stars behind long ago. In an open cluster there are about 3 times more stars per unit volume than there are in the present day solar vicinity.

Returning now to one of the final fragments of the cloud, the end point of its contraction depends strongly on its angular momentum. If there is very little, all the mass collects together to form a large single protostar and there is negligible nebulosity left behind. More angular momentum produces a less massive central star which is surrounded by a dense flattened nebula of gas and dust, this nebula having a radius of a few tens of astronomical units (1 AU is the mean distance between the Earth and the Sun, approximately 1.5×10^{13} cm). A rapidly rotating fragment will break up as it contracts and will form a binary or even a multiple star system. It is reasonable to suppose that each of these possibilities is equally likely. Planets are probably formed only in the second case. About half the stars are thought to be in binary or multiple systems. Of the remaining single stars, about 50% have planets in attendance.

THE PRODUCTION OF A SINGLE STAR

The best way to visualise the formation of a star like the Sun is by referring to a Hertzsprung–Russell diagram. This plots the logarithm of the temperature of the star's surface as a function of the logarithm of the star's luminosity (the energy it emits per unit time). An example is shown in Figure 1.2, starting with a protostar, one of the final fragments of the original cloud and ending up with a star of one solar mass, $M_\odot$. This final fragment, however, needs to have a mass of about $2M_\odot$. It is cold and has a position over on the right-hand side of the diagram (at point A in Figure

1.2). Fortunately, the initial contraction does not automatically lead to a large increase in temperature. Most of the thermal energy is removed by the collisional excitation of the radiation energy levels of the hydrogen molecules in the cloud, followed by the emission of 28 μm infra-red radiation. As the star moves across the H–R diagram it gets smaller. The steep decrease in luminosity just before point *B* is caused by the stellar

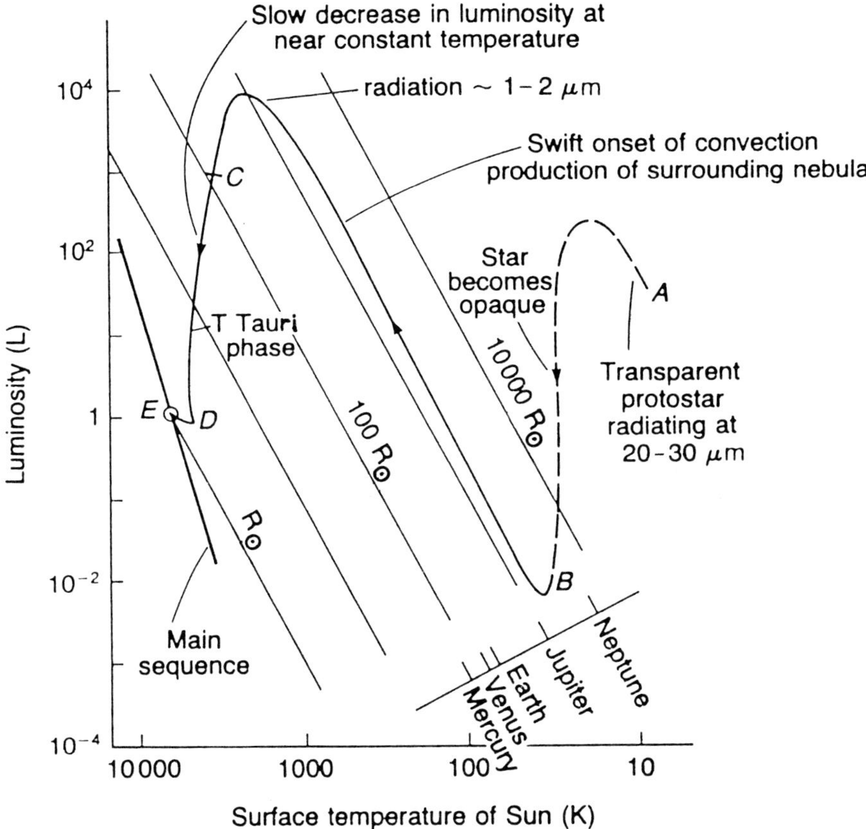

Figure 1.2. The evolutionary track of the Sun across a Hertzsprung–Russell diagram is shown for the first stage of the proto-Sun's stellar evolution. The H–R diagram is a logarithmic plot of stellar luminosity against surface temperature (T). The ordinate is in units of the solar luminosity, $L_\odot = 3.8 \times 10^{33}$ erg s^{-1}. The thin diagonal lines across the diagram are lines of equal stellar radii and are given in units of the present day solar radius, $R_\odot = 700\,000$ km. When the proto-Sun started to condense at *A*, it had a mass of about $2M_\odot$. In a period of rapid collapse between points *B* and *C* it left behind a large nebular cloud which subsequently condensed to form the planets. Now the Sun is near point *E*, on the main sequence, a line ($L \propto T^5$) on which all hydrogen burning stars are found. The Sun at present has a mass of $M_\odot = 2 \times 10^{33}$ g.

material becoming opaque and thus trapping the underlying radiation. This leads to an increase in the central temperature of the star and to a dissociation of the molecules, plus a subsequent ionisation of the atoms. The absorption of energy by these two processes causes a lowering of the pressure and a rapid decrease in size (between B and C in Figure 1.2).

There are two methods of energy transfer inside stars and these use either convective or radiative processes. Theoretical modelling of the stellar contraction in the first stage of stellar evolution shows that convection, the more effective and faster process, takes place. The onset of convection is accompanied by a sharp flare in the stellar luminosity, the star reaching a maximum luminosity, L_{max} given by

$$L_{max} = 10^4 \left(\frac{M_*}{M_\odot} \right)^2 L_\odot \quad (5)$$

where M_* is the mass of the star at that time, and $L_\odot$ is the present day main sequence luminosity of the Sun. This flare lasts for a few years (in the Sun's case the luminosity at that time would be about 10 000 times its present value). After the flare the star's radius decreases gradually while the surface temperature slowly increases from about 2500 K to 6000 K. The central temperature also rises and passes through a value of several hundred thousand degrees kelvin, a temperature at which the nuclear burning of the trace elements lithium, beryllium and boron takes place. Eventually the centre becomes hot enough for the nuclear burning of hydrogen. This produces a considerable amount of energy and a large increase in temperature. At this high temperature the gas and radiation pressure is sufficient to balance the gravitational force exerted by the contracting star and the star becomes stable. For the Sun this happens when the radius is about 700 000 km, the surface temperature is 5800 K and the luminosity is 3.8×10^{33} erg s^{-1}. Stars which are burning hydrogen into helium form the main sequence on the H–R diagram, a band which goes from the top left-hand corner to the lower right-hand corner as the stars become less massive, cooler and less luminous.

The main sequence evolution of a star is slow and gentle. As the mean molecular weight, μ, increases, the star gets larger (radius proportional to $\mu^{0.33}$), hotter (temperature proportional to μ^2) and more luminous (L proportional to μ^8). Eventually the helium core of the star collapses under the weight of the overlying hydrogen. The potential energy released by this collapse produces sufficient radiation pressure to blow out the surface of the star until it becomes a red giant, about 1000 times bigger than its main sequence size. (When this happens to the Sun most of the inner planets will be swallowed up into its tenuous atmosphere.) After the giant stage, the star finds itself without nuclear fuel and it then shrinks to become a white dwarf.

Early in the evolution of a star (Figure 1.2), two stages of speedy contraction occur and in both the stellar gas undergoes a free-fall in the gravitational field of the star. In the second (from point *B* to *C* in Figure 1.2) the radius of the protostar drops from a few thousand solar radii ($R_\odot$) to about $100R_\odot$. The speed of this collapse is shown in Figure 1.3. It is very fast in astronomical terms, taking only a few years. During this period the radius of the protosun drops from a value nearly the same as that of the radius of the orbit of the planet Neptune to a value that is close to the radius of the orbit of Mercury. In the penultimate stage (*C* to *D*) the Sun was a T Tauri star. These are irregular variables with massive and active outer chromospheres which are being blown away from the inner star by a stellar gale. The largest rate of mass loss occurs when the T Tauri star is young and still confined within a dark nebular cloud. The mass outflow decreases as the star approaches the main sequence. It is likely that the mass being lost is coupled to the stellar photosphere by strong magnetic fields, and this coupling will provide a mechanism whereby angular

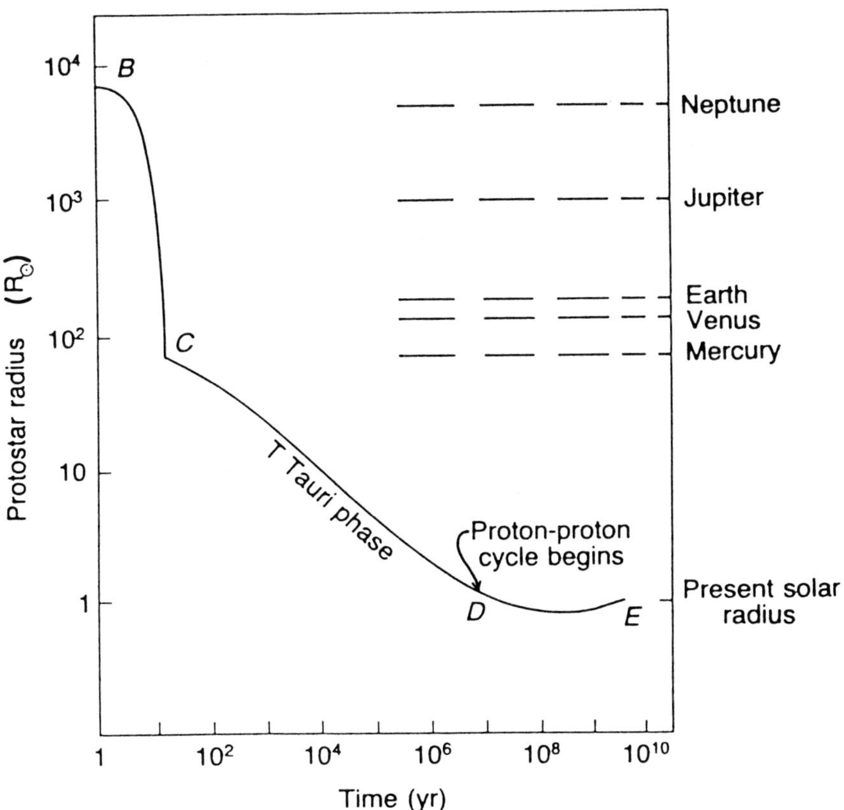

Figure 1.3. The radius of the protostar, during the first stage of stellar evolution is plotted as a function of time. The letters *B*, *C*, *D* and *E* correspond to points on Figure 1.2. $R_\odot$ is the present radius of the Sun, 700 000 km.

momentum can be drained from the star. During this time a low-mass star (less than 1.5 solar masses) will be despun until it has a rotation period of a few tens of days. At the height of T Tauri activity, the mass loss is in the range 10^{-6} to 10^{-5} solar masses per year. The Sun is at present losing mass at the rate of 10^{-14} solar masses per year; just prior to the onset of T Tauri activity, it had a mass of 1.3 $M_{\odot}$. The velocity of this stellar gale is several hundred kilometres per second, and the gale is probably sufficient, not only to halt the accretion of small planetesimals onto the protostar, but also to sweep away much of the gas from the circumstellar nebula.

In total it takes the Sun about 50 Myr to get from point *C* to point *D* on Figures 1.2 and 1.3. During this time it leaves behind a rounded, flattened, spinning, 'hatbox'-shaped, disc of gas and dust out of which the planets subsequently condense. This disc was opaque and isothermal during its early stages. It was also reasonably isothermal during the second phase of its evolution, at a time when the potential energy being released by its contraction was being used to evaporate molecular compounds such as H_2O, NH_3 and CH_4 from the surfaces of the dust grains, and subsequently to dissociate and ionise these molecules. As soon as these sinks for potential energy have been utilised, the temperature of the nebula, and concomitantly the gas pressure, increased sharply. This effectively put the brakes on the collapse, the nebula being at this stage about 40 AU in radius and at a temperature ranging from 10^4 K near the Sun to a few hundred degrees kelvin in the outer regions. As the Sun continued to contract (see Figure 1.3) the metallic elements in the nebula started to condense out. At greater distances from the infant Sun the temperature dropped even lower, and progressively dust grains of refractory and earthy materials and then snowflakes of volatiles such as the ices of water, ammonia and methane condensed out. In the outer regions many of the original interstellar grains retained their form.

The nebular disc is made up of particles that are orbiting the newborn Sun. These orbits have differing major axes, inclinations and eccentricities. The general velocity of the particles can, however, be resolved into two components, a general circular movement around the Sun and a much smaller chaotic relative velocity between individual particles. Collisions are frequent and as the particles are irregular and porous, these collisions are non-elastic and result in energy loss. The component of the velocity perpendicular to the mean plane of the orbits decreases more than the component in the plane, the result being that a disc of rotating dust, snow and gas becomes progressively flatter.

THE ORIGIN OF THE PLANETS

The conversion of this flattened, dense, slowly rotating, circumsolar disc (see above), which contains a myriad of millimetre- and smaller-sized particles, into the system of planets, asteroids and comets that orbit the

Sun today, must now be explained. The disc had a radius of around 40 AU. Perpendicular to the plane of the disc, only gas pressure balanced the gravitational force. This resulted in the thickness of the disc being about 1 AU. The mass is of the order of 10^{32} g, resulting in an average density of around 3×10^{-11} g cm^{-3}. The disc temperature ranges from around 2000 K down to 100 K. The Jeans criterion, using equation (4), can be rewritten

$$M_F > \frac{8.5 \times 10^{22}\, T^{3/2}}{\varrho^{1/2}} \text{ grams} \tag{6}$$

where ϱ is the density in g cm^{-3} and M_F is the mass of the smallest disc section that will condense. It can be seen that the initiation of condensation is rather touch-and-go. It requires the nebula to be cool, and the density to be considerably higher than the average value given above.

A second problem is the tendency for condensates to be tidally disrupted. Put simplistically, the Sun attracts the sunward side of the condensate more than the antisolar side and the resultant force tends to pull the condensate apart. In opposition is the gravitational attraction between the different regions of the condensate.

Condensation only occurs if

$$\frac{\beta M_F}{R_F} > \frac{M_\odot}{r^3} \tag{7}$$

where β is a dimensionless constant (depending on the density distribution) with a value between 1 and 3. M_F is the mass of the fragment, R_F is its radius and r is the distance between the fragment and the Sun (which has mass $M_\odot$). The minimum value of r at which condensation can take place, r_r, is known as the Roche limit and is given by

$$r_r = \left(\frac{0.12\, M_\odot}{\varrho}\right)^{1/3} \tag{8}$$

A density of 3×10^{-11} g cm^{-3} gives $r_r = 2 \times 10^{14}$ cm, i.e. 13 AU, which is between Saturn and Uranus.

The solutions that have been put forward to try and overcome this problem are all based on increasing the density in the equations (6) and (8) above. The first relies on the rotation of the disc to promote turbulence and a tendency for it to break up into whirlpool-like eddies. The absorption of radiant energy from the Sun pumps energy into the system and stops the eddies dying out. Between the eddies, gas and dust become concentrated. The second solution relies on the explosion of a nearby supernova. Considering that the Sun and its surrounding nebula are still in an open cluster, and that the massive stars in the cluster evolve much more quickly than the Sun, a nearby supernova is common and not unusual. The shock wave produced by this explosion will travel at u times the speed of sound in the nebula and will produce a density compression of $0.4 \times u^{6/5}$. A typical value of u is 5000 and this will drop the Jeans mass by a factor of 100 and reduce the Roche limit to about 0.8 AU. It will also seed the cloud with supernova remnants such as ^{26}Mg and ^{26}Al, and these have been found in meteorites such as Allende.

The third approach concentrates on the micrometre-sized dust particles which are unaffected by the Roche limit simply because they are small, and have a high density and significant strength. Collisions between these particles will be common and many will stick together because of the tar-like nature of the kerogen-type organic molecules that coat their surface. (Many tar-like substances are produced in laboratory experiments that simulate low temperature molecular cloud chemistry.) Thus dust accretes more dust.

The end result is the break-up of the equatorial disc and the production of a multitude of large particles usually known as planetesimals. Condensation, accretion and collision-induced fragmentation have all played a part. These planetesimals range in size from millimetres to tens of kilometres. They also have a range of compositions reflecting the change in temperature of the original nebula of gas and dust, a temperature that decreases considerably away from the Sun. The planetesimals are all in orbit around the Sun. They often collide with each other; some collisions are very gentle and result in an increase in the size of the more massive planetesimal, and others are more energetic and both masses fragment. Clearly growth triumphed in the end; the planets that today orbit the Sun are here to prove it. But the accumulation was an extremely wasteful process. From an original nebula of cosmic composition (see Table 1.1) and an initial mass probably in the 0.1 to 0.5 solar-mass region, a solar system of total mass 0.0013 times that of the Sun remains.

The collision rate in the disc slowly decreases until eventually all the planetesimals are moving in nearly circular orbits. At this stage accretion is governed by interplanetesimal gravitational forces. These are weak, but are still comparable to the *difference* between the solar gravitational forces acting on two neighbouring orbiting objects. In a few thousand years the metre-sized objects have accreted into planetesimals between 1 km and 100 km in size. Tens of thousands of years later these large planetesimals have grown even bigger. The process is illustrated in Table 1.2. Two things are clear: first, that the accretion process can be accompanied by considerable mass loss and this mass will escape from the solar system; second, that the process takes a long time. Neptune and Uranus took about 300 Myr to aggregate (or about 7% of the total age of the solar system). The inner terrestrial planets took around 2 Myr.

In the Uranus and Neptune region the planetesimals are made of a mixture of dust and ice, the ice being mainly H_2O but containing small percentages of CO_2, NH_3 and CH_4. This planetesimal swarm is essentially indistinguishable from a great circular stream of cometary nuclei. Many collected to form Uranus and Neptune, but many of the remainder were perturbed by these growing planets. Some of the planetesimals were

Table 1.2. *The aggregation of Uranus and Neptune and of the terrestrial planets from their specific regions of the nebula disc*

Mass of largest planetesimals (g)	*Number of bodies*	*Time scale of aggregation (yr)*	*Total mass of ring region (g)*
A. Uranus and Neptune			
10^6	10^{25}	10^1–10^2	10^{31}
10^{14}	10^{16}	10^3	10^{30}
5×10^{21}	4×10^7	10^6–10^7	2×10^{29}
4×10^{27}	50	3×10^8	2×10^{29}
1.0×10^{29} (Neptune) 8.7×10^{28} (Uranus)	2		2×10^{29}
B. The terrestrial planets			
10^6	10^{22}	10^2	10^{28}
6×10^{17}	2×10^{10}	10^4–10^5	10^{28}
6×10^{25}	200	2×10^6	10^{28}
			10^{28}
6.6×10^{27} (Earth–Mars) 5×10^{27} (Venus–Mercury)	2		10^{28}

ejected from the solar system and subsequently wandered off through the galactic disc. Perturbation-induced change in the energy of others moved them into typical short-period comet orbits. Here, after a few thousand passages of the Sun, they decayed, breaking up into dust which formed meteor streams and eventually fed the zodiacal cloud. A small percentage of the original planetesimals were perturbed into orbits which took them into the regions of space between the Sun and its nearby stellar neighbours. These 'dirty snowball' planetesimals were, and still are, effectively stored in deep freeze. Occasionally a passing star will perturb them into orbits which pass close to the Sun, where they will decay producing gaseous comae, dusty tails, and will be seen as comets. The present day collection of comets has a total mass of around one Earth mass, $M_\oplus$, and this is the exponential remnant of what was originally a much more massive cloud.

The cosmic material that made up the early solar system nebula can be divided into three types dependent on the melting temperature (Table 1.3). The planetary composition can also be divided into earthy, icy and gaseous groups (Table 1.4). Perturbations of the orbits of the spacecraft that have passed by Jupiter and Saturn indicate that both these planets have 'earthy' cores of about 20 Earth masses. If they were of cosmic composition the total mass of Jupiter and Saturn would be about 6000 $M_\oplus$ as opposed to their observed total mass of 413 $M_\oplus$.

This concept of having a cosmic mixture of material in each of the zones of the nebula, out of which the planets accreted, enables the original

Table 1.3. *Proportions of cosmic material making up the solar system nebula. Figures are based on combining the cosmic abundance of elements (Table 1.1) and crudely dividing them into metals, rocks, ices and gases (assuming, for example, that Fe is present both as a metal and as a rock, and that three atoms of oxygen combine with every two of Fe, Mg, Si, Al, etc., to form rock).*

	Earthy		*Icy*	*Gaseous*
	metal	*rock*	*(H_2O, CO_2 NH_3, CH_4)*	*(H, He)*
Mass (g) per tonne	910	4000	11 000	984 000
Ratio (by mass)	1		2.2	200
Melting temperature (K)	2000		270	10

nebula to be reconstructed. The present composition of each planet is used as a starting point. This is augmented to the cosmic composition by assuming that no iron was lost during the accretion process. This clearly only gives a lower limit, but is a reasonable indication because in a nebula in thermal equilibrium the heavy iron atoms will be moving very slowly and will rarely approach the escape velocity. Earth, for example, has an iron mass-fraction of 0.38. Bringing Earth up to cosmic composition would increase the mass to a total of 320 $M_\oplus$. This mass can then be spread over an annular zone of inner radius 0.86 AU (half way between Earth and Venus) and outer radius 1.26 AU (half way between Earth and Mars) giving a surface density for this part of the nebula of 3200 g cm^{-2}. The results of similar calculations for the other planets are shown in Figure 1.4.

Table 1.4. *Estimated composition of the planets* (per cent)*

	Earthy	*Icy*	*Gaseous*
Terrestrial planets	100	<1	0
Jupiter	6	~13	~81
Saturn	21	~45	~34
Uranus	~28	~62	~10
Neptune	~28	~62	~10
(Comets)	~31	~69	~ 0

*Jupiter and Saturn have rocky cores, both of about 20 $M_\oplus$. Uranus has a theoretically estimated rock and ice core of about 13 $M_\oplus$.

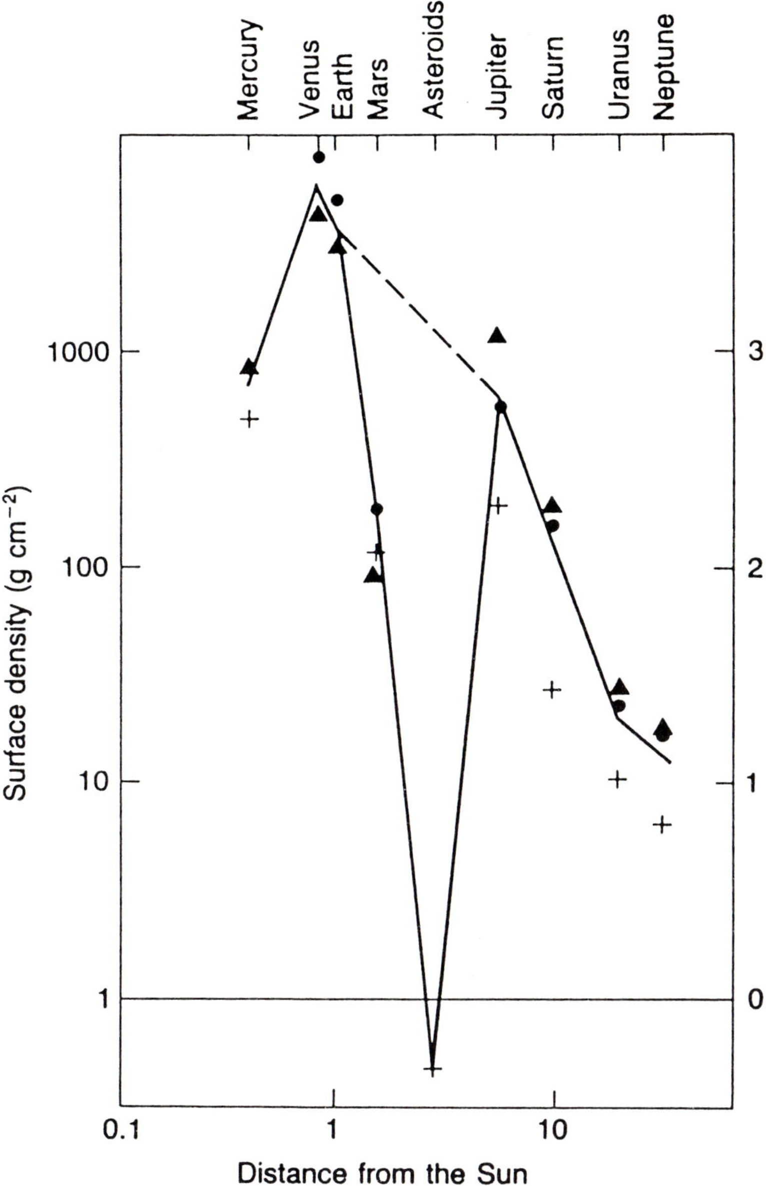

Figure 1.4. A logarithmic plot of the surface density of the preplanetary nebular disc as a function of distance from the Sun. The bold line represents an arithmetic mean of the data. The density estimates used are from Weidenschilling 1977 (triangles); Whipple 1972 (dots) and Harris 1978 (crosses).

It would be wrong to draw too sweeping a set of conclusions from this calculation, but four points are worth making. First, the decrease in the areal density beyond Jupiter is consistent with the termination of the planetary system at Neptune. There is too little material left over in the nebula for the formation of an additional 'Planet X'. It would also have taken too long to accrete (see Table 1.2). (Note that Pluto is not thought to be a planet but merely an escaped satellite; its mass is around 20% that of our Moon.) Second, moving to the inner solar system, it is highly probable that there has never been an intra-mercurial planet. Third, it is clear that something very odd happened at the asteroid belt. This is a 'Cassini division' in the solar system, and maybe the growing Jupiter had a profound gravitational effect on the rate of accretion in the 2.8 AU region. Fourth, although the solar system today has a mass of 1/745 that of the Sun, the nebula from which it condensed could easily have been 500 times more massive.

The early nebula is illustrated in Figure 1.5. The temperature of the nebula decreases considerably moving away from the Sun. The chemical and petrologic analysis of meteorites (which can be regarded as the crusts and cores of fragmented asteroids) indicates that the cooling nebula condensed under conditions of quasi-thermal equilibrium and at a pressure of around 10^{-4} atmospheres. At temperatures of about 1700 K, rocks and metals such as Al_2O_3, $CaTiO_3$, FeNiCo, Mg_2SiO_4 and $MgSiO_3$ condensed out. As the temperature cooled below 1000 K these were followed by FeS and FeO. The density of these condensates was in the 3.5 g cm^{-3} region. Upon cooling below 270 K, ices and shows such as H_2O, $NH_3 \cdot H_2O$ and $CH_4 \cdot 8H_2O$ formed around the rocky nucleation centres. Inert gases such as Ar and Ne condensed out at around 24 K (when the cumulative density of the condensate was around 1.4 g cm^{-3}). As the temperature dropped below 8 K and 1 K, H_2 and He finally condensed.

This sequence explains the variation in the observed planetary densities (Table 1.5). Other facts about the solar nebula can also be adduced from this table. There is a regularity in the disposition of the major axes of the planetary orbits (usually referred to as the Titius–Bode law). The planets all orbit in approximately the same nearly circular orbital plane. These regularities are simply the accumulated effects of 4570 Myr of interplanetary gravitational perturbation. The large majority of solar system objects are spinning directly, that is in the same sense as their orbital motion. This indicates that accretion started when the nebula had a near-constant density. Under these circumstances the orbital velocity increases in proportion to the distance from the central spin axis. Particles accreting onto the sunward side of a planetesimal would have a lower velocity than it had. Antisun particles would have still higher velocities. The resulting aggregate would spin in the same direction as it moved around its orbit. If the nebula, at the time of accretion, was diffuse, the planetesimals would have 'Keplerian' orbits (in which velocity is proportional to the inverse square root of the distance from the spin axis). The resultant aggregate would spin in a

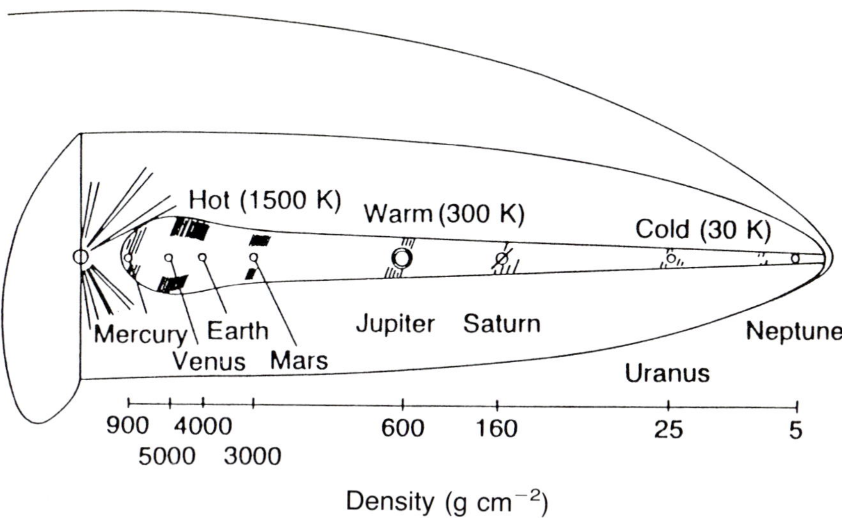

Figure 1.5. A cross-section through the nebular disc during the first stages of planetary accretion. The Sun is on the left of the diagram. The temperature decreases considerably away from the Sun. The lower scale represents the areal density of the disc (values from Figure 1.4).

retrograde fashion. As this is not the case, the nebula was dense rather than diffuse.

It can be seen that the orientations of the spin axes of the planets are not perpendicular to the ecliptic plane (i.e. the general plane of the planetary orbits). These deviations are probably due to the statistical nature of the final few accretion events. A grazing accretion can easily alter the spin axis direction of the major body by a few tens of degrees.

The spin period of most of the major planets and the asteroids is about 9 hours (see Figure 1.6), and their equatorial spin velocities are consequently about one-third their escape velocities. This indicates that accretion was the dominant factor in their formation. There are some exceptions among the terrestrial planets. Venus and Mercury have much less spin angular momentum than expected. Most of it has probably been lost due to tidal interactions with the Sun, at a time when their interiors were nearly molten, and when creep elastoviscosity could play a dominant role in dissipating energy. It is possible that Mercury was at one time in orbit around Venus. The Earth fits this general picture if the orbital angular momentum of the Moon is added to the Earth's spin angular momentum. The Moon is at present retreating from the Earth at a rate of 4 cm yr^{-1} and at the beginning of the solar system was very much closer, or even part of it.

The angular momentum distribution of the solar system has often been considered unusual. Put simply, the Sun has an angular momentum of 1.63×10^{48} g cm^{-2} s^{-1} whereas the orbital angular momentum of the

Table 1.5. *Present-day solar system parameters*

		Orbital semi major axis (AU)*	Inclination of orbit to ecliptic (deg)	Eccentricity	Mass (10^{27} g)	Density (g cm^{-3})	Uncompressed density (g cm^{-3})	Rotation period (h = hours; d = days)	Inclination of equator to orbital plane (deg)
	Sun	–	–	–	1 989 000	1.41		25.38d	[7.25]
Terrestrial	Mercury	0.39	7.00	0.206	0.33	5.4	5.4	58.65d	0
Terrestrial	Venus	0.72	3.40	0.007	4.87	5.2	~4.2	−243.01d	2
Terrestrial	Earth	1.00	–	0.017	5.98	5.518	~4.2	23.9345h	23.44
Terrestrial	Mars	1.52	1.85	0.093	0.642	3.95	3.3	24.6229h	23.98
	Asteroids	2.7	~9.5	~0.14	> 0.0048	~3.5	–	~8h	–
Major	Jupiter	5.20	1.30	0.049	1899.45	1.34	–	9.84h	3.08
Major	Saturn	9.54	2.48	0.056	568.64	0.70	–	10.23d	2.9
Major	Uranus	19.18	0.77	0.047	86.90	1.58	–	15.5h	97.92
Major	Neptune	30.06	1.77	0.009	102.97	2.30	–	(15.8)h	28.8
	Pluto	39.44	17.17	0.025	0.011	~0.5	–	6.39d	(≥50)

*(1AV = 1.5 × 10^3 cm)

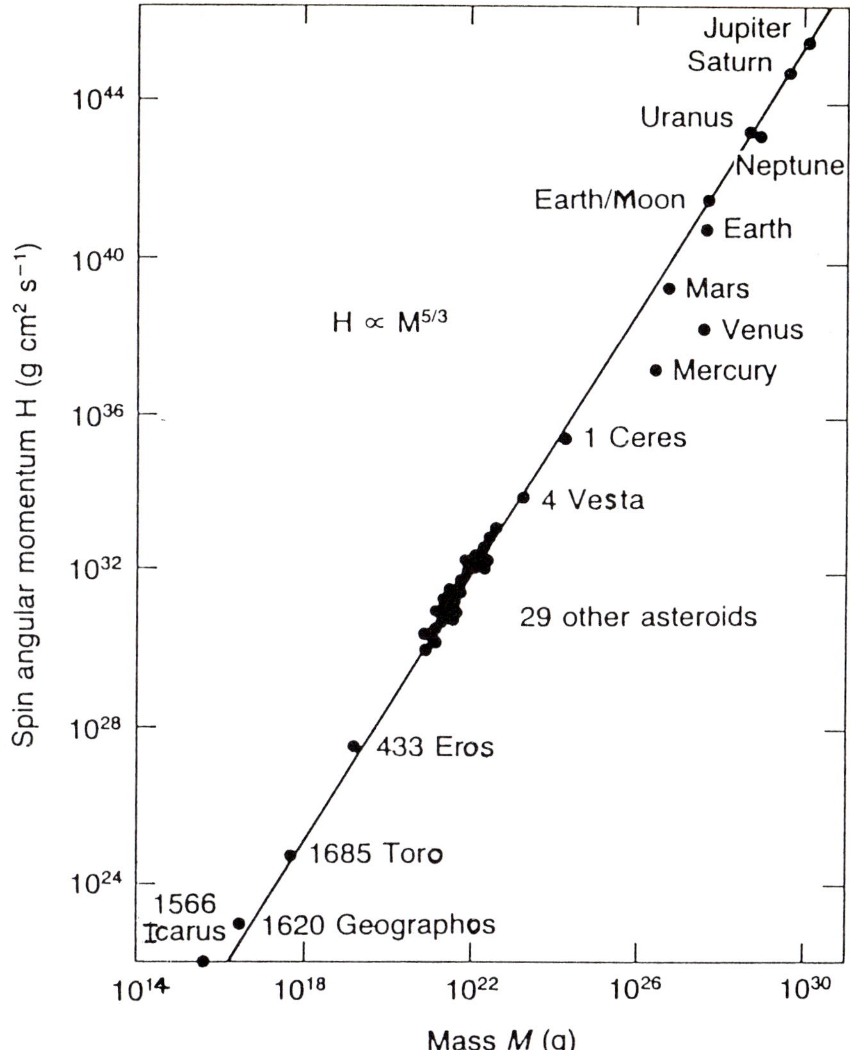

Figure 1.6. A logarithmic plot of the spin angular momenta, H, of the planets and asteroids of the solar system, as a function of their mass, M_p. The straight line indicates that $H \propto M_p^{5/3}$. The angular momentum, H, is equal to $2\pi I/T$ where T is the spin period and I the moment of inertia (which can be equated to $M_p R_p^2 \alpha$ where R_p is the raduis of the planet and α is a function of the density distribution and can be taken to be 0.4). Rearrangement gives

$$T = 4.3 \times 10^5 \, \alpha \, R_p^2 M_p^{-2/3}.$$

If the planet or asteroid is assumed to have a constant density of ϱ_p then the equation above leads to

$$T = 6.6 \times 10^4 \, \varrho_p^{-2/3} \text{ seconds}.$$

If it is assumed that all solar system objects have a density of 3 g cm^{-3} this results in a prediction that they all have a spin period of 8.8 hours.

planets is 3.15×10^{50} g cm^2 s^{-1}. Considering the Sun plus the planets, 61% of the angular momentum resides in Jupiter; Saturn, Uranus and Neptune account for a further 38%. Returning to a consideration of the condensation of the Sun, its rate of spin can be calculated. Once a condensate becomes controlled by its own gravitation, angular momentum is conserved, and its period of rotation is proportional to the square of its diameter. The Sun and solar nebula became independent at a diameter of about 10^{17} cm. As the Sun orbits the galaxy every 200 Myr a reasonable spin period for the initial independent nebula is 30 Myr. The Sun now has a diameter of 1.4×10^6 km and, according to the figures given above, should be spinning every 0.02 days. However, its present equatorial spin period is 25.3 days. Most of the solar angular momentum (and mass) loss is thought to have taken place during the T Tauri phase of its activity. The associated solar gale would also have swept away the material that was left over after planetesimal condensation. It is interesting to note that a similar calculation for objects near the orbit of Jupiter results in a period of a few tens of years which corresponds reasonably well with that planet's orbital period. The planets have therefore not speeded up, rather the Sun has slowed down.

The accreting planetesimals originally had the same temperature as the ambient nebula. During the growth of a protoplanet the potential energy of the incoming planetesimals is converted into thermal energy. The total energy available, V, is given by

$$V = 0.8\, G\pi \varrho_p R_p^2 M_p \tag{9}$$

where the protoplanet has mass M_p, radius R_p and density ϱ_p respectively. Assume, as a limiting case, that all this energy is used to heat up the protoplanet. The temperature rise, ΔT (per unit mass), generated by this process is given by

$$\Delta T = \frac{0.8\, G\pi \varrho_p R_p^2}{C_p} \tag{10}$$

where C_p is the specific heat of the accreting substance (~ 800 J kg^{-1} K^{-1} for rock). For Earth the resulting ΔT is 45000 K, and for an asteroid of radius 100 km the value is 11 K. This is a maximum value and thermal radiation will reduce it considerably. None the less, in the case of Earth, it is sufficient to make the rock viscous if not molten and this will enable the gravitational field to pull the planet into a spherical shape. The low value found in the case of the asteroid indicates that no melting or rock flow takes place. Small asteroids remain irregular objects.

THE EARTH

The protoplanet that formed earth was an accretion of cold, small planetesimals. As they came together the potential energy so released heated up the accumulated objects. Even with radiative heat losses, there was still ample energy to raise the temperature of the protoplanet to its melting point. As the hot fluid protoplanet condensed further, angular momentum would be conserved and, assuming that there was a considerable amount of angular momentum, the shape would change as it shrank. The initial large spherical protoplanet would form a smaller, egg-shaped object that would then progress to become pear-shaped. As condensation continued the pear would become smaller but also more elongated until eventually the 'stalk end' region would break off. This break would be messy, probably resulting in a major fragment and a few minor ones. The major fragment would take with it considerable angular momentum which, with tidal-induced interchange, would be in the orbital as opposed to the spin form. So the fragments could now settle down to solidify as spheres. Jeans (1919) studied this process and concluded that the break-up commonly led to a mass ratio of around 10 to 1 for the major fragments. It might be no coincidence that the mass ratio between Earth and Mars is 9 to 1 and that between Venus and Mercury is 15 to 1. The Moon has a mass of 1/81 that of Earth and was probably a droplet formed between Earth and Mars during the break up. The early protoplanet could have started to differentiate, the heavy metallic elements sinking to the centre and the lighter rocky crust floating to the surface. This could explain the differences in the mean densities of Earth, Moon and Mars. Both the Moon and Mars are much more 'rocky'.

The accretion time for a terrestrial planet, like Earth, would have been around 2 Myr (see above). Imagine a region of the pre-planetary nebula, with an inner radius about 0.2 AU (well inside the orbit of Mercury) and an outer radius of 2 AU (beyond the orbit of Mars). At this very early epoch, the conception time of the planets, the region in question contained about 700 Earth masses of cosmic material. One part of this was iron and rock, 2.2 parts ice and 200 parts gas (hydrogen and helium). The temperature was in the hundreds of degrees Celsius. The iron and rock in the nebula could condense; the rest could not and remained in the gaseous phase. Condensation took place around nucleation centres in a way rather similar to the formation of raindrops in clouds. The nebula was reasonably dense and interparticle collisions were frequent. At some of these collisions the impacting particles broke up. At others (the low-energy ones) the two colliding particles 'stuck' together, this accretion process obviously increasing the size of the largest particle. Accretion dominated over fragmentation. If it had not, there would be no planets today. Soon the 3.0 Earth masses of rock and iron in this region of the nebula, had formed a collection of some 10^{22} bodies, the largest being about 10^3 kg in mass. Some 100 years later the 10^{22} bodies had come together to form a mere 10^{10} bodies. Now, the diameter of the largest object was about 1 km and it had a mass of about 6×10^{17} g. Between 10 000 and 100 000 years later still the region contained only 200 bodies, the most massive being 6×10^{25} g. Finally, after another 2 Myr, these 200 bodies had come together to form the four planets Mercury,

Venus, Earth and Mars. The process involved considerable mass loss. All the ice and gas in the original nebula was blown away by the solar gales that raged throughout the Sun's T Tauri phase. Even some of the smaller, micrometre-sized, rocky particles were lost, being blown away from the solar system by radiation pressure.

The planetesimals that came together cooled to form Earth. But accretion produced considerable heat which melted the rock and drew it into a spherical shape. The heavier elements in the molten material, elements such as iron and nickel, sank under gravity to the centre of the planet forming a core. The less dense rocky material formed a huge layer of magma (a substance like volcanic lava) around the core. Gradually this magma cooled and solidified, forming a solid and ever-thickening crust around the planet. It is this crust that we live on.

The Sun's T Tauri stage and the associated solar gale stripped the Earth (and all the inner planets) of any possible primitive atmosphere. The atmosphere and oceans that surround our planet today are entirely secondary and have been derived by degassing the hot mantle layer as it solidified. Oxidising conditions during this stage led to the production of such gases as SO_2, NO_2 and CO_2. H_2 and N_2 molecules were also released.

Crustal minerals such as micas and amphiboles contain about 10% by weight of both H_2O and hydroxyl (OH) molecules. These are released by thermal 'cracking' when their temperatures are raised to between 800 and 1300 K. The primordial material contained about 1 part in 10 000 of H_2O. This was released by accretion-induced heating and radioactive heating about 3000 Myr ago. Water now covers about 70% of the surface of our planet. On a perfectly smooth globe this water would form a uniform layer 1.8 km deep. The total mass of H_2O is 1.4×10^{24} g compared to an Earth mass of 5.98×10^{27} g. Even if the frequency of volcanic eruption in the past did not exceed today's value, the extruded lavas would easily provide sufficient water to fill the oceans during geologic time.

The retention of an atmosphere depends on the mean velocity of the gaseous molecules being well below the escape velocity (i.e. the velocity needed to climb out of the gravitational potential well and leave the planet). This can be expressed as

$$\text{mean molecular velocity, } v_{\mathrm{m}} = \sqrt{\frac{3kT}{m}} \tag{11}$$

$$\text{escape velocity, } v_{\mathrm{e}}, = \sqrt{\frac{2M_{\mathrm{p}}G}{R_{\mathrm{p}}}} \tag{12}$$

where k is Boltzmann's constant, T is the absolute temperature of the gas and m is the mean mass of a gas molecule. The planet has a mass of M_{p}, a radius of R_{p}, and G is the gravitation constant. For the temperatures of Earth and Moon H_2 has a mean velocity of 1.8 km s^{-1}, N_2 moves at 0.49 km s^{-1} and O_2 at 0.46 km s^{-1}. Now molecular velocities in a gas in thermal equilibrium follow a Maxwellian distribution, and therefore some molecules are moving much faster than the mean value v_m. The escape velocity from the Earth's lower atmosphere is 11.2 km s^{-1}, whereas that from the surface of the Moon is 2.4 km s^{-1}. Statistically it is found that if $v_m < 0.33$ v_e, half the molecules are lost in a few weeks; if $v_m < 0.25$ v_e half are lost in 50 000 years and if $v_m < 0.20$ v_e half are lost in 100 Myr. The molecules most likely to be lost are those which have the largest velocity, that is, the smallest mass. For the Earth this means that hydrogen is continuously being lost from the top of the atmosphere.

Some 4400 Myr ago the atmosphere contained about 80% CO_2, 15% N_2 and 5% H_2. As time progressed CO_2 decreased and N_2 increased. With the evolution of plant life about 2500 Myr ago, O_2 began to enter the atmosphere. At present the atmosphere contains 78.084% N_2, 20.946% O_2, 0.934% Ar, 0.031% CO_2 and up to 1% water vapour. Volcanism indicates that the Earth is still an active planet. Movement of the mantle is evidenced by the gentle drifting of the plates that carry the continents.

ORIGIN OF LIFE

The way in which life originated on this planet is not well known and is still being actively researched. Living material contains proteins and nucleic acids, each of which is built up of a large number of smaller repeated units that are held together chemically. The proteins and nucleic acids existed on the surface of our planet before life originated.

Carbon, the twelfth most abundant element on Earth, is closely associated with the origin of life. Carbon chemistry took place both in the solid environment that was responsible for the formation of carbonaceous meteorites and also in the liquid environment of our early oceans. These oceans were initially hot and have often been thought of as an ever-thickening soup of chemicals. Initially the environment was non-oxidising. There was plenty of power available in the form of sunlight and atmospheric electricity (lightning). Spark discharges and ultraviolet radiation applied to vapours of water, ammonia and methane produce a whole range of organic compounds. Carbonaceous meteorites contain about 5% carbon. Most of this is in the form of complex molecules, alkanes, aromatics, benzoic acids, purines, alcohols, sugars, amino acids and porphyrins. All can be regarded as prebiological cellular aggregates; most are insoluble in water. If this material was present in the early Earth, or fell to its surface, it would float on top of the oceans, and be concentrated on the shorelines by sedimentary processes, and would have behaved like crude oil.

Opportunities for evolution were triggered as the atmospheric oxygen concentration passed certain critical values. It seems clear that to produce advanced forms of living organisms on other planets in the solar system, or on planets around other stars, the amount of oxygen in their atmospheres would have to be at least 1% of that in our present-day atmosphere. At less than that value, life is restricted to primitive unicellular forms. In addition, the planetary surface temperature must stay within a reasonable range and there must be suitable solvents present.

Biological evolution proceeds by a purely random process entailing successive mutations. The random nature of this process suggests that the evolutionary time-scale of life is reasonably constant and is independent of the type of parent star about which the living planet is orbiting. The habitable (living) zone about a star, however, is a function of the star's surface temperature. For the Sun, with its temperature of 5800 K, this zone stretches from just inside the orbit of Venus to just outside the orbit of Mars. Hotter stars have physically larger habitable zones, and cooler stars have much smaller ones. Unfortunately the expectation of having more 'living' planets around hotter (i.e. more massive) stars is not fulfilled simply because those stars evolve much faster than the Sun (see equation (1)) and so the surface temperature of an orbiting planet will change much too quickly to provide a stable environment for the slow development of life. The Sun is 'just right' for a central star, inasmuch as it has been stable for many thousands of millions of years, and the orbital distance of Earth is 'just right' for a 'living' planet.

Much has been written recently about possible extraterrestrial influences on the origin and development of life on Earth. Disease-ridden impacting comets and bug-infested meteorite stream particles are two of the favourite spectres. Both comets and meteorites contain carbonaceous material, and thus the wherewithal for very primitive life. But they contain no higher concentrations than the Earth itself; indeed they are made of similar material to the planetesimals that initially accreted to form the early Earth. So comets and meteorites are not necessary as harbingers of life's 'seeds'. Further, their orbits give them varying and unfavourable temperatures, and their small masses make them incapable of retaining any vestiges of an atmosphere, whether it included oxygen or not.

EPILOGUE

This chapter concentrates on one scenario for planetary formation, albeit the one thought to be the most likely. It regards the formation of planets to be a simple adjunct to the formation of single stars of medium initial angular momentum. It does not rely on rare collisions between stars, or on the Sun picking up a nebula cloud as it journeys around the galaxy. The hypothesis does indicate that planetary systems are reasonably common, planets being the companions of around 25% of all the stars in the galaxy. It is clear that time has played a considerable part in fashioning todays planets. They are now very close to equilibrium in the solar system. The next large upheaval will occur when the Sun runs short of its hydrogen nuclear fuel. This is estimated to be about 5000 Myr in the future. As the Sun will then switch to helium as a source of its energy its physical state will change. The central core will collapse and the consequent sharp increase in temperature will result in enhanced radiation pressure. This pressure will push out the surface layers of the Sun to such an extent that it will expand to about 1000 times its present size. Mercury, Venus, Earth and Mars will all be consumed in its outer envelope. The Sun will then be as big as the orbit of Jupiter and our planet will no longer exist. (For further reading on this chapter see the references which follow.)

REFERENCES

Cattermole, P. and Moore, P., 1985, *The story of the Earth*, Cambridge University Press, Cambridge.

Dermott, S.F. (ed.), 1978, *The origin of the solar system*, John Wiley and Sons, Chichester.

Harris, A.W., 1978, *Dynamics of Planetesimal Formation and Planetary Accretion*, In *The Origin of the Solar System*, S.F. Dermott (ed), John Wiley and Sons, Chichester.

Jeans, J., 1919, *Problems of cosmogony and stellar dynamics*, Cambridge University Press, Cambridge.

Meadows, A.J., 1967, *Stellar evolution*, Pergamon Press, Oxford.

Lewis, J.S. and Prinn, R.G., 1984, *Planets and their atmospheres, Origin and Evolution*, Academic Press, Orlando, FL.

Rowan-Robinson, M., 1977, *Cosmology*, Oxford University Press, Oxford.

Shklovskii, I.S., 1978, *Stars, their birth, life and death*, W.H. Freeman and Co., San Francisco.

Tayler, R.J., 1978, *Galaxies: structure and evolution*, Wykeham Publications Ltd. London.

Weidenschilling, S.J., 1977, *The distribution of mass in the planetary system and solar nebula*. Astrophysics and Space Science, **51**: 153–158.

Weinberg, S., 1977, *The first three minutes. A modern view of the origin of the universe*, André Deutsch, London.

Whipple, F.L., 1972. *Cometary Nuclei—Models*. In *Comets, Scientific Data and Missions*. G.P. Kuiper and E. Roemer (eds). NASA CR 129110: 4–15.

Williams, I.P., 1975, *The origin of the planets*, Adam Hilger, Bristol.

Wood, J.A., 1979, *The solar system*, Prentice Hall, New Jersey.

Chapter 2

CATASTROPHES IN THE HISTORY OF LIFE

George R. McGhee, Jr

RECOGNISING CATASTROPHES IN GEOLOGIC TIME

The ultimate measure of crises in the history of life is the loss of species diversity from one geologic time interval to another. Extinction is a normal part of the evolution of life and is inevitable. All species will eventually become extinct. Life continues, however, as species normally give rise to descendant lineages. Thus the diversity of species, while fluctuating, is maintained through the passage of geologic time.

At certain periods in the history of the Earth the diversity of life has decreased markedly. These catastrophic decreases are referred to as 'mass extinctions', to distinguish them from the normal extinctions which occur constantly but do not markedly affect the diversity of life, as the species which are lost are replaced by new species originations.

The simplest, and oldest, measure of mass extinction is a plot of the diversity of life as a function of geologic time. In 1860 John Phillips, then Professor of Geology at Oxford, published a rough estimate of the diversity of life (Figure 2.1) showing a relative increase in diversity with time. More importantly, he illustrated periods of decline in diversity as well, particularly the major losses in diversity which marked the close of his proposed Palaeozoic and Mesozoic Eras. Phillips's eras were in essence defined on the basis of mass extinction.

Much attention has been focused upon extinction itself as the mechanism for the loss of species diversity during periods of 'mass extinction', the equally important mechanism of species origination has in general been neglected. Clearly, one way in which diversity can fall abruptly is as a result of a substantial increase in the number of species which become extinct in a

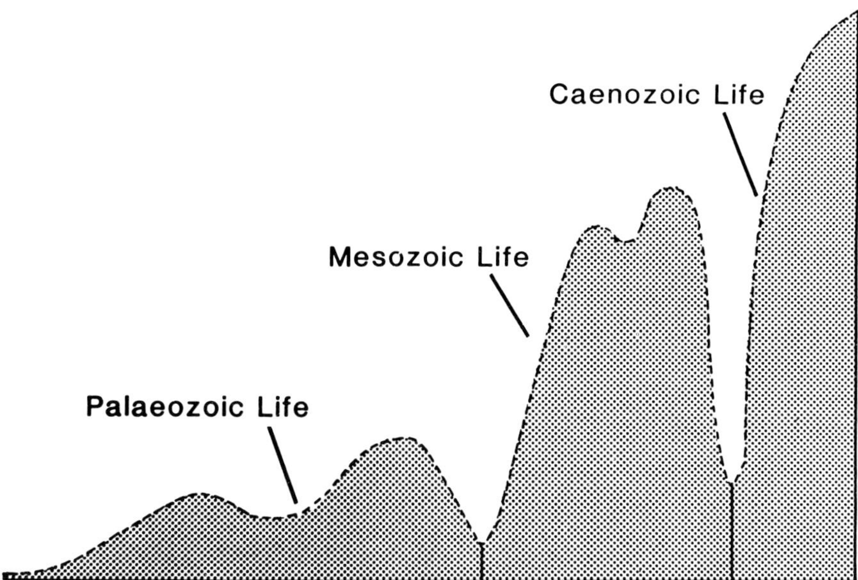

Figure 2.1. John Phillips's eras, proposed in 1860, were based on the recognition of two major periods of biotic crisis in the Earth's history. Note that he also recognised two smaller, less catastrophic, diversity declines within the Palaeozoic and Mesozoic. (Modified from Rudwick 1976.)

time interval. Alternatively, extinction can remain more or less constant and abrupt diversity loss can still occur if the number of species originations greatly decreases. Maximum diversity loss can be obtained by combining these two effects: greatly increasing the number of species extinctions while simultaneously decreasing the number of species originations. To date more attention has been given to the role of extinction in triggering the loss of species diversity on the Earth.

The quantification of extinction magnitudes is not a simple matter. Four different metrics for the study of extinction have been proposed:

1. The absolute number of species (or other taxa) which become extinct at some point in time, E.
2. The proportional number of extinctions, which is the number of extinctions scaled against the total number of species present (N) at that point in time, E/N.
3. The absolute rate of extinction, which is the number of extinctions in some time interval (Δt), usually taken to be a million years, $E/\Delta t$.
4. The proportional rate of extinction, which is the number of extinctions per time interval scaled relative to the number of species present in that interval, $E/(N\Delta t)$.

In the study of the role of species originations in triggering diversity decreases, the analogue of each of the above metrics can be used simply by substituting the number of originations (O) for the number of extinctions (E).

An alternative to plotting diversity as a function of time in order to pinpoint the intervals of diversity decrease, is to plot the rate of change of diversity per time interval, $\Delta N/\Delta t$. This metric, known also as the 'turnover rate', can be calculated as the derivative of the diversity function itself, or as the difference between the origination and extinction rates, $\Delta N/\Delta t = O/\Delta t - E/\Delta t$. Using this metric mass extinctions are identified as intervals characterised by negative turnover rates in geologic time.

THE MAJOR CRISES IN THE HISTORY OF LIFE

Five mass extinction episodes

Much more refined data are presently available than those used by Phillips over 120 years ago. Sepkoski (1982) compiled a massive amount of data on the diversity of life at the familial level for the past 600 Myr. Plotting familial diversity as a function of time reveals five major crisis periods in Earth history, when diversity decreased markedly (Figure 2.2 and Table 2.1). These crises occurred in the Late Ordovician, the Late Devonian, the Late Permian, the Late Triassic, and the Late Cretaceous. The mass extinction most popularly known is that of the Late Cretaceous, as it was then that the last of the dinosaurs perished. The Late Cretaceous event was not, however, either the first or even second most severe. The single most catastrophic event in the history of life on Earth was the Late Permian mass extinction.

In the Late Permian crisis over 50% of all of the animal families present in the Earth's oceans perished. As much as 83% of the marine genera were

Table 2.1. *Magnitudes of the major crises in the history of life on Earth. For each event the percentage of families of marine animals which perished is given. (Data from Sepkoski 1982).*

Mass extinction	*Marine families eliminated (%)*
I. Late Permian	50
II. Late Ordovician	22
III. Late Devonian	21
IV. Late Triassic	20
V. Late Cretaceous	15

eliminated (Sepkoski 1986), and it is estimated that 96% of all of the species in the oceans vanished (Raup 1979). Perhaps only 4% of the species present in the Late Permian world survived into the Triassic. The remaining mass extinctions were much less severe, based on the magnitude of familial diversity loss (Table 2.1). In all cases, however, a geologically significant amount of time elapsed before the numbers of new species originations were able to bring diversity back to previous levels (Figure 2.2).

A second quantitative aspect of mass extinction, in addition to magnitude, is the duration of the interval of diversity loss. The question of duration is of major importance in considering alternative hypotheses for the causes of mass extinction, as some models predict protracted periods of diversity decline while others call for geologically instantaneous diversity drops on a global scale.

Duration estimates are strongly influenced by the completeness of the stratigraphic record and the minimum interval of stratigraphic correlation.

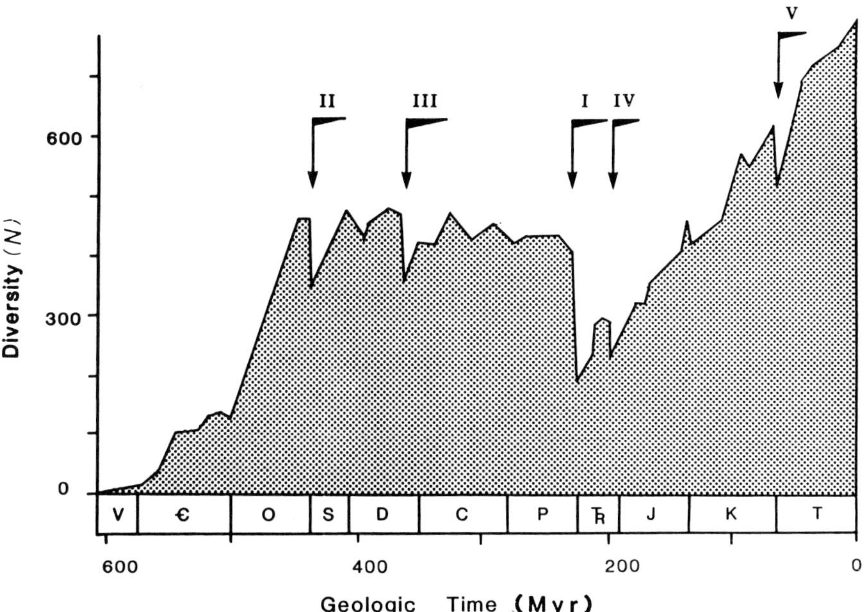

Figure 2.2. Five major periods of biotic crisis are recognised in the more recent data of Sepkoski (1982). The curve plots the diversity of animal families in the Earth's oceans as a function of geologic time. Major diversity losses, marked by arrows above, occurred in the Late Permian (I), the Late Ordovician (II), the Late Devonian (III), the Late Triassic (IV), and the Late Cretaceous (V). Following each crisis, a geologically significant span of time elapsed before marine life recovered to previous levels of diversity (delineated by the bars from the arrows above). (Modified from Erwin *et al.* 1987.)

Species range terminations are artificially smeared backwards in time during periods where the record is fragmentary or incomplete, which is often the case in the marine record during regression phases. Correlation to the level of stage is the most detailed possible for much of the geologic record on a global scale, yet most stages are several millions of years in duration. Both of these factors bias duration estimates towards maximum figures, and may conceal intervals of very short-term diversity loss.

Lastly, it can be shown in some cases that the interval of diversity loss does not correspond to the interval of maximum extinction. In the Late Devonian event, brachiopods experienced high rates of extinction (Figure 2.3A) which spanned some 4 Myr of the Frasnian Age, with peak extinction occurring 2 Myr before the end of the Frasnian. Turnover rates reveal a quite different picture (Figure 2.3B). The diversity loss occurred at the very end of the Frasnian, and was of much shorter duration. The actual 'mass extinction', at least for the Late Devonian, was more a function of a drop in origination rate rather than a direct result of an increase in extinction rate. Unfortunately, the role of origination decline in most biotic crises has yet to be fully explored.

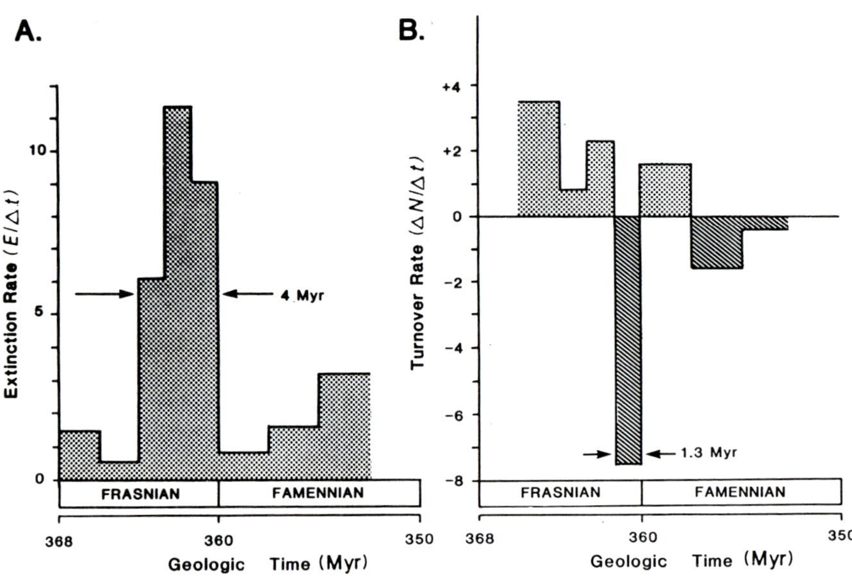

Figure 2.3. Two different measures of the timing and duration of the Late Devonian biotic crisis, which reveal different patterns. A: plots the absolute rate of extinction of brachiopods during the Late Devonian. B: plots the rate of change in brachiopod diversity (the 'turnover rate'). While extinction rates were very high for a period of 4 Myr during the Frasnian Age (A), diversity loss was very abrupt and concentrated in a single sharp pulse at the very end of the Frasnian (B). (Modified from McGhee 1988.)

Of the five major biotic crises, the Late Permian is generally considered to have been a long-term event. Extinction rates (see Figure 2.4 for two different metrics) increased in the Leonardian Age, reached a maximum in the Guadalupian, and declined slightly in the terminal age of the Permian, the Tatarian. Thus the event appears to have spanned a minimum of 10 Myr, using extinction rates as a measure of the duration of the crisis. This pattern may be artificial, however, as the major marine regressions which occurred at the end of the Permian resulted in the loss of much of this interval of time from the stratigraphic record. Many apparent Guadalupian, and perhaps Leonardian, extinctions may actually have occurred in the Tatarian and were recorded in strata which no longer exist.

The Late Triassic and Late Devonian events likewise remain problematical. Extinction rates were at maximum in the Norian Age of the Late Triassic, and declined but remained high in the terminal Rhaetian. Just as for the Permian, this may indicate a protracted event (15 Myr) with maximum extinction occurring before the very end of the period, or may represent artificial smearing backwards of species' last occurrences into older strata, as Late Triassic marine strata are also scarce.

Unlike the Triassic or Permian, abundant stratigraphic sequences exist for the Late Devonian interval. Rather than stratigraphic completeness, the problem now is correlation resolution. At the stage level a single broad extinction peak spans the Middle and Late Devonian, some 25 Myr. Depending upon the metric used, maximum extinction occurred in either the Frasnian Age of the Late Devonian (Figure 2.4A) or the Givetian Age of the Middle Devonian (Figure 2.4B). Substage data suggest that the single broad extinction peak is an artifact of the level of resolution, and that in fact numerous smaller extinction events occurred throughout this interval (House, 1985) with maximum extinction magnitudes occurring in the latter half of the Frasnian (Figure 2.3A). Note again, however, that the actual period of diversity loss (Figure 2.3B) was much shorter in duration than estimated by extinction rates alone (Figure 2.3A).

The Late Ordovician and Late Cretaceous events both appear to have been short-term events. In both cases, maximum extinction rates occurred within the later half of the terminal ages of these periods, the Ashgillian (Ordovician) and Maastrichtian (Cretaceous).

Other significant crisis intervals

Several other extinction peaks appear in Figure 2.4, and are of variable magnitude depending upon the metric used. Caution is required in using extinction magnitudes alone as a measure of biological crises, as discussed above. High extinction rates may also reflect high evolutionary turnover, in that origination rates may be equally high and no major change in standing diversity occurs. High apparent extinction rates are also caused by unusual fossil preservation (the 'Lagerstätten effect'). In areas of exceptional

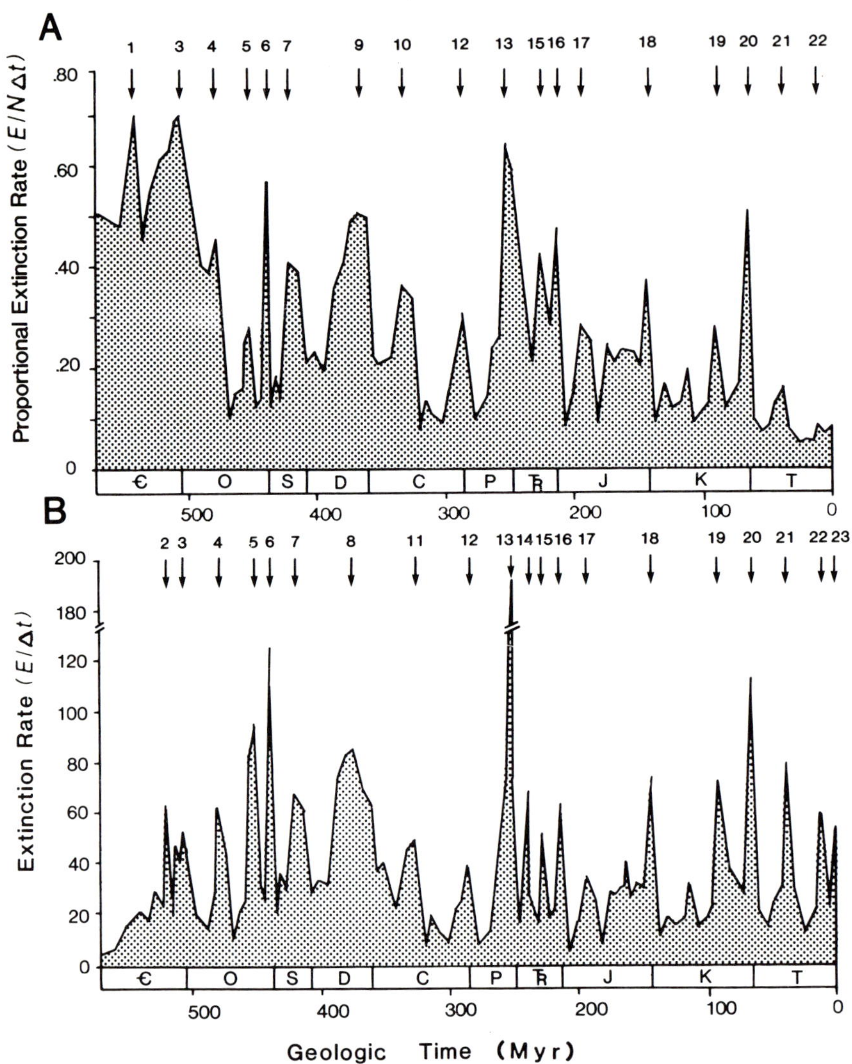

Figure 2.4. Two different extinction metrics for genera of marine animals, which reveal 23 peaks in extinction (marked by arrows) during the Phanerozoic. A: plots the proportional rate of extinction as a function of geologic time. B: plots the absolute rate of extinction. While 23 peaks are recognisable, not all correspond to actual periods of biotic crisis, as discussed in the text. Ages of the peaks are as follows: 1. Botomian; 2. Dresbachian; 3. Trempealeauan; 4. Arenigian; 5. Caradocian; 6. Ashgillian; 7. Wenlockian; 8. Givetian; 9. Frasnian; 10. Visean; 11. Serpukhovian; 12. Stephanian; 13. Guadalupian; 14. Olenekian; 15. Carnian; 16. Norian; 17. Pliensbachian; 18. Tithonian; 19. Cenomanian; 20. Maastrichtian; 21. Late Eocene; 22. Middle Miocene; 23. Pliocene. (Modified from Sepkoski 1986.)

preservation, such as the Mazonian Delta Complex (Carboniferous) or Solnhofen Limestone (Jurassic), species appear that are known from no other time interval, and technically become extinct as well.

Lastly, to qualify as a true biosphere crisis it is generally considered that a significant proportion of members of the ecosystem must be affected, and that the geographic scope of the event must be global and not just regional. Thus the elimination of all species of one or two families of restricted geographic distribution, while catastrophic, is not considered a 'mass' extinction.

Late Precambrian

Evidence suggests that biotic crises have struck life even in its simplest phases, before the Phanerozoic diversification of metazoans and macrophytes. Crises in the Precambrian world are difficult to quantify, and thus do not appear in Figures 2.2 or 2.4. However, a major loss in acritarch diversity appears to have occurred 650 Myr ago, with the decimation of perhaps 70% of the flora of that time (Vidal and Knoll 1982). The peculiar Ediacara biota also appears to have vanished abruptly shortly before the beginning of the Phanerozoic, and this may record yet another extinction event in the latest Precambrian (Seilacher 1984).

Botomian, Dresbachian, Trempealeauan

At the dawn of the Phanerozoic, the entire Cambrian Period and the Early Ordovician were characterised by high extinction magnitudes (Figure 2.4A). The Botomian and Trempealeauan Ages stand out above this high background extinction (Figure 2.4A), and when absolute extinction rates are considered, the Dresbachian Age emerges as the maximum. These three periods were times of mass extinctions within the global trilobite fauna, which were the dominant elements in Cambrian marine ecosystems. Considering the trilobites alone, five pulses of extinction ('biomere events' of Palmer 1984) have been recognised within the Cambrian, and the Botomian, Dresbachian, and Trempealeauan events appear to have been the largest of these.

Arenigian

The Arenigian Age within the Early Ordovician is identified as a peak using both absolute and proportional extinction metrics (Figure 2.4). It is of relatively low magnitude, but did have significant ecological impact on the structure of marine ecosystems in the Early Ordovician (Boucot 1983).

Caradocian

The Caradocian Age at the beginning of the Late Ordovician exhibits an absolute extinction rate (Figure 2.4B) approaching that of the first major mass extinction, the terminal Ordovician (Ashgillian) event. It has been suggested, however, that this high rate is artificial or a Lagerstätten effect of rich Caradocian fossil deposits (Sepkoski 1986), and thus may be balanced by an equally high apparent origination rate with no major loss of species diversity.

Wenlockian/Ludlovian
A broad but low peak in extinction appears spread through the Wenlockian and Ludlovian Ages in the middle of the Silurian. This peak may represent a series of smaller extinctions at the substage level, which may also have extended into the terminal Pridolian Age of the Silurian although their record is temporally smeared backwards due to the regressive nature of latest Silurian strata.

Visean–Serpukhovian
A broad peak, similar to that in the Silurian, extends through the Visean and Serpukhovian Ages in the Lower Carboniferous. This event may also represent a series of smaller sequential extinctions spread over perhaps as much as 15 Myr.

Stephanian
The terminal Stephanian Age of the Carboniferous stands out as a small peak using both extinction metrics (Figure 2.4). This event did affect a variety of genera, though in a modest way, and can be detected at the familial level as well (Sepkoski and Raup 1986).

Olenekian and Carnian
Extinction peaks within the Triassic, particularly the Olenekian (Figure 2.4B) and Carnian (Figures 2.4A and 2.4B), have been argued to be the result of sampling artefacts in the data, and thus not real (Sepkoski, 1986). Alternatively, these peaks may represent high rates of turnover, particularly among ammonites, and may thus not be periods of major diversity loss.

Pliensbachian–Toarcian
The Pliensbachian, and later Toarcian, Ages in the Early Jurassic stand out as a broad peak (Figure 2.4). The timing of this event is uncertain. It is possible that the Pliensbachian point reflects a backward smearing of generic ranges from the Toarcian, and thus the event may have been of shorter duration. Hallam (1986) argued that this event was confined geographically to Europe, and thus does not represent a global biotic crisis.

Tithonian
The Tithonian Age of the terminal Jurassic was a time of extinction for many faunal groups, and represents the largest extinction event in the Mesozoic between the Late Triassic and Late Cretaceous mass extinctions. An estimated 37% of marine genera became extinct (Sepkoski 1986), although Hallam (1986) argued that these, like the earlier Jurassic extinctions, were strongly regional in nature, and not global.

Cenomanian
The Cenomanian event, in the beginning of the Late Cretaceous, displays only a slightly lower extinction-rate magnitude than the Tithonian event. Detailed stratigraphic analysis reveals that this event occurred over a 2–3 Myr interval, in a series of sequential extinction pulses (Hut *et al.* 1987).

Late Eocene
A sharp peak in absolute extinction rate occurred in the Late Eocene (Figure 2.4B). This event particularly affected the marine plankton, and it has been argued that it represents a series of sequential extinction pulses, distributed over 3–4 Myr (Hut *et al.* 1987).

Middle Miocene and Pliocene
Two low peaks, the Middle Miocene and Pliocene, stand out in a plot of absolute extinction rates during the Neogene (Figure 2.4B). Though two separate peaks are resolvable within approximately 10 Myr, the two events may be causally related and thus represent a single phenomenon, the onset of the current ice age. The Pliocene event was highly regional, affecting mostly the North Atlantic marine fauna, and thus may not qualify as a global event (Stanley and Campbell, 1981).

It is possible that a continuum in extinction magnitudes exists, with no quantitatively definable boundary between 'mass' and 'background' extinction. Several of the 23 intervals discussed above (Figure 2.4), while exhibiting high extinction magnitudes, have nevertheless been considered to be intervals in which global losses in diversity did not occur. At smaller extinction magnitudes, a total of 29 possible 'mass' extinction intervals exist (Sepkoski 1986).

The question of periodicity

A decade ago, Fischer and Arthur (1977) made an intriguing suggestion—that extinctions may have occurred regularly every 32 Myr, at least for the past 250 Myr or so. This suggestion was made on the basis of apparently periodic bursts of extinction in ammonites and foraminifera during the Mesozoic and Cenozoic. The much more comprehensive data of Sepkoski (1982) have also been used to support the occurrence of periodic extinction since the time of the great Late Permian event (Figure 2.5), though these data suggest a 26 Myr periodicity (Sepkoski and Raup 1986).

The immediate implication of periodic extinction is that there exists a single causal mechanism of extinction, rather than many independent and unique causes. Curiously, following Sepkoski and Raup's (1986) analysis, attention was directed to possible extraterrestrial mechanisms, rather than terrestrial. This may be due to the fact that another intriguing proposal, that the Late Cretaceous event was triggered by an asteroid impact, was also under intensive debate in scientific circles. That the Earth is occasionally impacted by asteroidal debris or comets is a reasonable and perhaps expected prediction; that the Earth is impacted with clock-like regularity is another matter. No known periodic mechanism for perturbing the asteroidal belt between Mars and Jupiter exists, thus the extraterrestrial search has been directed further out to the very fringes of the solar system. There a large reservoir of cometary debris and comets is believed to exist in the Oort Cloud. Speculative mechanisms for perturbing the Oort Cloud have included the possible existence of an undetected tenth planet ('Planet X'),

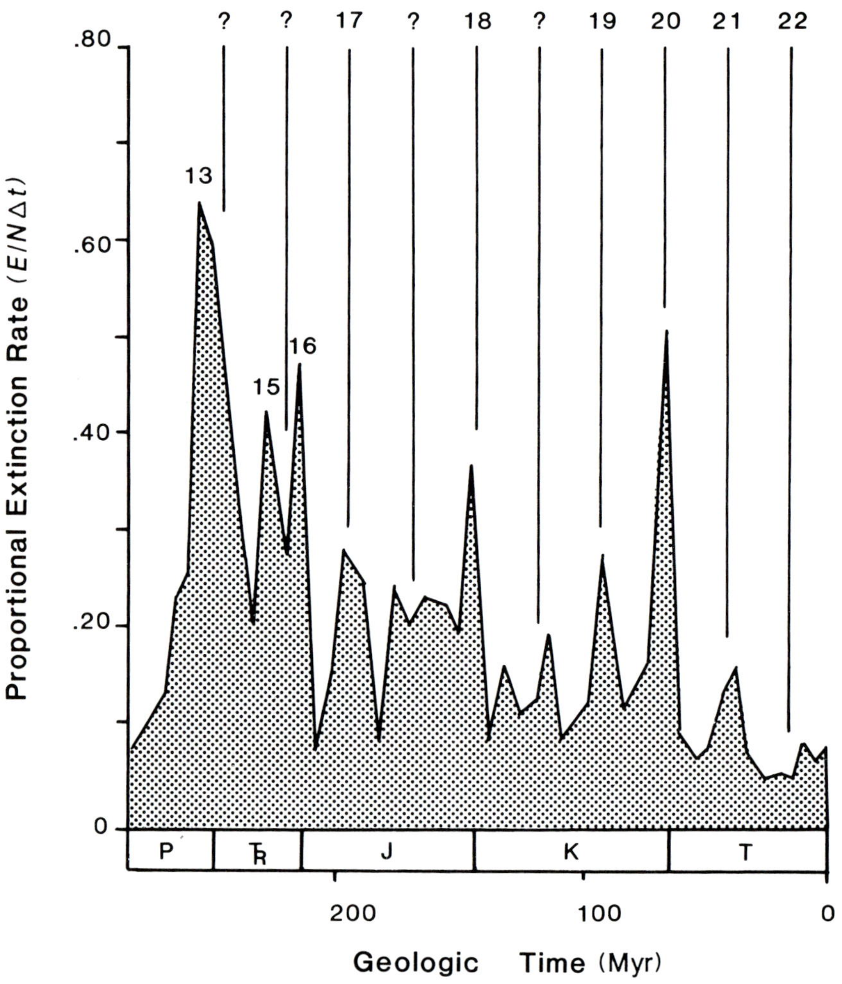

Figure 2.5. The proposed 26 Myr periodicity in extinction (marked by vertical lines) during the Mesozoic and Cenozoic, compared with the actual proportional rate of extinction of marine genera (from Figure 2.4A). Ages of peaks as in Figure 2.4.

with a highly eccentric orbit beyond that of Pluto, and the existence of an undetected binary solar companion (variously known as the 'Death Star', 'Nemesis', or 'Siva'), in a highly elliptical orbit ranging from 0.3 to 3 light years away from the Sun. The gravitational perturbation of either body, passing through or near the Oort Cloud, would have to be adequate to drive a sufficient number of comets and debris into the inner solar system to ensure a collision with the Earth every 26 Myr.

Needless to say, such proposals have generated heated debate. Both are *ad hoc*, as their predictions are based entirely on the suggested 26 Myr periodicity in extinctions. Analyses of the ages of terrestrial impact craters have suggested a 28 Myr periodicity, though this result is also under debate (Alvarez and Muller 1984; Kerr 1985). If correct, such a result would provide important support both for the postulated periodicity in extinctions and for extraterrestrial impacts as their causal mechanism.

Many problems remain with the original suggestion of periodic extinctions itself. One serious problem is the lack of exact correspondence between predicted and observed extinctions (Figure 2.5). Two predicted peaks are missing: the Bathonian in the Jurassic and the Aptian in the Cretaceous, though the Aptian is included in the 29 smaller events listed by Sepkoski (1986). Five peaks occur which are not predicted: the Guadalupian, Rhaetian, Olenekian, Carnian, and Pliocene. The Late Permian and Late Triassic 'double peaks' (Guadalupian–Tatarian and Norian–Rhaetian), as well as the two additional peaks within the Triassic (Olenekian and Carnian), may be artificial, however. It can be argued that the unpredicted Pliocene peak was not global in nature, but the same applies to the predicted Pliensbachian and Tithonian peaks. Other problems include the imprecise and variable dating of the geologic time-scale used in computing the temporal spacing of extinction events, and continuing statistical debates concerning both the 'events' to be used or discarded in periodicity computations, and the results of those computations (Raup and Sepkoski 1988; Stigler and Wagner 1988). A related problem is the apparent lack of periodicity in Palaeozoic extinctions which, if real, would require the initiation of a single and periodic causal mechanism relatively late in the Earth's history.

COMPARATIVE ECOLOGY OF TIMES OF BIOTIC CRISIS

Latitudinal effects

One of the most pervasive effects during times of biotic crisis is the massive disruption of tropical and low-latitude ecosystems, and the relative non-disturbance of high-latitude and polar ecosystems. Tropical reef ecosystems are consistently decimated, even though the biotic composition of the reefs themselves has changed markedly through geologic time. Thus there is a sequential destruction of primitive archaeocyathid reef structures in the Botomian (Cambrian), bryozoan bioherms in the Late Ordovician, massive stromatoporoid and tabulate coral reefs in the Late Devonian, diverse sponge, algae, bryozoan, brachiopod reefal communities in the Late Permian, hexacoral reefs in the Late Triassic, and rudistid bivalve reef structures in the Late Cretaceous.

Biogeographic selectivity in extinction is also seen in higher taxa with cosmopolitan or wide-ranging species distributions. Brachiopods in the Late Ordovician, Late Devonian, and Late Permian crises exhibited disproportionate survival in higher latitudes than in lower, and a similar latitudinal selectivity was exhibited by bivalves, gastropods, and dinoflagellates in the Late Cretaceous event.

Lastly, it can be shown that high-latitude fauna and flora migrated into lower-latitude regions during or immediately after some of the extinction events. These fine-scale biogeographic shifts are difficult to document in the fossil record, but have been demonstrated for graptolites and brachiopods in the Late Ordovician event, foraminifera in the Late Devonian, marine gastropods and foraminifera in the Late Cretaceous, and calcareous nannoplankton in the Late Eocene.

Bathymetric effects

Another pervasive effect of times of biotic crisis is the decimation of the marine plankton. In all major mass extinctions benthic organisms also suffered extinction, though for some extinction events it can be shown that organisms which inhabited shallow-water regions were more strongly affected than those which inhabited the deeps.

As for reefs, planktonic ecosystems were repeatedly disrupted, though the biotic composition of the plankton had changed greatly through time. The first extinction event currently recognised is recorded by the decimation of the acritarchs in the Late Precambrian. Acritarchs were repeatedly decimated in later Palaeozoic events, as were zooplanktonic trilobites in the Cambrian events, graptolites in the Late Ordovician, tentaculitids and pelagic conodonts in the Late Devonian, planktonic foraminifera in the Late Cretaceous and calcareous nannoplankton in the Late Eocene.

Shallow water epicontinental and continental shelf benthic ecosystems suffered disproportionately relative to deep-water oceanic ecosystems in all extinction events. In some events, decimated shelf faunas were later replenished by deep-water species (as in the Cambrian biomere events). In the Late Devonian event, deep-water glass sponge faunas migrated into shallow-water regions during and immediately following the Frasnian pulse of extinction. On the other hand, many of the very near-shore species are geographically widespread and tolerant of environmental variability, and have tended to survive periods of mass extinction (as discussed below).

Marine versus terrestrial habitat effects

The terrestrial ecosystem has been well developed only since mid-Palaeozoic time. In all major extinction events since the mid-Palaeozoic, however, both marine and terrestrial ecosystems have been affected, thus demonstrating that mass extinction is truly a global phenomenon.

The terrestrial flora appears to be particularly resistant to mass extinction (Knoll 1984). In some events, such as the late Cretaceous, it can be shown that plant populations were decimated yet quickly recovered after the event with only minor species losses. While relatively extinction-resistant, the evolution of the angiosperms does provide important information concerning climatic change for the latest Mesozoic and following Cenozoic. Modern species having leaves with entire margins are found in warm climates, and the proportion of species having leaves with complicated margins increases in lower-temperature regions. The percentage of angiosperm species with entire leaf margins dropped markedly in both the Late Cretaceous and Late Eocene events, indicating global cooling.

Large land animals appear particularly prone to extinction. Smaller animals tend to have larger population sizes and higher birth rates, in addition to lower nutritional needs, all of which may contribute to their relative survival in times of crisis.

Terrestrial climates and environments are much more variable than marine. Though extinction rates are higher in terrestrial ecosystems, the adaptation of many terrestrial species to large seasonal fluctuations in climate may tend to make these species more extinction-resistant in times of global environmental deterioration, relative to marine species. Fish species inhabiting rivers and lakes experienced greater survival than marine species during the Late Devonian crisis, relatively early in the development of terrestrial ecosystems.

Demonstrating the precise timing of terrestrial, relative to marine, extinctions is often problematical. It is clear that both regions are affected, but there may be a significant time-lag in their response to periods of crisis. Extinction on land and in the sea appears to have been diachronous in the Late Permian and Late Triassic (beginning on land first, then spreading to the seas), and synchronous in the Late Cretaceous. Diachroneity in the former two events may be artificial (due to sampling incompleteness in the generally poor terrestrial record; see Knoll 1984), whereas in the latter it is still not possible to demonstrate simultaneity, in that even magnetostratigraphic correlation between land and sea has an average resolution in the hundred-thousand-year range.

Biogeographic effects

Jablonski (1986a; 1986b) argued cogently that characteristic biogeographic patterns of selectivity exist for periods of mass extinction. Taxa which are geographically widespread and have broad environmental tolerances typically suffer fewer losses during mass extinctions than taxa which have restricted geographic ranges and environmental tolerances. For many marine organisms, these traits are directly related to reproductive mode. Taxa which possess planktotrophic modes of life tend to have high larval dispersal capabilities and environmental tolerance, in contrast to taxa with non-planktotrophic modes of life.

Paradoxically, planktotrophic organisms predominate in regions most susceptible to the effects of extinction. In the modern marine ecosystem, planktotrophic modes of life are typically found in tropical and subtropical latitudes, the same regions which are adversely affected during periods of mass extinction. A bathymetric gradient in modes of life occurs today such that planktotrophic species dominate near-shore environments, and the frequency of non-planktotrophic species increases with water depth (Jablonski and Bottjer 1983). As discussed above, shallow-water ecosystems are characteristically more severely affected by extinction intervals than are deep-water regions, hence once again planktotrophic species occur in high-risk environments. Indeed, it has been suggested that many Palaeozoic brachiopod and crinoid groups possessed planktotrophic modes of life, and that these groups were among the many low-latitude, shallow sublittoral groups selectively exterminated in the Late Permian event, leaving only non-planktotrophic survivors in these taxa today (Valentine and Jablonski 1983).

CAUSES OF GLOBAL ECOSYSTEM DETERIORATION

Earthbound mechanisms

Temperature

One cause of mass extinction which is most compatible with the observed ecological patterns of selectivity is lethal temperature decline. As Stanley (1987) pointed out, temperature is a pervasive environmental variable, one which affects the entire planet, terrestrial and marine. Global decline in temperature would decimate reefal and tropical ecosystems, as species which inhabit these regions have no refuge against cold. Higher-latitude ecosystems would be less affected, as many species could simply migrate to lower latitudes and thus maintain their tolerable temperature ranges. Indeed, expansion of temperate and polar biota into the lower latitudes can be documented for several of the major mass extinction episodes.

Global cooling would most adversely affect the warm-water, temperature-buffered, epicontinental sea benthos. Deeper-water faunas are generally adapted to cooler-water conditions, and thus would be expected to survive differentially. Shallow oceanic planktonic ecosystems would also be decimated, with greater survival of the plankton in high-latitude regions.

Terrestrial ecosystems are more resistant to temperature fluctuations than are marine. On the land, however, temperate to polar ecosystems are adapted to seasonal temperature fluctuations or prolonged periods of cold, whereas tropical ecosystems are not.

Many of these expected patterns of ecological selectivity have been documented for the major mass extinctions, as discussed in the previous section. Thus lethal temperature decline is a prime candidate as a common cause for global biotic crises. A problem arises in determining the ultimate cause of global temperature decline. In fact there may be no single ultimate cause, in that the same temperature effect may be produced by a variety of mechanisms, both earthbound and extraterrestrial.

The most obvious correlate with global cooling is continental glaciation. Yet even here it can be debated whether the onset of glaciation has an extraterrestrial trigger (cycles in orbital eccentricities of the Earth, or possible variations in the solar constant) or whether the trigger is a function of plate tectonics (positioning of a continental landmass over a pole), or both.

The Late Precambrian, Late Ordovician, and Pliocene extinctions are clearly associated with major glaciation phases, the Visean–Serpukhovian, Stephanian, and Middle Miocene extinctions are probably associated, and the Late Devonian and Late Permian may be. The Late Devonian extinction maximum in the Frasnian appears to pre-date the onset of glaciation in Gondwana by 7–13 Myr, however, and the Late Permian maximum appears to post-date Gondwana glaciation by 5–8 Myr. The glacial correlation breaks down for the remaining major and minor extinction phases. It has been argued that the existence of sea ice is associated with some of these events; on the other hand, the Earth has never been totally ice-free.

Marine regression

Sea-level fall is also associated with many periods of biotic crisis. As marine regression is produced by ice volume build-up, separating the effects of temperature decline from regression can be difficult. In geologic periods where regression occurred without demonstrable glaciation, the causal link between lowered sea levels and elevated extinction levels is still problematical. Lastly, many marine regressions have occurred with no obvious biological effect. This last observation may be expected, however, as discussed below.

An oft-cited causal link between regression and extinction is the species–area effect. Marine regression clearly results in the reduction of habitable shelf area on the continents, leading to decreased population sizes and eventual extinction of many species (following the predictions of the equilibrium theory of island biogeography). The most serious crisis in the history of life on Earth, the Late Permian event, has been argued to be due solely to the habitat-reduction effect of the massive marine regressions and low sea-level stands which characterise that interval (Schopf 1974; Simberloff 1974). Regressions also occurred during the Late Ordovician, Late

Triassic, and Late Cretaceous events, and perhaps the Late Devonian, although regressions during the late Frasnian Age appear to have been minor in scale.

However, there are many problems with the species–area scenario. First, while regression produces habitat reduction and destruction in the marine realm, the areal extent and habitat space of the terrestrial realm increases. Yet, as discussed above, terrestrial ecosystems also suffer during mass extinction intervals. Second, even within the marine realm itself, regression produces habitat expansion around oceanic islands, many of which harbour high species diversities today (Stanley 1987). Lastly, even given the highly provincial nature of marine biota today, Jablonski (1986a) argued that only 13% of Recent families would become extinct if all of the modern shelf biota were eliminated.

Further complications arise concerning the extent of continental inundation and regression. Given a sea-level drop of equal magnitude, species–area effects would be most severe during intervals when the continents are covered with large epeiric seas than during intervals when absolute sea-level stand is low, with only narrow continental shelf zones. Thus the expected loss of species diversity due to the species–area effect is not a simple function of regression magnitude, but also of the pre-regression condition of the marine biosphere.

A more promising causal link between regression and extinction may exist in climatic effects (Jablonski 1986a), though again the pre-regression configuration of land and sea is of major importance. Global climatic extremes may be buffered during intervals when the continents are covered with extensive shallow seas, which may simultaneously warm oceanic waters and dampen seasonal temperature fluctuations on the land. Regression would destroy this ameliorating effect, producing sharp climatic gradients betwen land and sea. Oceanic waters would become colder in the absence of warm-water influx from epicontinental seas, and seasonal temperature fluctuations on land would become extreme in the absence of the dampening effect of large shallow-water masses. An increase in exposed land area would heighten the albedo of the Earth, and would reduce the carbon dioxide content of the atmosphere, due to increased weathering. Both effects would produce overall global cooling, and temperature decline is a consistent ecological signal for times of biotic crisis.

Explosive volcanism

Magnitude and duration are key factors in considering volcanism as a trigger for mass extinction. Massive and explosive volcanism, concentrated in a short period of time, would produce many of the same climatic effects as those proposed for the impact of a large asteroid. The injection of large amounts of volcanic dust into the atmosphere would increase the albedo of the Earth and simultaneously decrease the amount of sunlight reaching the surface, both of which would produce a sharp drop in global temperature.

Several biotic crises do occur during tectonic phases of orogenesis and volcanism, such as the Late Ordovician, Late Devonian, and Late Cretaceous, but others do not. In addition, most volcanic phases are spread over long intervals of time in comparison to the duration of most mass extinctions. A threshold effect can be invoked here, however, in that extended environmental stress produced by long-term periods of volcanism may eventually trigger global ecosystem collapse. Volcanism that is simultaneously explosive and deep enough to tap the Earth's mantle could produce many of the features associated with the impact of an asteroid (discussed below). Thus, it has been argued that the volcanism which produced the Deccan Traps in the Late Cretaceous may not only have triggered the Late Cretaceous extinction event, but also produced the anomalous geochemical and metamorphic shock features associated with sediments of latest Maastrichtian Age (Officer *et al.* 1987). However, the Deccan volcanics extended over a considerable interval of time and occurred in multiple phases, whereas the anomalous geologic features apparently are restricted to a single time horizon and thus record a single short-term event.

Oceanographic effects

Changes in oceanic circulation patterns, stratification, oxygen distribution, salinity composition, and other chemical factors have also been advanced as possible causes of mass extinction. However, the duration and timing of known oceanographic events in the Phanerozoic do not appear to correlate well with those of periods of biotic crisis. Also, a major difficulty arises in demonstrating a causal link between oceanic events and terrestrial extinction. One interesting theory, that of 'pulsation tectonics', has been advanced which seeks to link changes in ocean chemistry, atmospheric composition, volcanism, climate, and the magnetic field of the Earth, all driven by episodic convective events in the mantle (Sheridan 1987). This theory is interesting in that it also proposes a periodic mechanism of change. The period length proposed is approximately 80 Myr for the Palaeozoic, slowing to approximately 65 Myr for the later Mesozoic and Cenozoic. However, neither cycle corresponds to the proposed 26 Myr periodicity in Mesozoic–Cenozoic extinctions, nor the lack of periodicity in Palaeozoic extinction events.

Extraterrestrial mechanisms

Intra-solar

Perhaps the most interesting and controversial proposal made in palaeobiology in the past decade is the proposal that the Late Cretaceous mass extinction was triggered by the impact of a large asteroid (Alvarez *et al.* 1980), perhaps half the size of the Martian moon Phobos (Figure 2.6), itself believed to have originated in the asteroid belt between Mars and Jupiter. There is an entire class of asteroid-sized bodies, the Apollo objects, which

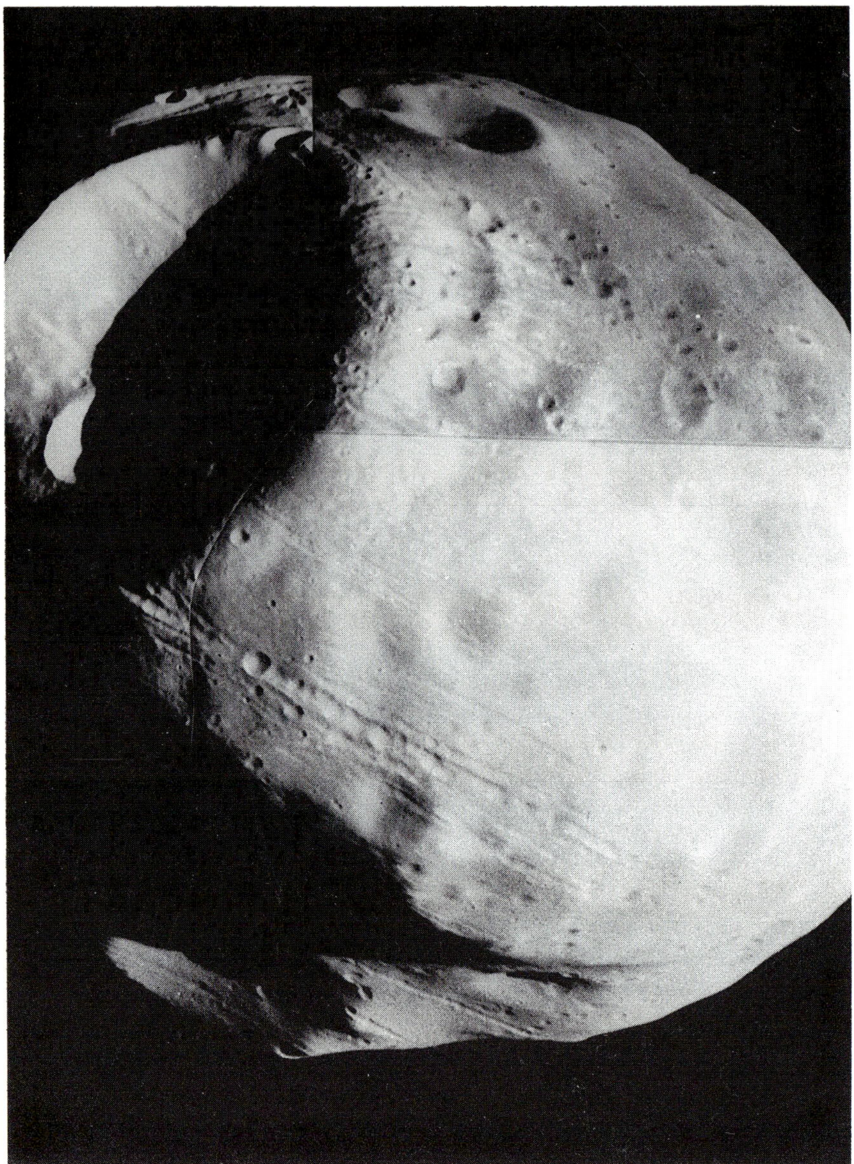

Figure 2.6. The Martian moon Phobos, photographed from 611 km away by Viking Orbiter I. Phobos is most likely an asteroid, captured by Mars from the asteroid belt, and the surface shown here (19 km by 22 km) is approximately twice the size of the asteroid which is proposed to have impacted the Earth at the end of the Cretaceous. (Photograph courtesy of NASA.)

have orbits which cross that of the Earth and it is only a matter of time before a collision takes place. Thus it is entirely reasonable to propose that such an event has taken place in the past, and that the catastrophic results of such a collision could trigger global ecosystem collapse. In retrospect, it is the very catastrophic and geologically instantaneous nature of such an impact which has led to rejection of the idea by many, who see the proposal as a return to nineteenth-century catastrophism. On the other hand, others have gone to the extreme of maintaining that *all* mass extinctions are triggered by impacts, and that these impacts occur with clock-like regularity.

An impressive body of evidence has been amassed which supports the idea that an asteroid did indeed impact the Earth at the end of the Maastrichtian Age in the Late Cretaceous. The initial evidence was geochemical—the discovery of anomalously high concentrations of the element iridium in sediments of latest Maastrichtian Age. Iridium is depleted in crustal rocks of the Earth, but enriched in meteoritic materials. The anomalous iridium layer has now been discovered at sites all over the globe, and in both marine and terrestrial sediments. Thus it appears that the Earth was enveloped in a global dust cloud, anomalously enriched in iridium, at the end of the Cretaceous.

Other evidence includes the discovery of minerals with shock features, characteristic of impacts, in sediments of latest Maastrichtian Age in widely separated regions of the globe. Small spherules, argued to be impact-melt-derived, have been discovered in these same sediments as well as carbon concentrations, interpreted as the soot fallout of massive wildfires triggered by the impact. Geochemical anomalies, in addition to iridium, have also been interpreted as extraterrestrial in origin.

The climatic effects of an asteroid or cometary impact are difficult to predict. Large amounts of debris from the vaporised bolide and impact site would be injected into the atmosphere. Meteorological models suggest that such a global dust cloud would result in the total blockage of sunlight from the Earth's surface for weeks, and light intensities too low to support photosynthesis for several months. Surface temperatures would quickly drop to and remain below freezing for a period of several months. Other proposed effects include poisoning of the atmosphere with nitrous oxides and cyanide, global acid rain, and destruction of the ozone layer.

Certainly such a climatic upheaval would have a major effect on the biosphere, and global temperature decline is consistent with the ecological signature of most mass extinctions. One problem with the impact hypothesis is that the effect of such a collision should be instantaneous but short-term. Most mass extinctions appear not to have been geologically instantaneous, however; diversity losses are spread over hundreds of thousands to millions of years. Last appearances of species in an extinction interval often are not simultaneous but stepwise, with one group of species disappearing first, followed by another some thousands to tens of thousands of years later, and yet another later in time.

It has been suggested that the stepwise or protracted pattern of species extinction seen in the fossil record is an artifact. Many extinction intervals also correlate with regressions, and marine regression has a strong distorting effect on the stratigraphic record. The progressive loss of sampling area produced by regressing seas, combined with the erosion and destruction of previously deposited strata, can smear last appearances of species backwards in time and transform even an instantaneous diversity drop into a progressive diversity decline (Signor and Lipps 1982). The magnitude of this distortion can be estimated to some extent by noting the number of 'Lazarus species' (species which disappear during the extinction interval but are known to survive, as they reappear later) associated with the event (Jablonski 1986a). If the Lazarus effect is strong for a particular event, then it is not possible to reject the hypothesis that the event was sudden and only artificially appears protracted. Jablonski (1986a) noted that the Lazarus effect appears worst for the Late Permian event, and raised the interesting possibility that this most severe of all crises in Earth history may have been more abrupt than is normally considered.

The stratigraphic record of many crisis intervals is sufficiently complete to negate the possibility of simultaneous extinction of all species which disappear during the event. Accepting the stepwise extinction pattern as real, an alternative impact hypothesis has been proposed (Hut *et al.* 1987) in which a series of smaller impacts, distributed over a geologically significant period of time, is argued to have produced the sequential extinctions!

Lastly, it should be noted that the climatic predictions of the impact hypothesis are strongly model-dependent. The climatic effects of a large-body collision could be more long-term than currently predicted, and may produce global changes on a time scale of tens of thousands of years. Such long-term climatic effects would be more in concert with the pattern of species extinctions seen in the fossil record.

To date, a strong case cannot be made for large impacts during crisis intervals other than the Late Cretaceous and Late Eocene events (Donovan 1987; Kerr 1987), and perhaps a small impact in the Pliocene (Kyte *et al.* 1988). The Late Eocene is particularly interesting, in that an iridium anomaly as well as microtektites have been found, clearly indicating extra-terrestrial impact. The microtektites, however, occur in several horizons of Late Eocene and Early Oligocene age, thus evidencing a series of impacts over this time interval. Curiously, none of these multiple horizons appears to correspond exactly to known extinction horizons (Keller *et al.* 1983), thus providing no strong causal link between the extinctions and the impacts. Small iridium anomalies have been reported for the Cenomanian and Middle Miocene events, but an impact origin for these has yet to be conclusively demonstrated. On the other hand, iridium-enrichment horizons have been discovered which were produced by sediment geochemical and biological processes. Thus anomalous concentrations of iridium can no longer be taken solely as direct evidence for extraterrestrial impacts.

Other possible intra-solar mechanisms to trigger terrestrial crises include possible fluctuations in the luminosity of the Sun, and known orbital variations in the distance of the Earth from the Sun. The possibility that the solar 'constant' varies is interesting, but difficult to demonstrate on the basis of terrestrial data alone. Obviously any increase or decrease in energy from the Sun reaching the Earth would have major effects on global climate. Orbital changes in Earth–Sun relationships can also cause variations in solar energy reaching the Earth. Such changes are of particular interest in that they are periodic, but the various fluctuations in the Earth's orbital (Milankovitch) periods fall in the range of thousands to hundreds of thousands of years. Geologic periodicities of the order of 2–7 Myr, which may be orbit-related, have been proposed but not confirmed (Rampino and Stothers 1987). None match the proposed 26 Myr periodicity in Mesozoic–Cenozoic extinctions.

Extra-solar

One of the longest-standing suggestions for an extraterrestrial trigger of mass extinction is the explosion of a nearby star or supernova. The consequent burst of high-energy radiation reaching the Earth was initially proposed as the killing agent. Radiation influx would adversely affect terrestrial ecosystems (particularly large animals) and shallow-depth plankton in the oceans, but the great majority of marine organisms would be shielded by oceanic waters. Thus the expected pattern of extinction and survival does not match what is actually observed in mass extinctions. More elaborate versions of the supernova hypothesis combine radiation with climatic effects. Here the high-radiation flux heats the upper levels of the Earth's atmosphere intensely, initiating massive convection and bringing low-level water vapour into upper levels of the atmosphere where it freezes, producing a global ice-cloud layer. Such a cloud layer would greatly increase the albedo of the Earth, and result in sharp drops in temperature at the surface. The end result (global cooling) is very similar to that envisaged as a consequence of asteroid impacts or explosive volcanism, though without any of the associated geochemical and geologic signatures.

Additional extra-solar hypotheses for mass extinction are framed with reference to the position of the solar system relative to the galaxy, and are particularly interesting in that they entail periodic processes. The solar system oscillates about the galactic plane as it orbits the galaxy, and reaches an extremum above or below the galactic plane approximately every 33 Myr. It has been proposed that the Earth is exposed to high-energy radiation flux emanating from the core of the galaxy at these extrema, and that global climatic changes are induced as a result. Alternative hypotheses argue for gravitational perturbation of the Oort Cloud surrounding the solar system, sending comet showers into the inner solar system which may impact the Earth. Such perturbation might occur as the solar system crosses the higher-density regions of the galactic plane itself (thus also having a 33 Myr periodicity), or as the solar system passes through the spiral arms of the galaxy (having an approximately 50 Myr periodicity). Neither of these two periodicities match the proposed 26 Myr period length for Mesozoic and Cenozoic extinctions, however.

IMPLICATIONS OF MASS EXTINCTION FOR EVOLUTIONARY THEORY

The known biotic crises in the history of life on Earth are usually viewed as destructive, and as aberrations from the normal process of evolution. The disappearance of large numbers of species can accurately be called destructive, and any selection which chiefly eliminates and seemingly does not allow the evolution of resistance to mass extinction is clearly coarse-grained.

Major perturbations in the course of evolution, however, may have played a creative role in the development of life on Earth (Sepkoski 1985; Jablonski 1986b). Repeated periods of diversity loss may have prevented life from ever reaching a state of equilibrium or saturation, and have kept the biosphere in a state of continual flux. Major ecosystem changes, such as the replacement of the dinosaurs by the mammals, might never have occurred without the magnitude of a perturbation associated with mass extinction. Mass extinctions may have repeatedly 'reset the clock' of evolution, thus reversing the process of adaptation of species to ever more subtle ecological interactions and increasingly intense biotic pressure. Extinctions may have allowed the repeated origination of new and less fit evolutionary innovations through the course of geologic time. Thus biotic crises, while decreasing diversity in the short term, may actually have operated to maintain the ecological diversity of life on a geologic time-scale.

REFERENCES

Alvarez, L.W., Alvarez, W., Asaro, F. and Michel, H.V., 1980, Extraterrestrial cause for the Cretaceous–Tertiary extinction, *Science*, 208: 1095–1108.

Alvarez, W. and Muller, R.A., 1984, Evidence from crater ages for periodic impacts on the earth, *Nature*, **308**: 718–20.

Boucot, A.J., 1983, Does evolution take place in an ecological vacuum?, *Journal of Paleontology*, **57** (1): 1–30.

Donovan, S.K., 1987, Iridium anomalous no longer?, *Nature*, **326**: 331–2.

Erwin, D.H., Valentine, J.W. and Sepkoski, J.J., Jr, 1987, A comparative study of diversification events: the early Paleozoic versus the Mesozoic, *Evolution*, **41** (6): 1177–86.

Fischer, A.G. and Arthur, M.A., 1977, Secular variations in the pelagic realm, *Society of Economic Paleontologists and Mineralogists Special Publication*, **25**: 19–50.

Hallam, A., 1986, The Pliensbachian and Tithonian extinction events, *Nature*, **319**: 765–8.

House, M.R., 1985, Correlation of Mid-Palaeozoic ammonoid evolutionary events with global sedimentary perturbations, *Nature*, **313**: 17–22.

Hut, P., Alvarez, W., Elder, W.P., Hansen, T., Kauffman, E.G., Keller, G., Schoemaker, E. and Weissman, P.R., 1987, Comet showers as a cause of mass extinctions, *Nature*, **329**: 118–26.

Jablonski, D., 1986a, Causes and consequences of mass extinctions: a comparative approach. In D.K. Elliott (ed.), *Dynamics of Extinction*, John Wiley & Sons, Inc., New York, pp. 183–229.

Jablonski, D., 1986b, Evolutionary consequences of mass extinction. In D.M. Raup and D. Jablonski (eds), *Patterns and processes in the history of life*, Springer-Verlag, Berlin, pp. 313–29.

Jablonski, D. and Bottjer, D.J., 1983, Soft-bottom epifaunal suspension-feeding assemblages in the Late Cretaceous: implications for the evolution of benthic paleocommunities. In M.J.S. Tevesz and P.L. McCall (eds), *Biotic interactions in recent and fossil benthic communities*, Plenum Press, New York, pp. 747–812.

Keller, G., d'Hondt, S. and Vallier, T.L., 1983, Multiple microtektite horizons in Upper Eocene marine sediments: no evidence for mass extinctions, *Science*, **221**: 150–2.

Kerr, R.A., 1985, Periodic extinctions and impacts challenged, *Science*, **227**: 1451–3.

Kerr, R.A., 1987, Beyond the K–T boundary, *Science*, **236**: 667.

Knoll, A.H., 1984, Patterns of extinction in the fossil record of vascular plants. In M.H. Nitecki (ed.), *Extinctions*, University of Chicago Press, Chicago, pp. 21–68.

Kyte, F.T., Zhou, L. and Wasson, J.T., 1988, New evidence on the size and possible effects of a Late Pliocene oceanic asteroid impact, *Science*, **241**: 63–5.

McGhee, G.R., Jr, 1988, The Late Devonian extinction event: evidence for abrupt ecosystem collapse, *Paleobiology*, **14** (3): 250–7.

Officer, C.B., Hallam, A., Drake, C.L. and Devine, J.D., 1987, Late Cretaceous and paroxysmal Cretaceous/Tertiary extinctions, *Nature*, **326**: 143–9.

Palmer, A.R., 1984, The biomere problem: evolution of an idea, *Journal of Paleontology*, **58**: 599–611.

Phillips, J., 1860, *Life on Earth: its origin and succession*. Cambridge and London.

Rampino, M.R. and Stothers, R.B., 1987, Episodic nature of the Cenozoic marine record, *Paleoceanography*, **2**: 255–8.

Raup, D.M., 1979, Size of the Permo-Triassic bottleneck and its evolutionary implications, *Science*, **206**: 217–18.

Raup, D.M. and Sepkoski, J.J., Jr, 1982, Mass extinctions in the marine fossil record, *Science*, **215**: 1501–3.

Raup, D.M., and Sepkoski, J.J., Jr, 1988, Testing for periodicity of extinction, *Science*, **241**: 94–6.

Rudwick, M.J.S., 1976, *The meaning of fossils*, Neale Watson Academic Publications, Inc., New York.

Schopf, T.J.M., 1974, Permo-Triassic extinction: relation to sea-floor spreading, *Journal of Geology*, **82**: 129–43.

Seilacher, A., 1984, Late Precambrian and Early Cambrian metazoa: preservational or real extinctions? In H.D. Holland and A.F. Trendall (eds), *Patterns of change in earth evolution*, Springer-Verlag, Berlin, pp. 159–68.

Sepkoski, J.J., Jr, 1982, Mass extinctions in the Phanerozoic oceans: a review, *Geological Society of America Special Paper*, **190**: 283–9.

Sepkoski, J.J., Jr, 1985, Some implications of mass extinction for the evolution of complex life. In M.D. Papagiannis (ed.), *The search for extraterrestrial life: recent developments*, D. Reidel Publishing Co., Dordrecht, pp. 223–32.

Sepkoski, J.J., Jr, 1986, Phanerozoic overview of mass extinction. In D.M. Raup and D. Jablonski (eds), *Pattern and process in the history of life*, Springer-Verlag, Berlin, pp. 277–95.

Sepkoski, J.J., Jr and Raup, D.M., 1986, Periodicity in marine extinction events. In D.K. Elliott (ed.), *Dynamics of extinction*, John Wiley & Sons, Inc., New York, pp. 3–36.

Sheridan, R.E., 1987, Pulsation tectonics as the control of long-term stratigraphic cycles, *Paleoceanography*, **2** (2): 97–118.

Signor, P.W. and Lipps, J.H., 1982, Sampling bias, gradual extinction patterns, and catastrophes in the fossil record, *Geological Society of America Special Paper*, **190**: 291–6.

Simberloff, D., 1974, Permo-Triassic extinctions: effects of area on biotic equilibrium, *Journal of Geology*, **82**: 267–74.

Stanley, S.M., 1987, *Extinction*, Scientific American Books, Inc., New York.

Stanley, S.M. and Campbell, L.D., 1981, Neogene mass extinction of western Atlantic molluscs, *Nature*, **293**: 457–9.

Stigler, S.M. and Wagner, M.J., 1988, Testing for periodicity of extinction: response, *Science*, **241**: 96–9.

Valentine, J.W. and Jablonski, D., 1983, Larval adaptations and patterns of brachiopod diversity in space and time, *Evolution*, **37**: 1052–61.

Vidal, G. and Knoll, A.H., 1982, Radiations and extinctions of plankton in the late Proterozoic and early Cambrian, *Nature*, **297**: 57–60.

Chapter 3

PRE-METAZOAN LIFE

Paul K. Strother

At the cellular level, most of the major steps in the evolution of life took place during the Precambrian. During the period of time between the beginnings of a sedimentary rock record (about 3800–3900 Myr ago) and the radiation of shelly invertebrates near the base of the Phanerozoic (about 570 Myr ago), life itself began, the genetic code became established, the basic types of prokaryotic cell structures evolved, and the composite, eukaryotic cell originated. The origin of the Metazoa is a Precambrian phenomenon as well. Of all the major groups of organisms (at the Kingdom level) which have a fossil record, only the Kingdom Plantae (and possibly the Kingdom Fungi) has an origin after the end of the Precambrian. Thus, the Cambrian 'explosion' represents not only the beginning of the evolution of the macroscopic world of animals and plants but also the culmination of an evolutionary history of the cell throughout the Precambrian.

Although the evolutionary processes which occurred during the Precambrian interval are of major significance, the discussion of events during this time is severely constrained by the vagaries of preservation which provide only a limited fossil record of cellular evolution. The taphonomic processes involved in the preservation of cells as fossils are as yet poorly understood. We are dependent upon numerous indirect lines of geochemical and mineralogical evidence for the presence of various types of organism. The recovery and description of buried organic carbon, carbon isotope data, and temporal aspects of the mineralogical history of the Earth are all necessary adjuncts to the more traditional approaches to the study of evolution through comparative morphology of fossils.

In spite of taphonomic constraints which limit the record of Precambrian fossils, three classes of morphological fossils occur during this time interval:

microorganisms preserved in three dimensions in chert; stromatolites; and organic compressions found in shales and siltstones. This last class contains both macroscopic compressions and microscopic vesicles generally known as acritarchs. The record of these three fossil types yields information about organic diversity over time (systematics) and places constraints on the depositional systems in which they are found (palaeoecology). Because of the antiquity of Precambrian sediments, even meagre fossil data may have a considerable effect on our understanding of fundamental evolutionary processes.

The serious study of life during the Precambrian is a relatively recent undertaking, becoming coherent during the 1950s and existing today as a loose collective of phycology, microbial ecology, micropalaeontology, sedimentary geology, biogeochemistry and evolutionary biology. The recognition that a long history of cellular-level evolution preceded the Cambrian explosion, has prompted the study of Recent microbial communities with the hope of gaining a better understanding of Precambrian micro-ecosystems. This interdisciplinary approach is exemplified in the study of stromatolites which result from the *interaction* between micro-organisms and their sedimentary environment.

THE ORIGINS OF LIFE ON EARTH

In 1924, A.I. Oparin (see Oparin 1957) opened up the way to serious thought about the origin of life on Earth by outlining a series of possible steps to abiogenesis (the generation of life from matter not previously living). Crucial to his arguments was the premise that the atmosphere of the early Earth was non-oxidizing, allowing for a build-up of various abiogenic organic compounds. The hypothetical synthesis of complex organic polymers was followed by their self-assembly into replicating cell-like systems which began to evolve via the standard pathways of Darwinian evolution. The belief in this general mechanism crystallised after experiments by Miller (1953) demonstrated the feasibility of the initial steps outlined in Oparin's theory.

In order for the Oparin hypothesis to work, certain conditions must hold for the early Earth. The Earth's atmosphere must have been non-oxidising; there could not have been excess free oxygen available to combine with abiogenic organic compounds. The rise of atmospheric O_2 throughout the Precambrian to its present level of 20.8% can be seen in the subsequent geologic record of detrital uranitite, banded iron formations, red beds and evaporites. The Earth's general temperature must have been between 0° and 100°C, allowing for an aqueous chemistry to exist. Life must have originated after the crust had cooled to below 100°C and probably not before the end of the meteoritic bombardment 4000 Myr ago. There is a consensus among palaeontologists that isolated shallow marine to perhaps freshwater ponds or tidal pools were a necessary ingredient for the origins of life. This may be due to the need to provide conditions which allowed polymerisation reactions involving dehydration syntheses to take place.

One of the accomplishments of geologists studying the earliest sedimentary record has been to demonstrate the existence of shallow-water, evaporitic environments on the Earth's surface during Archaean time. The recognition of iron formation and metaquartzite in the supracrustals from Isua, western Greenland, clearly indicate that the hydrological cycle existed 3800–3900 Myr ago. Sedimentary rocks from the Pilbara Block of Western Australia clearly indicate the existence of shallow-water habitats 3450 Myr ago (Groves *et al.* 1981). The evidence consists of barite pseudomorphs after gypsum, found within a primary sedimentary context of laminated muds and possible evaporites. A restricted marine embayment associated with tidal flats within a volcano-sedimentary setting is the palaeoenvironmental reconstruction preferred for the Pilbara sediments. These deposits also contain evidence for the existence of life at this time.

An understanding of the structural components common to all living systems should, in theory, point to a minimum set of conditions and materials necessary for life's origins. The requirement of elemental and energetic exchange between organisms and the 'non-living' or physical environment implies that biology, particularly at its initiation, is not a separate system from a general cosmic evolution of matter and energy. Palaeontology has the role of providing constraints on the temporal and, to some extent, environmental components of scenarios for the origin of life. While there is no fossil evidence of the morphological steps involved in the process of abiogenesis, indications of the origin of *biological activity* are preserved in the sedimentary record. The first occurrence of fossils provides a minimum age for the origin of life; however, the evidence which constitutes that first occurrence is still the subject of some debate.

THE EARLIEST RECORD OF LIFE

The Isuasphaera *problem*

When the supracrustal sequence was first discovered at Isua, western Greenland, rock samples of quartzites and argillites were examined for traces of life. Extractions of organic compounds too complex to be abiogenic were later shown to be contamination from encrusting lichens. Yeast-like microfossils were described from thin sections of Isua quartzites and designated *Isuasphaera* (Pflug 1978; Pflug and Jaeschke-Boyer 1979). These were later shown to be limonite-stained carbonate and possibly silicate grains which were embedded in the metaquartz and adjacent to quartz grain boundaries (Roedder 1981).

The significance of the claim that the *Isuasphaera* structures are biogenic is threefold: In the first place, it points to the need for critical empirical assessment of proposed microbial fossils, especially those of greatest antiquity. Further, the claim that the earliest organisms found on Earth were yeasts, a form of eukaryote which, on the basis of numerous other lines of evidence, did not evolve until at least 2000 or 3000 Myr later, was used to support pseudoscientific theories of both panspermia and creationism. Finally, the description of *Isuasphaera* led to the erection of additional pseudotaxa based on the misinterpretation of populations of minerals embedded in quartz (Robbins, 1987).

The current consensus is that there is no evidence for the existence of living organisms during Isua time 3800–3900 Myr ago. However, there is serious debate over the biogenicity of organic microspheres, stromatolites and isotopically light carbon in sediments from greenstone belts.

The Barberton Mountainland and the North Pole Dome

Organic-rich cherts from the Swartkoppie zone and related silicified volcaniclastic sediments from the Swaziland Supergroup in eastern Transvaal, South Africa, contain numerous samples of populations of organic spheres embedded in a fabric of primary laminae. Originally designated *Archaeosphaeroides barbertonensis* (Schopf and Barghoorn 1967), these microspheres are spherical vesicles composed of highly carbonised and granular organic matter. Size-frequency distributions of populations of *Archaeosphaeroides* are similar in form to those of other coccoid cyanobacteria, both living and fossil. Subsequent descriptions of microfossils from the Swaziland Supergroup revealed populations of microspheres differing in size (Knoll and Barghoorn 1977; Muir and Grant 1976) in addition to filaments (Walsh and Low 1985) (see Figure 3.1). The morphology of filaments is more complex than that of simple spherical structures, and thus their presence, in combination with the known populations of microspheres, strongly supports the claim that living organisms existed 3400 Myr ago.

In addition to the recovery of possible microorganisms, Reimer *et al.* (1979) showed that the preservation of buried organic carbon from several formations in the Swaziland sequence was comparable to that found in modern sedimentary basins. The weighted average carbon percentage of the 81 samples they examined was 0.46% by weight. This value is comparable to later buried organic carbon values which range from 0.18 to 0.76% (Reimer *et al.* 1979, Table II). Thus, at an 'ecosystem' level, the operation of the carbon cycle in respect of sedimentary processes was on a similar scale to that in later biotic systems.

Hays *et al.* (1983) compiled $\delta^{13}C_{PDB}$(‰) values for the 'Swaziland Sequence' and found an average value of −27‰ for 31 samples. Such negative $\delta^{13}C$ values are consistent with a biogenic origin for this carbon. They are comparable with values from later sequences where the biogenicity of carbon is not in question. Deines (1980) obtained a value of −25‰ in a survey of modern marine sediment values, very close to the value obtained by Hays *et al.* (1983). Biological chemical pathways are thought to bias fractionation toward the lighter isotope, generating negative $\delta^{13}C$ values as compared to the standard PDB carbon which was formed in equilibrium with seawater to produce a $\delta^{13}C$ of 0.

Dunlop *et al.* in (1978) reported the discovery of microfossils both in thin section and maceration from the Warrawoona Group at North Pole, Western Australia. The unit has an age of around 3450 Myr. In association with presumed microspheres were stromatolitic (oncolitic) cherts replacing carbonate, and barite pseudomorphs after gypsum. Primary sedimentary structures and fabrics indicating shallow-water evaporitic conditions occur in conjunction with probable stromatolites (Lowe 1980; Walter *et al.* 1980; Groves *et al.* 1981), but these stromatolites are all void of *in situ* organic-walled microfossils.

Populations of microspheres from the Warrawoona sediments have yet to be well characterised. However, several different types of filaments found in the cherts have been described (Awramik *et al.* 1983). These include non-septate sheath-like filaments (*Archaeotrichion contortum* Schopf and *Siphonophycus antiquus* Schopf) with clear affinities to Proterozoic fossils, a stellate form (*Warrawoonella radiata* Schopf), and a possible septate filament (*Primaevifilum septatum* Schopff). In addition, there is a doublet of *Archarosphaeroides pilbarensis* Schopf similar in character to *Archaeosphaeroides* from the Swaziland Supergroup of roughly equivalent age in South Africa.

In combination, the silicified evaporitic carbonates and volcaniclastics from the Warrawoona Group and the Swaziland Supergroup contain four lines of evidence which support the presence of an active ecosystem on the

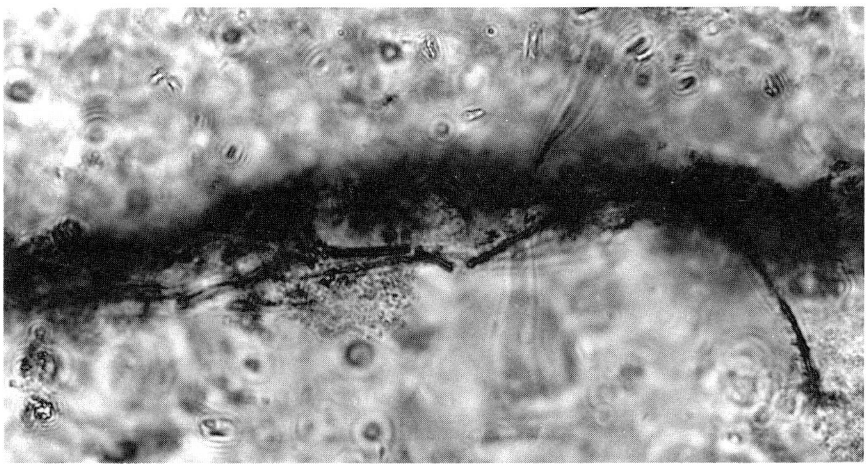

Figure 3.1. Carbonaceous filament from the Kromberg Formation, Barberton Mountainland, South Africa. The filaments are about 1.2 μm wide. They have an age of 3400 Myr (Photo courtesy of M. Walsh).

Earth 3400 Myr ago. The first is the occurrence of populations of spheroidal organic microspheres found embedded in cherts with fine-scale organic laminae. The second is the occurrence on both continents of isolated organic filaments, some of which may be septate. These structures are likely to be biogenic because the filamentous habit represents a step above simple spheres in the hierarchy of biological complexity (Awramik *et al.* 1983). The third line of evidence is that of the stromatolitic sediments found in the Warrawoona Group. Lastly, the organic geochemical signal of negative $\delta^{13}C$ values in conjunction with absolute buried carbon abundances in some greenstone sequences favours the hypothesis that life was established on the Earth's surface by 3400 Myr ago.

Taken as a whole, these lines of evidence suggest that cellular life on Earth originated during a 500 Myr window between 3900 and 3400 Myr ago. The systematic position of these earliest known morphologically preserved cells is unknown. Spherical and filamentous shapes generally require the possession of a somewhat rigid cell wall; stromatolite formation is usually due to phototrophic organisms. It is generally felt that these simple morphologies represent the remains of prokaryotes, some of which may have been cyanobacteria.

Well-preserved stromatolites in carbonates younger than 3000 Myr are indistinguishable from those found in the upper Proterozoic in association with cyanobacterial cells preserved in cherts. It seems likely that some of the stromatolites from the upper Proterozoic were built by O_2-producing cyanobacteria. In spite of the record of carbonate stromatolites, however, the record of microfossils found as structurally preserved cells in cherts is poor until the occurrence of the 2000-Myr-old Gunflint biota.

PROTEROZOIC MICROBIOTAS

The Gunflint microbiota

With the discovery of a microbial assemblage in organic-rich cherts of the 2000-Myr-old Gunflint Iron Formation (southern Ontario, Canada) in the early 1950s by Tyler and Barghoorn (see Barghoorn and Tyler, 1965), a new era in the understanding of Precambrian life began. For the first time, entire populations of a mixed assemblage of micro-organisms were found *in situ*. The enclosing cherty matrix from this deposit preserves the structural integrity of individual cells in addition to their three-dimensional position within the rock fabric. The systematic relationships of most of the microfossils from the Gunflint remain unclear; hence the evolutionary value of the assemblage is not specific to particular phylogenetic lineages. Instead, a descriptive account of the Gunflint microbiota raises many of the major questions which have faced researchers in Precambrian palaeobiology since the 1950s.

The Gunflint microbiota consists of two very distinct biofacies both of which formed as microbenthos at the sediment–water interface. The first is

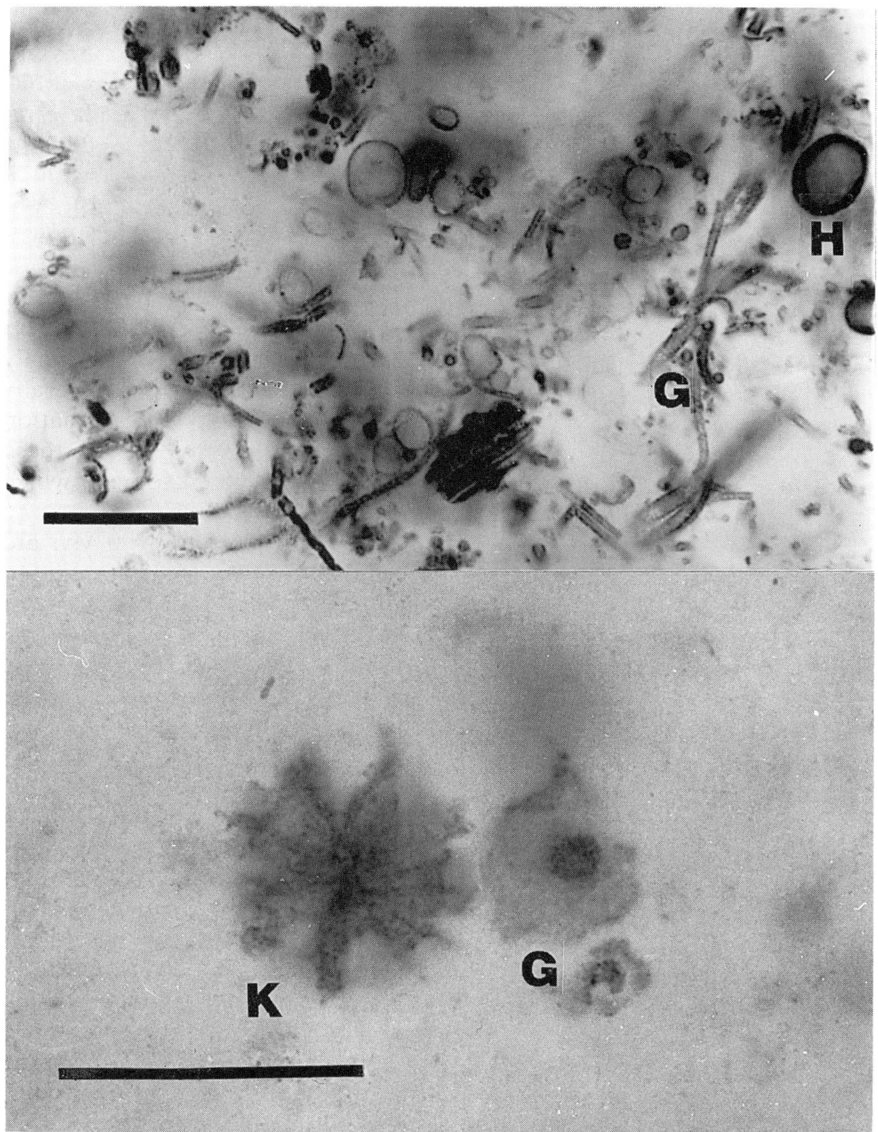

Figure 3.2. Microfossils from the Gunflint Chert. A: *Huroniospora* (H) and *Gunflintia* (G) dominated assemblage from the 'stromatolitic' biofacies. Scale bar equals 20 μm. B: *Kakabekia* (K) and *Galaxiopsis* (G) dominated non-stromatoliotic biofacies. Scale bar equals 20 μm.

a pseudostromatolitic to oncolitic cryptalgal laminite which is dominated by *Gunflintia minuta*, a non-septate tubular sheath, and by an enigmatic, spherical spore-like vesicle placed in the genus *Huroniospora* (Figure 3.2A). Although the convex-upward laminae which define the stromatolitic form are largely composed of tubular sheaths of *G. minuta*, the preserved condition of the cells and their miscellaneous orientation with respect to bedding are unlike subsequently discovered cyanobacterial stromatolites (Knoll and Golubic 1979). In addition, the microscopic appearance of the Gunflint 'stromatolitic' laminae is similar to extant geyserite deposits in Yellowstone (Walter 1976). The second biofacies is unique in consisting of a three-dimensional array of uniformly spaced cells classified as *Kakabekia umbellata* and *Galaxiopsis melanocentra* (Figure 3.2B). These three-dimensional clusters of cells occur at varying densities up to 10^7 individuals per cubic centimetre, which is comparable to that for living populations of bacteria (Knoll *et al.* 1978).

Both of these cherty iron formation biofacies include rarer species which represent planktonic cells (e.g., *Leptoteichos golubicii*), filamentous (septate) trichomes of possible cyanobacteria (*Gunflintia grandis*), and sheaths of possible oscillatoriacean cyanobacteria (*Animikiea septata*). The systematic affinities of the dominant types, however, are still a matter of some debate. If the organic laminae of the Gunflint cherts contain the remains of their mat-building microbial communities, then *G. minuta*, which dominates the assemblage, should represent the remains of cyanobacterial sheaths. However, the assemblage more closely resembles natural colonies of the sheath-forming bacteria, *Sphaerotilus* and *Leptothrix*; the latter is distinguished by the precipitation of iron and manganese oxides on its sheath.

The *Kakabekia–Galaxiopsis* biofacies contains *Kakabekia* and *Eoastrion*, both of which resemble the manganese-oxidising bacterium, *Metallogenium personatum*. The micro-organisms from this biofacies are not cyanobacteria, and it is now clear that the Gunflint microflora is atypical of most Precambrian microbiotas preserved in cherts which are dominated by cyanobacteria.

The dominant organisms of the Gunflint assemblage (*Gunflintia*, *Huroniospora* and *Kakabekia*) have also been found in cherts of similar age from the Sokoman Iron Formation from the Labrador Trough (Knoll and Simonson 1981), Minnesota (Cloud and Licari 1968), the 2000-Myr-old Duck Creek Dolomite (Knoll *et al.* 1988) and the 1700-Myr-old Frere Formation (Walter *et al.* 1976) of Western Australia. More poorly preserved examples of haematite-replaced microfossils of *Gunflintia–Huroniospora–Kakabekia*, all around 2000 Myr old, are fairly widespread.

The chert-carbonate microbiotas

Patterns of Precambrian taxonomic diversity which have a bearing on evolutionary questions can only be discerned on the basis of microbiotas from chert-carbonate facies. Fossiliferous cherts found in association with carbonate stromatolites and 'algal' laminates are dominated by the remains of cyanobacteria. On the basis of morphological comparison, some of these fossil taxa are virtually indistinguishable from modern cyanobacteria at the generic, and perhaps even specific, level. The association of cyanobacteria with mineralogical and sedimentological indicators of shallow carbonate and evaporitic environments has enabled excellent palaeoecological circumscription of the Precambrian chert-carbonate microbiotas. Thus, within an ecologically bounded chert-carbonate biofacies, elements of a phylogenetic lineage spanning 2000 Myr can be recognised.

The first monographic treatment of a chert-carbonate microbiota described a stromatolitic/algal-laminite assemblage from the Bitter Springs Formation of Southern Australia (Schopf 1968). The shallow marine to supratidal sequence of carbonates and cherts is approximately 850 Myr old. Schopf was able to make direct comparisons between the filamentous and spheroidal microfossils from the Bitter Springs cherts and species and genera of Recent cyanobacteria. This allowed the unambiguous placement of Precambrian taxa into an evolutionary systematic classification for the first time. There is no doubt that the cyanobacterial families Oscillatoriaceae and Chroococcaceae are represented in the Bitter Springs microflora. This was a clear advancement on the status of the microbial systematics of the earlier Gunflint microbiota.

Perhaps even more valuable was the impact of the discovery of the Bitter Springs microbial assemblage on palaeoecological interpretations of Precambrian microbiotas (Knoll and Golubic 1979). The association of evaporitic sediments and stromatolites with chert formation prompted the study of the ecology of Recent microbial mat-building communities thought to be analogues to those in the Precambrian. The dominance of cyanobacteria in the Bitter Springs and other chert-carbonate microbiotas enabled direct comparisons with subtidal stromatolite-building and inter- to supratidal mat-building communities. It is now possible to describe cyanobacterial communities from Proterozoic cherts in terms of subdivisions of Recent cyanobacterial communities from evaporitic and carbonate coasts and shallow platforms.

Several taxa from the Bitter Springs were originally described as eukaryotes: *Eomycetopsis* was thought to be fungal, *Gloeodiniopsis* and *Zosterosphaera* were tentatively placed in the Pyrrophyta, *Gleonbotrydion* and *Globophycus* were considered to be chlorophytes. In the late 1960s and early 1970s this was compatible with the idea that the iron formation microbiotas might be more primitive that the younger Bitter Springs type, and that eukaryotes (in the Bitter Springs chert) probably had evolved betwen 1900 and 900 Myr ago. It seemed logical that the fossil record of Precambrian cherty microbiotas would show an evolutionary pathway of increasing biological complexity.

Hofmann (1976) described an assemblage of fossilised cyanobacteria from the Belcher Island Supergroup which, although contemporaneous with the Gunflint, contained a microflora more similar to that from the much younger Bitter Springs Formation. This discovery was completely at odds with the concept of a progressive evolutionary sequence from a prokaryotic iron-formation microbiota to a mixed prokaryotic–eukaryotic chert-carbonate microbiota. It led to the realisation that the silicified microbial assemblages represent samples which are wide-ranging temporally, but narrowly defined ecologically.

The restricted nature of the benthic carbonate microbial mat-building ecology has proved particulary significant in seeking early Precambrian eukaryotic organisms. Recent microbial mats and stromatolite-building communities are essentially prokaryotic in composition. Many such communities exist today only in areas defined by the ecological exclusion of eukaryotes. Benthic microbial mats in evaporitic sequences may never have been colonised by eukaryotes, therefore their inclusion in fossil assemblages from this habitat is likely. Planktonic eukaryotes which occur in several formations from the Precambrian of Svalbard are thought to have been rafted into subtidal benthic prokaryotic mats (Knoll 1984). There were clearly ecological restrictions to the distribution of early planktonic eukaryotes. Evidence for the timing of the origin and subsequent evolution of the eukaryotes occurs in another lithofacies: normal marine organic-rich shales and siltstones.

Precambrian palynofacies

In contrast to the ecology and taphonomy of both the iron-formation and chert-carbonate microbiotas, the microfossils extracted from organic-rich shales (e.g. Vidal, 1976) yield a completely different view of Precambrian life. Although the historical approach to the study of organic remains from shales and cherts led to two separate schools of research, they are now seen to complement each other.

The modern study of Precambrian microfossils from shales began at about the same time as that of the cherty microbiotas with Timofeev (1959). Timofeev described an entire taxonomy of acritarchs, extracted from shales and siltstones using palynological preparation techniques involving acid maceration of the rock samples. This method frees intact organic structures from the enclosing mineral matrix, allowing them to be studied microscopically.

The silicification process leads to the preservation of cellular remains in cherts in three-dimensional detail. In addition, the spatial relationships between microfossils and bedding are preserved within a benthic microbial community in three dimensions. In contrast to this, acritarchs in non-silicified sediments are usually flattened, folded and crushed. Taphonomically, they have acted as sedimentary particles which have fallen through the water column accumulating along with mineral clasts to form muds and silts. Hence acritarchs are thought to represent an allochthonous biotic component, the resting or sexual cysts of planktonic algae. Since no Recent cyanobacteria are known to produce these cysts, it is generally assumed that acritarchs from Precambrian shales represent the remains of eukaryotes.

The record of compressed organic structures from Precambrian shales represents a much more ecologically diverse sample than that from the chert-carbonate biofacies. In addition, the number of individual fossiliferous sample sites within any given rock section is almost always greater when shales are analysed. This is not a surprising result; fossil preservation through silicification and chert formation is a rare geological occurrence compared to the formation of organic-rich shale. Although sampling density may be greater, however, the quality of biological information available from the study of acritarchs is often less than that from cherty microbiotas. Acritarchs represent only the encystment stages of an algal life cycle; the vegetative portions remain unknown. Morphological characteristics such as vesicle surface texture, wall thickness and colour, and perhaps even diameter, are all subject to diagenetic variation. Many Precambrian acritarchs have a simple spheroidal shape, and these sphaeromorph acritarchs exhibit few distinguishing characters with biological specificity which are useful in taxonomy.

The Precambrian acritarch record does not reveal much about evolutionary relationships among the protist and algal groups. It does, however, provide some data of evolutionary interest. The first concerns the timing of the origin of the eukaryotes. If acritarchs represent the cysts of marine algae (and no prokaryotes are known to produce such cysts), then the earliest occurrence of acritarchs with resistant organic walls should record the upper age limit for the prokaryotic–eukaryotic transition. Currently this limit stands at 1400 Myr ago, based primarily on the record of sphaeromorph acritarchs from the lower Belt Supergroup in Montana, USA (Horodyski 1980).

Although the construction of biologically meaningful taxonomies for acritarchs is difficult, there is enough morphological variation (e.g. in shape, vesicle construction, wall thickness, sculpture etc), to allow a comparison of acritarch diversity through the Upper Proterozoic. Based on a simple tabulation of acritarch taxa, Vidal and Knoll (1983) constructed a curve of late Proterozoic diversity which shows a hypothesised gradual rise from the earliest records 1400 Myr ago to about 30 taxa by 750 Myr ago (Figure 3.3). Immediately following this peak, possibly due to the effects of glaciation during the Varangerian, there is a drop in diversity to about ten taxa. This is followed by an exponential increase in taxa due to the initial radiation of the much better-documented Palaeozoic plankton.

This diversity pattern is correlated with generalised changes in taxonomic composition. The earliest acritarch assemblages contain only sphaeromorph acritarchs; *Leiosphaeridia*, *Kildinella* and *Trachysphaeridium* are typical genera of this type. These groups are very wide-ranging stratigraphically. This low-diversity assemblage gradually gives way to a higher-diversity uppermost Riphean assemblage which includes cysts with poly-

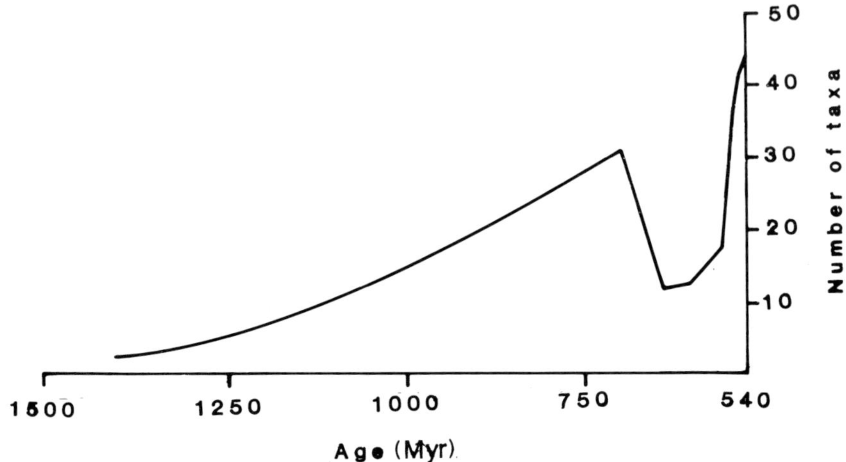

Figure 3.3. Acritarch diversity in the later Proterozoic and earliest Cambrian. (Redrawn from Vidal and Knoll 1983, used with permission.)

gonal geometries, multiple vesicles, equatorial flanges and other morphological differences. The subsequent low-diversity Varangerian assemblages (Figure 3.3) characterised by the presence of *Bavlinella faveolata*, which is thought to be a planktonic cyanobacterium. Classic acathomorphic (with spines or processes) acritarchs dominate among the distinctive morphologies which appeared during the radiation which continued from latest Precambrian into the Palaeozoic.

Many acritarch-dominated assemblages include organic-walled algal filaments of various types. Some filaments appear to have crosswalls, giving them a distinct oscillatoriacean appearance. Most, however, are featureless, leading to the conclusion that they are probably the resistant sheaths of cyanobacteria and sheath-forming bacteria. The crosswall patterns in some specimens may be impressions made by filamentous trichomes upon the inner sheath wall. It appears unlikely that such filaments were a major component of the Precambrian plankton; they may be the remains of an autochthonous benthic microbiota released during the maceration process. The recovery of entangled mats of such sheaths supports this hypothesis. Isolated individual sheaths, viewed either in parallel-to-bedding thin sections or in maceration, may be autochthonous or allochthonous. The systematics and stratigraphic occurrence of the filamentous component of Precambrian palynofacies is an important area which has yet to be addressed.

If cyanobacterial sheaths of filamentous species can survive diagenesis in clastic sediments, in addition to the acid maceration process, then sheaths of spherical cyanobacteria may be present in acritarch assemblages. Thus, a component of the Precambrian acritarchs may be of prokaryotic origin (as suggested for *Bavlinella*). This weakens the argument that the earliest acritarchs represent the first remains of eukaryotes.

Robust, ornamented, obviously eukaryotic vesicles also occur in chert-carbonate facies, for example the Hunnberg Formation of Svalbard (Knoll 1984). In these late Precambrian cherts, the distinction between eukaryotic acritarchs and spherical prokaryotic sheaths is ambiguous only for simple-walled forms; otherwise the superior three-dimensional preservation usually allows unequivocal identification. Acritarchs have thick, robust walls which are often multilayered and, in these latest Precambrian sequences, spines and processes on the cyst are common.

The occurrence of acritarchs and other organic remains in a wide variety of palaeoenvironmental settings attests to the universality of the microplanktonic life mode during the Proterozoic. This, in turn, implies a certain level of evolutionary complexity. Cyst formation may have evolved in response to changing nutrient supply, seasonal conditions, reproductive necessity, or as the outcome of the accumulation of secondary biosynthetic products. Regardless of how it evolved, however, the co-ordination of cyst formation with long-term reproductive survival represents a significant evolutionary step.

Vendotaenids and other macroscopic algae

Some of the acritarchs, such as *Chuaria*, are large enough to be clearly visible to the naked eye. There is little evidence that the larger examples represent anything other than enormous cysts of single-celled eukaryotic plankton. The occurrence of macroscopic alga-like compressions with more complex morphology in later Proterozoic shales does seem to be evidence of multicellularity. An example is *Vendotaenia*, macroscopic striated organic sheets of Vendian age described by Gnilovskaya (1971) as the probable remains of algae.

The vendotaenids are generally known from latest Precambrian shale assemblages. They usually appear as fragments of elongate, strap-shaped thalli covered by roughly parallel striations. Since the subsequent record of algal thalli in compression is so poor, there is very little material other than the Palaeozoic 'fucoids' with which to compare the vendotaenids. The vendotaenid sheet could, however, have been produced by bacteria, thus their presence in shales need not be evidence of eukaryotes.

'U-shaped' and elongate compressions occur in the Little Dal Group (1100–800 Myr old) in Canada (Hofmann and Aitken 1979). These are comparable with compressions from the Belt Supergroup which include definite frond-like organic films on bedding planes (Walter *et al.* 1976). Most are simple elongate structures, but a few palmate forms also occur. There is little doubt that such compressions are the remains of algal thalli of possible chlorophyte or phaeophyte affinity. Thus there is fossil evidence that macroscopic algae inhabited the late Proterozoic seas, but the systematic position and evolutionary history of these groups remains essentially unknown.

THE EVOLUTION OF THE CYANOBACTERIA DURING THE PROTEROZOIC

Of the organisms with a Precambrian fossil record, only cyanobacterial species reveal enough morphological diversity to allow taxonomic assignment within the current framework of evolutionary systematics. This does not diminish the evolutionary importance of problematic or '*incertae sedis*' micro-organisms from the Precambrian. To date, most systematic assignments of Precambrian taxa have been speculative; witness the earlier assignment of taxa to the fungi and higher algae which are now regarded as prokaryotic (e.g. *Eomycetopsis*). Many well-preserved specimens with complex morphology have no known homologues among Recent microbial groups. The importance of the cyanobacteria in the study of Precambrian evolution lies in their preservation potential, robust size, and relatively complex morphology which together permit their classification within an existing taxonomic framework.

The cyanobacteria are the primary builders of stromatolites and 'algal'-laminites in carbonate-rich sedimentary environments. This seems to be due to a combination of their phototropic habit, gliding motility, and the production of profuse amounts of extracellular mucopolysaccharide sheath which traps detritus. These characteristics, combined with the chemical properties of early silicification (the chert-carbonate system) during the Precambrian, seem to have favoured the preservation of cyanobacterial cells and sheaths. In addition, cyanobacteria are fairly large prokaryotes which exhibit a wide degree of morphological variation, colonial morphology and cell-division patterns. All these features are used to help establish the systematic placement of fossilised forms.

The oscillatoriacean genera from the Bitter Springs Formation were probably the first Precambrian fossils to be assigned unambiguously to the Recent cyanobacteria (Schopf 1968; Schopf and Blacic 1971). Cell size and shape, terminal cell shapes and overall habit allowed the erection of genera such as *Paleolingbya* and *Oscillatoriopsis* as 'equivalent' to the Recent *Lingbya* and *Oscillatoria*. The difference is that the Precambrian taxa are fossilised and of great antiquity; on the basis of morphology and habit alone, they are indistinguishable. This leads to the confident assignment of the fossil material at the family level.

Empty sheaths which are hollow and have smooth to granular wall-surface textures are the most common type of 'cyanobacterial' filament known from the Precambrian. In cases of exceptional preservation, the remains of trichomes (filaments of cells) which originally produced the sheaths are sometimes found. This supports the hypothesis that many of the smooth-walled 'filaments' are, indeed, the sheaths of oscillatoriacean cyanobacteria. Since such simple morphology is found among a wide group of bacteria and algae, the taxonomic treatment of smooth to granular sheaths is treated phenetically, on the basis of diameter. *Tenuofilum* refers to filamentous diameters of less than 1 μm, *Eomycetopsis* ranges from 2 to 3 μm, and *Siphonophycus* encompasses forms which are 4 μm and greater in diameter (see Knoll 1982). (*Gunflintia* overlaps with *Eomycetopsis* in size. The former, however, is always composed of straight, unorientated sheaths; whereas the latter is generally sinuous, forming orientated mats of intertwined sheaths.) Without the morphological details available from trichomes themselves, further taxonomic treatment of sheaths becomes very difficult.

Most researchers assign the Gunflint genus *Animikiea* to the Oscillatoriaceae; this then represents the oldest occurrence of that family, and the order Hormogonales. An early example of the Hormogonales is a specimen of *Halythrix* from the 1900-Myr-old Kasegalik Formation (Hofmann 1976). The recognition of *Halythrix* as oscillatorian is based on the appearance of cells preserved in the trichome. Taken together, these findings clearly indicate that the oscillatoriacean cyanobacteria have a fossil record extending back 2000 Myr.

Except for the increase in variety of cell shapes (and sizes) and the addition of multiple sheaths, the subsequent Precambrian evolution of the filamentous cyanobacteria shows very little tendency toward increasing morphological complexity. Claims for the presence of heterocysts in Precambrian cyanobacterial filaments can in every case be interpreted as the degradational remnants of normal cells, and it is remarkable (given the fine level of preservation in many Precambrian cherts) that cell shapes and patterns of distinctly Nostocacean, Rivulariacean, and Scytonematacean types of cyanobacteria have yet to be discovered (Knoll 1985).

Solitary coccoid cells are much more difficult to classify. In order to be confident of taxonomic assignment, cell-division patterns apparent in sheath morphology or in population structure must mimic patterns known to have taxonomic significance in modern taxa. *Synechococcus* cells stick together producing linked chains of two, four and eight subspherical to rod-shaped cells. The divisional patterns produced by this property enable the recognition of the taxon in the Precambrian record. Hofmann (1976), for example, erected the genus *Eosynechococcus* based on fossils from the 1900-Myr-old Belcher Island microflora. Additional species of *Eosynechococcus* have subsequently been described from the Bitter Springs Formation (850 Myr old), the Narssarssuk Formation (ca 700 Myr old), and the Draken Conglomerate (700–800 Myr old), extending its temporal range throughout the Proterozoic.

In addition to expanding the stratigraphic range of undisputed cyanobacteria, Hofmann (1976) described a taxon from the Belcher Island microbiota with clear affinities to the Recent genus *Eoentophysalis*. As with *Eosynechococcus*, this taxon was later recognised in the Bitter Springs Formation itself, and has been found in numerous microbial assemblages throughout the Proterozoic. *Eoentophysalis* is characterised by successive divisions of changing orientation in co-ordination with continual sheath production. Thus three-dimensional packets of cells enclosed in multiple sheaths are formed. This morphology is clearly recognisable in Recent and fossil examples of the genus (Golubic and Hofmann 1976).

Of all the morphological transitions in the cyanobacteria, the construction of colonial chroococcalean genera is most clearly documented in the Late Precambrian record. *Eucapsis* exhibits an extremely regular division pattern which produces packets of 2^n cells arranged orthogonally in three dimensions (Licari *et al.* 1969). This pattern is unlike similar ones found in bacteria, and the resemblance of the 1600 Myr fossil to its Recent counterpart is unquestioned. The cherts of the 700–800-Myr-old Draken Conglomerate contain a spherical colony, *Synodophycus*, which is packed with up to 150 individual or paired cells which are not arranged in any regular order. This form was considered by Knoll (1982) to be planktonic. The division of paired cells attached to stalks which become aligned along the surface of a sphere is characteristic of the extant cyanobacterial genus, *Gomphosphaeria*. Its exact morphological counterpart has been found in the *c.* 750-Myr-old Narssarssuk Formation in a form treated under the genus *Gyalosphaera*. In this instance, the pattern of cell division remains clearly preserved in the spatial relationship of paired cells in surface view.

Two other basic chroococcalean colonial forms also occur in the silicified intertidal pools of the Narssarssuk carbonates (Strother *et al.* 1983). *Gyalosphaera* is the morphological equivalent of the Recent genus *Coelosphaerium* which contains various numbers of individual cells, uniformly distributed on the outer surface of a sphere. The number of individual cells increases with increasing diameter of the spherical colony. The fossil genus *Coleogleba* is similar to species of the living *Microcystis* which is a large, somewhat irregular colony, heavily gelatinised, with cells distributed throughout the colony interior. Both of these forms are found today in freshwater plankton, indicating the probable use of colonial form as an adaptation to the planktonic life-mode.

With so few points on the time-scale it is difficult to prove progressive complexity of form in the evolutionary diversification of the Chroococcaceae through the Proterozoic. As documented so far, the record does show a progression from solitary and irregularly associated cells to colonies with highly determined cell positions (*Eucapsis*) and determinate colonial structure (*Gyalosphaera*).

An undisputed example of the genus *Pleurocapsa*, designated *Paleopleurocapsa*, occurs in the 1000–1200-Myr-old Skillogalee Dolomite of South Australia (Knoll *et al.* 1975). A pattern of three-dimensional branching, cell shape, cell size and the production of endospores unites the fossil and Recent forms. A baeocyte-producing member of the Dermocarpaceae from 700–800-Myr-old cherts from east Greenland, which is indistinguishable from an undescribed cyanobacterium from the Bahama Banks, provides additional evidence of pleurocapsalean cyanobacteria (Green *et al.*

1987). These findings firmly establish the existence of all three orders of cyanobacteria in the Precambrian: Chroococcales, Pleurocapsales, and Hormogonales. If the three cyanobacterial orders are monophyletic, then it is clear that fundamental diversification in the cyanobacteria had occurred by 1000 Myr ago.

TAPHONOMY AND PALAEOECOLOGY IN PRECAMBRIAN PALAEOBIOLOGY

There are several general properties of Precambrian fossils which make their study fundamentally different from conventional palaeontological endeavour. Temporal resolution is often poor within Precambrian sequences; it is not uncommon for the age uncertainty of a Precambrian fossiliferous horizon to be half the entire Phanerozoic. Ironically, the preservation of Precambrian fossils in chert represents some of the most exquisite known in the fossil record. Thus morphological details are often available at the cellular level, prompting debate even about subcellular detail. In an effort to maximise information from a very small number of fossiliferous occurrences, two kinds of specialised study have characterised our approach to Precambrian fossils: investigations of taphonomy and of the ecology of living microbial mat-forming communities. Both methods are based on a strong uniformitarianist approach to the interpretation of the Precambrian record.

The study of microbial mat-forming communities is still in its infancy. Gross comparisons between fossil algal laminates and the buried surface layers of Recent intertidal and subtidal microbial mats can be made (Monty 1976). But at present clear connections between original organismal composition and resultant rock fabric cannot be made. Microbial mat-forming communities represent a highly dynamic ecosystem in which chemically and taxonomically distinct phototrophic layers are continually replaced, leaving behind only degraded remnants of the principal organisms responsible for mat production.

Microbial mats usually occur today in environmentally 'stressed' sites, with high salinity, insolation or odd tidal flow. These conditions are thought to exclude many of the eukaryotes which normally graze on benthic microbial growth. Those locations where subtidal stromatolites occur (Bahama Banks, Shark Bay in Western Australia) are valued as study sites for microhabitats which have persisted throughout the entire history of the Earth. Although stromatolites peaked in diversity during the Late Precambrian, there is still no consensus on the correspondence between the taxonomic composition of a stromatolite-building community and the resultant stromatolite form. The surrounding physical conditions of deposition must be taken into account when describing stromatolite construction.

A Precambrian history of taxonomic turnover, extinction and radiation in stromatolites or associated microbiotas has yet to be revealed. Rather, the fossil record shows a long-standing reign of various cyanobacteria, with little evolutionary change, persisting in certain habitats.

The interpretation of cellular morphology in its preserved state is fundamental to studies of taxonomy, ecology and the evolution of life in the Precambrian. Palaeobiologists, therefore, have been concerned with the effects of silica preservation on cell morphology. In addition, an understanding of the morphological changes which cells undergo during degradation prior to silicification is crucial to the correct interpretation of morphology. It is clear, for example, that the degradation of trichomes proceeds much faster than that of enclosing sheaths. Empty sheaths are the most common form of preservation found among the filamentous cyanobacteria. The timing of silica emplacement during burial and diagenesis has been a matter of much debate. It is generally accepted, however, that the emplacement of silica often occurs very early in the diagenetic process, prior to sediment compaction for example; and that the cells which are preserved can range from completely undegraded to fully decomposed. The recognition of this variation in preservational quality is essential for the accurate assessment of fossil species diversity and ecology.

Cells preserved in cherts in varying stages of degradation may appear completely different in terms of shape, surface texture, and interior cross-wall formation. The silicification process is capable of arresting morphology in various states of decay. Thus, morphological variation within a series of similar fossil forms may be due simply to differences in the timing of silicification. In species of coccoid cyanobacteria, degradation is characterised by centripetal shrinkage of the cell membrane and cytoplasm away from the relatively rigid sheath. Intermediate stages result in the retention of interior 'spots' of condensed organic matter. This process has been observed in the laboratory (Knoll and Barghoorn 1977) and in buried cyanobacterial mats from the Arabian Penninsula (Golubic and Barghoorn 1977).

Condensed interior 'spots' of organic matter from cells preserved in Precambrian cherts have been interpreted as the preserved nucleii of eukaryotic cells (Schopf 1968). However, experiments have shown that simple prokaryotic cells can degrade to produce structures which superficially resemble nucleii (Francis *et al.* 1978; Knoll and Barghoorn 1977). Therefore, proof of eukaryotic affinity must be based on supracellular structure; the preservation of ultrastructural detail as a criterion for the recognition of eukaryotic affinity is an unrealistic expectation.

An understanding of the taphonomic processes which produce morphological variants is needed in order to assess assemblage diversity, and ultimately, global organic diversity during the Precambrian. A reasonable estimate of species abundance is difficult to achieve on the basis of morphology alone, without taking into account degradational and preservation-induced morphologies. Schopf (1968) originally recognized some 30 species from the Bitter Springs assemblage, and others were added later

(Schopf and Blacic 1971). Subsequent examination of the Bitter Springs cherts showed that 12 coccoid species from eight (fossil) genera were indistinguishable from the cells derived from one degraded culture of *Chroococcus* (Knoll and Barghoorn 1977). A similar situation occurs in oscillatoriacean taxa which have been split taxonomically on the basis of terminal cell shape. These cells are also subject to changes during decay.

Several of the Bitter Springs taxa originally thought to have evolutionary systematic significance have instead provided the basis for an essentially phenetic system of classification of Precambrian microbes. *Sphaerophycus parvum* refers to small (less than 5 μm in diameter), spherical cells, with or without an internal membrane. *Myxococcoides* refers to roughly spherical cells which are somewhat larger (9–18 μm in diameter). *Eomycetopsis*

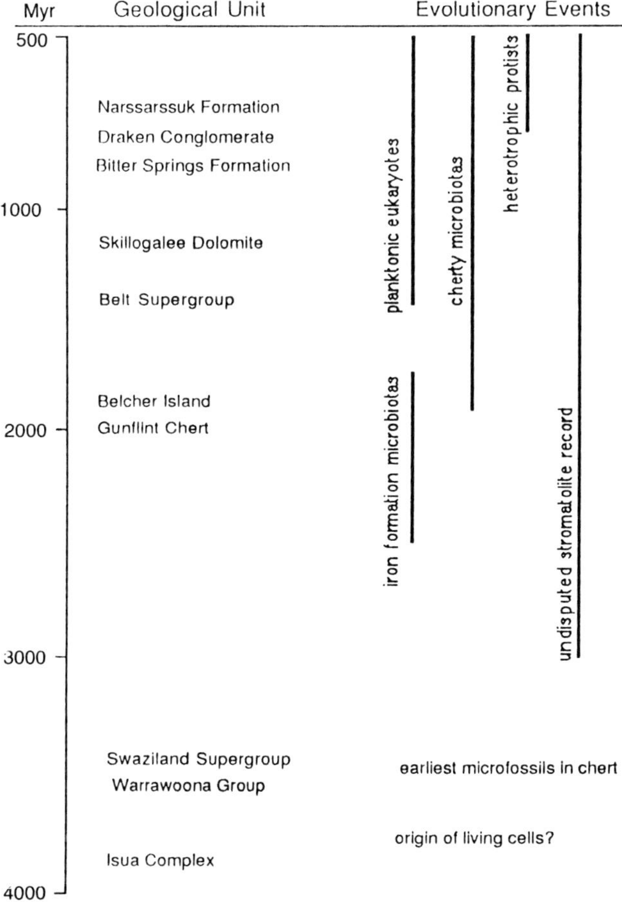

Figure 3.4. Stratigraphic chart showing the positions of fossiliferous units discussed in the text. Time is expressed in Myr, or millions or years before present. Evolutionary events include the ranges of major microbial assemblage types.

robusta is an extremely common sinuous tubular sheath of about 3 μm in diameter, originally thought to be of fungal origin. *Siphonophycus* is the generic name given to larger diameter sheaths which presumably housed trichome bundles. Taxa like these, which lack the morphological details needed for placement into a phylogenetic classification, have nevertheless been an essential basis for comparisons with subsequently described Precambrian assemblages. However, the quantitative relationship between form taxa and biological species is not well understood.

Estimates of organic diversity during the Precambrian are controlled by several mutually opposing factors. Most importantly, the facies which are capable of preserving cells of this age are rather restricted, both in stratigraphic and in palaeoenvironmental range. In all probability, many of the earlier described taxa are degradational variants of fewer biological species. On the other hand, long-ranging, morphologically simple form taxa, like *Eomycetopsis* and *Sphaerophycus*, could easily represent the sheaths of numerous different biological species. With the addition of shale facies microbiotas, a wider sampling of palaeoenvironments is achieved. Although the systematics of Precambrian palynology suffers from many of the problems of the taxonomy of cherty microbiotas, in combination they indicate that microbial life was adapted to a wide range of habitats by mid-Proterozoic time (Knoll 1985).

REFERENCES

Awramik, S.A. and Barghoorn, 1977, The Gunflint microbiota, *Precambrian Research*, **5**: 121–42.

Awramik, S.A., Schopf, J.W. and Walter, M.R., 1983, Filamentous fossil bacteria from the Archean of Western Australia, *Precambrian Research*, **20**: 357–74.

Barghoorn, E.S. and Tyler, S.A., 1965, Microorganisms from the Gunflint Chert, *Science*, **147**: 563–77.

Cloud, P.E. and Licari, G.R., 1968, Microbiotas of the banded iron formations, *Proceedings of the National Academy of Sciences*, **61**: 779–86.

Deines, P., 1980, The isotopic composition of reduced carbon. In P. Fritz and J.Ch. Fontes (eds), *Handbook of Environmental Isotope Geochemistry*, Elsevier, Amsterdam, pp. 329–406.

Dunlop, J.S.R., Muir, M.D., Milne, V.A. and Groves, D.I., 1978, A new microfossil assemblage from the Archaean of Western Australia, *Nature*, **274**: 676–8.

Francis, S., Margulis, L. and Barghoorn, E.S., 1978, On the experimental silicification of microorganisms II. On the time of appearance of eukaryotic organisms in the fossil record, *Precambrian Research*, **6**: 65–100.

Gnilovskaya, M.B., 1971, The oldest aquatic plants from the Vendian of the Russian Platform (late Precambrian), *Paleontological Journal*, **3**: 101–7 (in Russian).

Golubic, S. and Barghoorn, E.S., 1977, Interpretation of microbial fossils with special reference to the Precambrian. In E. Flugel (ed.), *Fossil Algae*, Springer-Verlag, Berlin, pp. 1–14.

Golubic, S. and Hofmann, H.J., 1976, Comparison of Holocene and mid-Precambrian Entophysalidaceae (Cyanophyta) in stromatolitic algal mats: cell division and degradation, *Journal of Paleontology*, **50** (6): 1074–82.

Green, J.W., Knoll, A.H., Golubic, S. and Swett, K., 1987, Paleobiology of distinctive benthic microfossils from the upper Proterozoic Limestone-dolomite 'Series', central east Greenland, *American Journal of Botany*, **74** (6): 928–40.

Groves, D.I., Dunlop, J.S.R. and Buick, R., 1981, An early habitat of life, *Scientific American*, **245** (4): 64–73.

Hays, J.M., Kaplan, I.R. and Wedeking, K.M., 1983, Precambrian organic geochemistry, preservation of the record. In J.W. Schopf (ed.), *Earth's Earliest Biosphere*, Princeton University Press, Princeton, NJ, pp. 93–134.

Hofmann, H.J., 1976, Precambrian microflora, Belcher Islands, Canada: significance and systematics, *Journal of Paleontology*, **50** (6): 1040–73.

Hofmann, H.J. and Aitken, J.D., 1979, Precambrian biota from the Little Dal Group, Mackenzie Mountains, northwestern Canada, *Canadian Journal of Earth Sciences*, **16** (1): 150–66.

Horodyski, R.J., 1980, Middle Proterozoic shale-facies microbiota from the lower Belt Supergroup, Little Belt Mountains, Montana, *Journal of Paleontology*, **54** (4): 649–63.

Knoll, A.H., 1982, Microfossils from the late Precambrian Draken Conglomerate, Ny Friesland, Svaldbard, *Journal of Paleontology*, **56** (3): 755–90.

Knoll, A.H., 1984, Microbiotas of the late Precambrian Hunnberg Formation, Nordauslandet, Svaldbard, *Journal of Paleontology*, **58** (1): 131–62.

Knoll, A.H., 1985, The distribution and evolution of microbial life in the late Proterozoic Era, *Annual Review of Microbiology*, **39**: 391–417.

Knoll, A.H. and Barghoorn, E.S., 1977, Archean microfossils showing cell division from the Swaziland system of South Africa, *Science*, **198**: 396–8.

Knoll, A.H., Barghoorn, E.S. and Awramik, S.A., 1978, New microorganisms from the Aphebian Gunflint Iron Formation, Ontario, *Journal of Paleontology*, **52** (5): 976–92.

Knoll, A.H., Barghoorn, E.S. and Golubic, S., 1975, *Paleopleurocapsa wopfnerii gen. et sp. nov.*: A late Precambrian alga and its modern counterpart, *Proceedings of the National Academy of Sciences*, **72** (7): 2488–92.

Knoll, A.H. and Golubic, S., 1979, Anatomy and taphonomy of a Precambrian stromatolite, *Precambrian Research*, **10**: 115–51.

Knoll, A.H. and Simonson, B., 1981, Early Proterozoic microfossils and penecontemporaneous quartz cementation in the Sokoman Iron Formation, Canada, *Science*, **211**: 478–80.

Knoll, A.H. and Strother, P.K., Rossi, S., 1988, Distribution and diagenesis of microfossils from the lower Proterozoic Duck Creek Dolomite, Western Australia, *Precambrian Research*, **38**: 257–79.

Lowe, D.R., 1980, Stromatolites 3,400 Myr old from the Archean of Western Australia, *Nature*, **284**: 441–2.

Licari, G.R., Cloud, P.E. and Smith, W.D., 1969, A new chroococcaean alga from the Proterozoic of Queensland, *Proceedings of the National Academy of Sciences*, **62** (1): 56–62.

Miller, S.L., 1953, A production of amino acids under possible primitive earth conditions, *Science*, **117**: 258–9.

Monty, C.L.V., 1976, The origin and development of cryptalgal fabrics. In M.R. Walter (ed.), *Stromatolites*, Elsevier, Amsterdam, pp. 193–249.

Muir, M.D. and Grant, P.R., 1976, Micropaleontological evidence from the Onverwacht Group, South Africa. In B.F. Windley (ed.), *The Early History of the Earth*, John Wiley, London, pp. 595–604.

Oparin, A.I., 1957, *The Origin of Life on the Earth*, Oliver and Boyd, Edinburgh.

Pflug, H.D., 1978, Yeast-like microfossils detected in the oldest sediments of the earth, *Naturwissenschaften*, **65**: 611–15.

Pflug, H.D. and Jaeschke-Boyer, H., 1979, Combined structural and chemical analysis of 3,800-Myr-old microfossils, *Nature*, **280**: 483–6.

Reimer, T.O., Barghoorn, E.S. and Margulis, L. 1979, Primary productivity in an early Archean microbial ecosystem, *Precambrian Research*, **9**: 93–104.

Roedder, E., 1981, Are the 3,800-Myr-old Isua objects microfossils, limonite-stained fluid inclusions, or neither?, *Nature*, **239**: 459–62.

Robbins, E.I., 1987, *Appelella ferrifera*, a possible new iron-coated microfossil in the Isua iron-formation, southwestern Greenland. In P.W.U. Appel and G.L. LaBerge (eds), *Precambrian Iron-formations*, Theophrastus Publications, Athens, pp. 141–54.

Schopf, J.W., 1968, Microflora of the Bitter Springs Formation, Late Precambrian, central Australia, *Journal of Paleontology*, **42** (3): 651–88.

Schopf, J.W. and Barghoorn, E.S., 1967, Alga-like fossils from the early Precambrian of South Africa, *Science*, **156**: 508–12.

Schopf, J.W. and Blacic, J.M., 1971, New microorganisms from the Bitter Springs Formation (late Precambrian) of the north-central Amadeus Basin, Australia, *Journal of Paleontology*, **45** (6): 925–60.

Strother, P.K., Knoll, A.H. and Barghoorn, E.S., 1983, Micro-organisms from the late Precambrian Narssarssuk Formation, north-west Greenland, *Palaeontology*, **26** (1): 1–32.

Timofeev, B.V., 1959, *The Ancient Flora of the Baltic region and its stratigraphic significance*, Trudi VNIGRI, 129, Lenningrad (in Russian).

Vidal, G., 1976, Late Precambrian microfossils from the Visingso Beds in southern Sweden, *Fossil and Strata*, 9.

Vidal, G., and Knoll, A.H., 1983, Proterozoic plankton. In L.G. Medaris, C.W. Byers, D.M. Mickelson and W.C. Shanks (eds), *Proterozoic Geology: Selected Papers from an International Proterozoic Symposium*, Geological Society of America, Memoir 161, pp. 265–77.

Walsh, M.M. and Lowe, D.R., 1985, Filamentous microfossils from the 3,500-Myr-old Onverwacht Group, Barberton Mountain Land, South Africa, *Nature*, **314**: 530–2.

Walter, M.R., 1976, Geyserites of Yellowstone National Park: an example of abiogenic 'stromatolites'. In M.R. Walter (ed.), *Stromatolites*, Elsevier, Amsterdam, pp. 87–112.

Walter, M.R., Buick, R. and Dunlop, J.S.R., 1980, Stromatolites 3,400–3,500 Myr old from the North Pole area, Western Australia, *Nature*, **284**: 443–5.

Walter, M.R., Goode, A.D.T. and Hall, W.D.M., 1976, Microfossils from a newly discovered Precambrian stromatolitic iron formation in Western Australia, *Nature*, **261**: 221–3.

Walter, M.R., Oehler, J.H. and Oehler, D.Z., 1976, Megascopic algae 1300 million years old from the Belt Supergroup, Montana: a reinterpretation of Walcott's Helminthoidichnites, *Journal of Paleontology*, **50** (5): 872–81.

Chapter 4

THE ORIGINS AND RADIATION OF THE EARLY METAZOA

Mark A.S. McMenamin

THE CAMBRIAN EXPLOSION

Animals have existed for only about 15–20% of the history of life, and their origin has posed one of the most enduring problems in palaeontology. When Charles Darwin was writing *The Origin of Species* (1859), the oldest fossils known to him included members of a number of living phyla: arthropods, molluscs and brachiopods. These fossils posed a perplexing question: where were the ancestral forms that had given rise to the earliest animals? The sudden appearance or 'explosion' of animal fossils at the base of the Cambrian and the absence of fossils in older rocks was, by Darwin's admission, a major stumbling block to his theory of natural selection. He was forced to conclude that there was an enormous gap in the fossil record immediately below the Cambrian.

This distinction between the fossiliferous sediments of Cambrian and younger age and the apparently lifeless strata below them resulted in the cardinal division of the geologic time-scale, the Precambrian–Cambrian boundary. Since the first recognition of this boundary, a number of stratigraphic sections have been located throughout the world in which Precambrian sediments grade continuously upward into the Cambrian, so that Darwin's postulated major world-wide gap does not exist. Precambrian rocks are not entirely devoid of animal fossils; over the past 40 years a number of areas have produced Precambrian body fossils and trace fossils. Precambrian animal or animal-like fossils, and trace fossils formed by the peristaltic movement of arthropod and annelid-like organisms, are now relatively well known. These fossils constitute an assemblage of soft-bodied organisms known as the Ediacaran fauna. The rarity of these

Precambrian body fossils compared to body fossils of Cambrian and later faunas is at least partly explained by the unlikelihood of preserving soft-bodied organisms in most marine depositional environments.

Discoveries of multicellular animals with systems of internal organs (metazoans) in Precambrian sediments do not, however, muffle the Cambrian explosion. A large number of major metazoan groups first appear at or near the base of the Cambrian, strongly suggesting that a major diversification occurred at this time.

In addition to familiar groups such as trilobites and brachiopods, many strange types of animal are known from Cambrian faunas. The bizarre echinoderm *Helicoplacus* (Figure 4.1) has a spindle-shaped, plated skeleton which, at least in external form, is completely unlike that of any living echinoderm (Durham and Caster 1963). It is a good example of an early 'experimental' echinoderm because of its unusual body plan and because its stratigraphic range is limited to a very short interval of the Early Cambrian. Short-lived higher taxa, many with unusual morphologies, seem to be characteristic of the Cambrian. Some Cambrian forms, such as *Anomalocaris* (Figure 4.2; known from Lower and Middle Cambrian strata), are so unusual as to defy assignment to any known phylum.

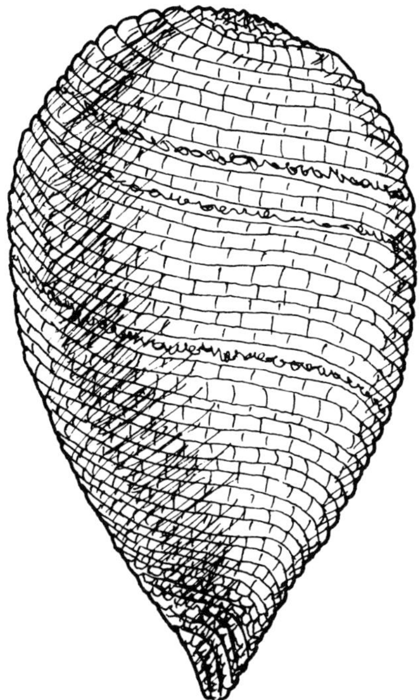

Figure 4.1. An Early Cambrian helicoplacoid, a bizarre Early echinoderm. Helicoplacoids are known only from the Cordillera of western North America. The spindle-shaped test was able to expand and contract. Length 5 cm.

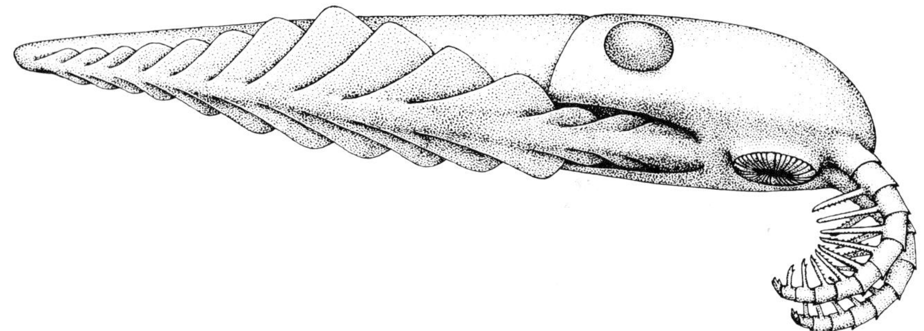

Figure 4.2. *Anomalocaris*, thought to have been a top carnivore during the Early and Middle Cambrian. *Anomalocaris* does not belong to any known phylum. Note the large eyes, the circular mouth, and the prey-grasping anterior appendages. Length 45 cm. (From 'The Emergence of Animals' by M.A.S. McMenamin. Copyright © 1987 by Scientific American, Inc. All rights reserved.)

The study of these strange groups is making it clear that modern phyla do not fully cover the spectrum of body plans possible for animals Bengtson 1986). Modern phyla are merely the descendants of the lineages that happened to survive the millions of years that have elapsed since their first appearances. Less than half of the animal phyla which inhabitated the Cambrian seas have living representatives. As many as 100 animal phyla existed during the Cambrian; the modern fauna, by contrast, consists of only 25–35 phyla (Pearse *et al.* 1987).

DATING THE PRECAMBRIAN–CAMBRIAN TRANSITION

The dating of geologic and biotic events during the Precambrian–Cambrian transition has advanced greatly in recent years, even though reliable radiometric dates for this interval are scarce. At present, the best control for the relative age dating of these sediments is the first and last appearance of various fossil groups. The four main thresholds of the transition interval are: the appearance of the Ediacaran fauna and the first shelly fossils; the appearance of phosphatic shelly fossils and sudden increase in the numbers and types of trace fossils; moderate diversity shelly faunas and the first archaeocyathans; and the appearance of high-diversity shelly faunas and the first trilobite shelly fossils.

These thresholds have been used and misused as biostratigraphic boundaries. On a relatively local scale, such as within a single basin or along a contiguous continental margin, these thresholds have important biostratigraphic utility, and indeed they form the basis for local biostratigraphic zonation. For example, the Tommotian Stage of the Siberian Platform is recognised by the first appearance of archaeocyathans (the third threshold in the scheme presented above). The Tommotian Stage has in the past, however, been applied in sections outside Siberia as a globally synchronous stage that begins with the initial appearance of shelly fossils. This is an inappropriate usage for three reasons. First, shelly fossils are known from pre-Tommotian strata in Siberia, so even the type Tommotian Stage is not defined by the first appearance of shelly fossils. Second, there are no fossils known to occur both within and outside Siberia that are restricted to the Tommotian Stage. Third, confident stage-level biostratigraphic correlations cannot yet be made between the Tommotian strata of Siberia and Lower Cambrian sediments elsewhere.

The absolute age of the beginning of the Cambrian is a contentious topic. A commonly accepted estimate is 570 Myr, based on radiometric age determinations, for the base of the moderate-diversity shelly fossil fauna. There is, however, an 80 Myr spread between the least and greatest radiometric ages for this boundary! Recent estimates seem to indicate an absolute age for the Precambrian–Cambrian boundary that is closer to 540 Myr than to 570 Myr (Landing *et al.* 1988). A recent concordant uranium-lead date on euhedral zircons recovered from a tuff bed slightly upsection from the Mistaken Point Ediacaran biota locality in eastern Newfoundland provides an age of 565 ± 3 Myr (A. Benus *in* Narbonne 1987a). The first shelly fossils in the Newfoundland section occur hundreds of metres above the date-yielding tuff bed, suggesting that the age of the base of the Cambrian is substantially less than 570 Myr.

THE EDIACARAN FAUNA AND EARLY SHELLY FOSSILS

The first appearance of the soft-bodied Ediacaran fauna occurs below typical Cambrian shelly metazoan communities and always above late Precambrian tillites. These tillites record several drawn-out episodes of late Precambrian glaciation throughout the world, and the only known evidence of animals below the tillites is a few simple trace fossils. Not all experts agree that these pre-tillite traces are indeed ancient burrows; convincingly biological examples are rare and these structures may be pseudofossils.

The Ediacaran biota consists of frond- and disc-shaped soft-bodied fossils, plus associated animal tracks and burrows (Glaessner 1985). In most stratigraphic sections where they occur, the Ediacaran fossils are found below the appearance of skeletons. *Pteridinium* (Figure 4.3) is a widespread, typical member of the Ediacaran fauna. Its body consists of elongate tubes that are joined together in a manner reminiscent of the partitions in an air mattress (Seilacher 1984). The body architecture in this and other Ediacaran fossils is so unusual that Seilacher (1984; 1985) suggested that the members of the Ediacaran fauna are unrelated to Cambrian and subsequent animals. Seilacher (1985) proposed the term 'Vendozoa' as a systematic term of high (phylum- or possibly even kingdom-level) taxonomic rank to classify the Ediacaran organisms.

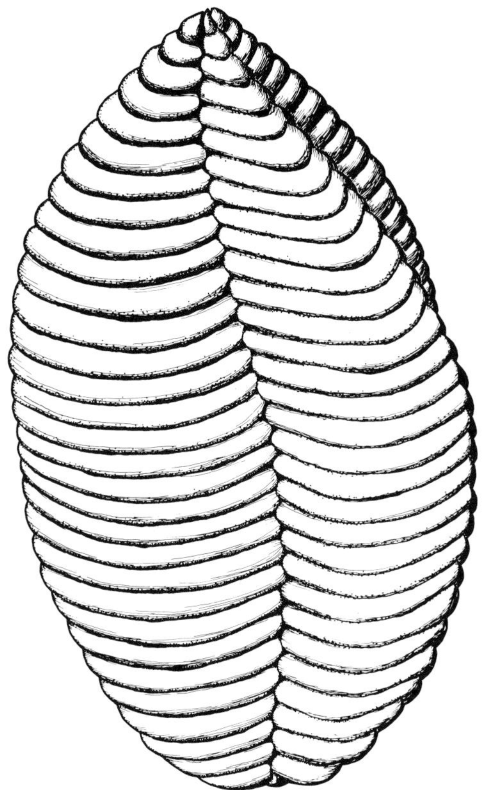

Figure 4.3. *Pteridinium*, a frond-like fossil from the Ediacaran fauna. The organism consisted of three vanes extending from a common medial axis. This reconstruction is based on specimens found in the Carolina Slate Belt in the United States. Length 7 cm.

Although Seilacher's taxonomy is disputed by some experts, it does emphasise the strangeness of the Ediacaran fauna.

A handful of mineralised fossils occur below diverse Cambrian communities as contemporaries of the Ediacaran fauna. An enigmatic tubular fossil, *Cloudina*, known with certainty only from Namibia, has a shell composed of calcium carbonate. *Cloudina* occurs in the Nama Group, a group of formations which also contains typical Ediacaran soft-bodied organisms such as *Pteridinium*. A similar calcareous tubular fossil, *Sinotubulites* (Figure 4.4), occurs in late Precambrian and earliest Cambrian strata in China, Mexico and the western United States.

Organisms with *phosphatic* hard parts do not appear in the fossil record until well after the first appearances of the Ediacaran fauna and Precambrian tubular fossils. The earliest phosphatic fossils are all a few millimetres or less in length. They form low-diversity faunas (less than about

Figure 4.4. *Sinotubulites*, a shelly fossil known from latest Precambrian and earliest Cambrian sediments. This specimen, now silicified but presumably originally calcareous, is from the earliest Cambrian of northwestern Sonora, Mexico. Length 12 mm.

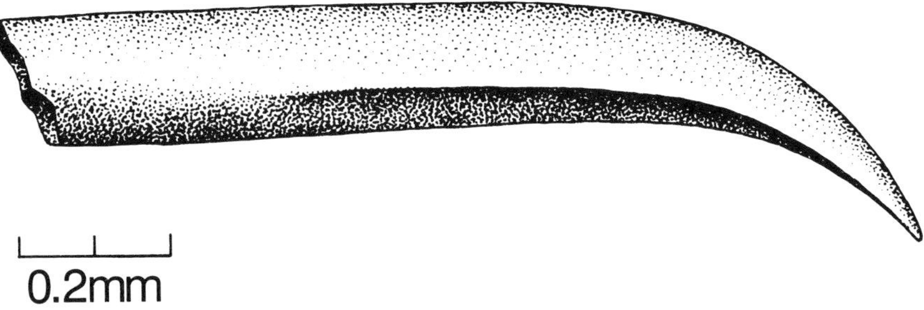

Figure 4.5. *Protohertzina*, a phosphatic shelly fossil from sediments which span the Precambrian–Cambrian boundary. *Protohertzina* is thought to have been the grasping spine of an early, possibly chaetognath-like predator. (From 'The Emergence of Animals' by M.A.S. McMenamin. Copyright © 1987 by Scientific American, Inc. All rights reserved.)

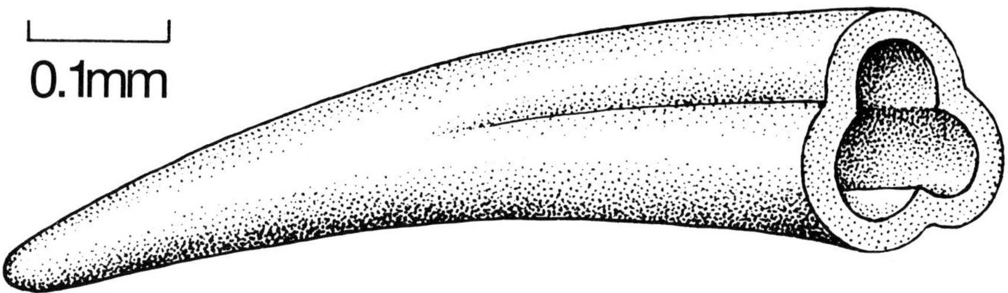

Figure 4.6. *Anabarites*, an Early Cambrian organism with a calcareous, possibly aragonitic shell. The shell has a distinctive trilobed cross-section, and originally housed a tubular animal with triradiate symmetry. (From 'The Emergence of Animals by M.A.S. McMenamin. Copyright © 1987 by Scientific American, Inc. all rights reserved.)

five species) and include the tiny tusk-shaped fossil *Protohertzina* (Figure 4.5). *Protohertzina* has strong microstructural similarities to the grasping spines of modern chaetognaths or arrow worms (Szaniawski 1982). Chaetognaths are tiny but voracious marine predators, so *Protohertzina* is the first fossil which can be confidently identified as a predatory metazoan.

Calcium carbonate shells also occur alongside the first phosphatic shells. *Anabarites* is a tubular fossil with a distinctive three-lobed transverse section (Figure 4.6). Its original shell mineralogy is not known with certainty, but it may have been composed of aragonite, a calcium carbonate mineral. Original shell mineralogy can be difficult to ascertain in these early shelly fossils, since, during sediment lithification, calcium phosphate can replace calcium carbonate shell matter with such fidelity that the microstructural details are preserved.

TRACE FOSSILS

At about the same time as the appearance of the first phosphatic shelly faunas (although the exact stratigraphic position of this point is unclear at some localities), there was a tremendous increase in the abundance and diversity of trace fossils. Trace fossils became more abundant and diverse, and underwent what has been called 'modernisation' (Landing *et al*. 1988). Deep vertical burrows such as *Skolithos* appeared in appreciable numbers for the first time. *Phycodes pedum* (Figure 4.7), a probing trace with significant vertical relief, is used to delineate the base of the Cambrian in some sections. The Ediacaran fauna, with a few possible exceptions, apparently died out at or below this level. Some palaeontologists, however, attribute the loss of Ediacaran-style soft-bodied fossils to the increased activities of burrowing metazoans, whose scavenging behaviour

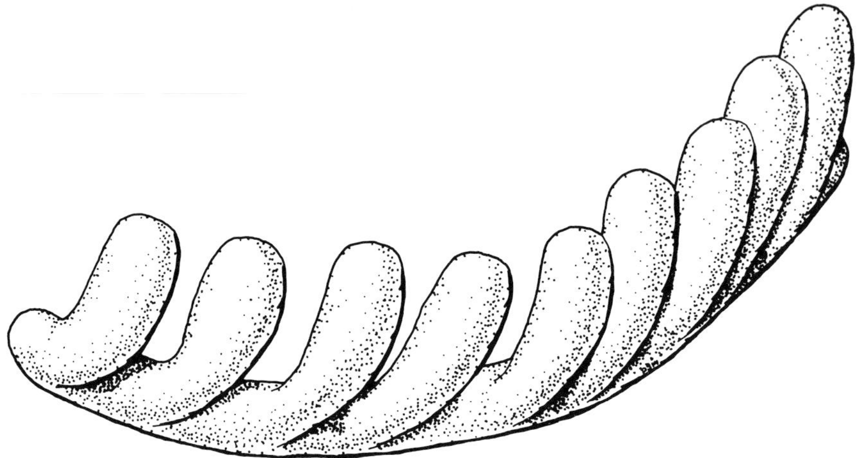

Figure 4.7. *Phycodes pedum*, a trace fossil made by a probing deposit feeder during the Early Cambrian. Each lobe of the trace fossil represents an upward excursion into the sediment of a burrowing animal searching for food. Complex feeding traces such as this one, with a significant vertical component, are unknown before the Cambrian. Total length of trace approximately 8 cm.

would have decreased the preservation potential of the soft-bodied corpses. Fedonkin (1987) argued that microbial decomposition in sediments may have become more pervasive towards the end of the Precambrian, further decreasing the likelihood of preserving soft-bodied impressions.

In most areas of the world, the first moderate-diversity shelly faunas (consisting of five to about fifteen species) occur immediately above the earliest phosphatic shelly fossils. In the Tommotian strata of the Siberian Platform, the moderate-diversity faunas are contemporaneous with the appearance of the earliest archaeocyathans.

Archaeocyathans (Figure 4.8) are cup- or vase-shaped fossils with double-walled, porous skeletons composed of calcium carbonate. Somewhat resembling corals and sponges, they bear no close relationship to any living group and are currently placed in their own phylum. Archaeocyathans and calcareous algae, which first became abundant during the Precambrian–Cambrian transition, formed wave-resistant reefs in the Cambrian by growing together in mounded accumulations of calcareous skeletal material known as bioherms.

Higher-diversity shelly faunas (often more than 15 skeletalised species) followed the moderate-diversity faunas. The appearance of these faunas defines the stratigraphic interval known as the Atdabanian Stage on the Siberian Platform and the Qiongzhusian Stage in South China, two areas where high-diversity shelly faunas have been well studied. The first trilobite carapaces appeared at this time, and archaeocyathans greatly expanded

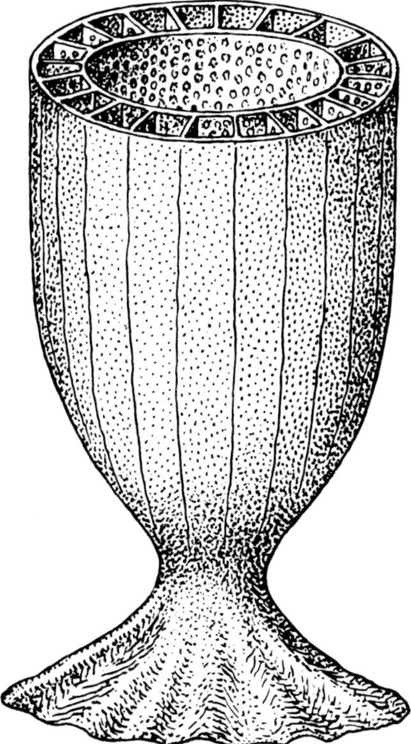

Figure 4.8. An archaeocyathan from the Early Cambrian. The biological affinities of archaeocyathans are unknown; their calcitic skeletons are not composed of spicules as are the tests of sponges. Archaeocyathans, together with early calcareous algae, formed the first wave-resistant reefs in the Palaeozoic. The majority range from 10 mm to 25 mm in diameter. (From 'The Emergence of Animals' by M.A.S. McMenamin. Copyright © 1987 by Scientific American, Inc. All rights reserved.)

their distribution, reaching many areas outside the Siberian Platform. At this stratigraphic position, relatively confident stage-level correlations can be made between Cambrian stratigraphic sections on different continents.

CORRELATION AND PALAEOGEOGRAPHY

Locating the base of the Cambrian was easy for nineteenth-century geologists since there was a hiatus between Precambrian and Cambrian rocks in most of the areas they studied. Our expanded data base has actually complicated the task of precisely locating the Precambrian–Cambrian boundary because many relatively continuous boundary sections are now known. Some stratigraphers prefer to place the boundary between the first phosphatic shelly fossils and the first moderate-diversity shelly faunas. Others would place it at the first appearance of trilobites, that is to say, at the beginning of high-diversity shelly faunas. A more recent view advocates placing the boundary at the appearance of distinctive and complex trace fossils such as *Phycodes pedum* (Narbonne *et al.* 1987b). Geographically separated Precambrian–Cambrian stratigraphic sections have proved difficult to correlate because of the limited number of fossil species available, and because many of the earliest shelly fossils have long stratigraphic ranges—and are thus not very useful for biostratigraphic subdivision of stratal sequences. For example, it is unclear whether or not the first moderate-diversity shelly faunas of South China are the same age as those of the Siberian Platform, because most of the species known from the moderate-diversity stage are also found higher up in the stratigraphic section with the high-diversity shelly faunas. This type of uncertainty hinders long-range correlation of the Precambrian–Cambrian boundary.

Many of the earliest animal (or animal-like) fossils have wide geographic distributions. Frond-shaped (e.g. *Charniodiscus* and *Pteridinium*) and/or discoid members (e.g. *Cyclomedusa*) of the Ediacaran fauna occur in nearly every region where fossils of this biota are known. Certain distinctive members of the Ediacaran fauna, such as *Dickinsonia*, *Tribrachidium*, and *Spriggina*, have a more restricted distribution, and are known only from the Baltic platform and continents (such as Australia and Africa) which once formed the ancient supercontinent Gondwana. Because of the similarities in their Ediacaran fossils, Gondwana and the Baltic Platform may have been closer together during the late Precambrian than during any subsequent time (McMenamin 1982; Donovan 1987).

The shelly fossils which first appear during the low-diversity interval are surprisingly widespread. For example, *Anabarites* is known from Siberia, China, Mongolia, Kazakhstan, India, Australia, western and eastern North America, and Iran.

The major reason for the wide distribution of some elements of the Ediacaran fauna and early shelly fossils is the equatorial position of most continents during the Late Precambrian and Early Cambrian (Cloud and Glaessner 1982). Equatorial currents must have facilitated the dispersal of these organisms between adjacent continents. Many small shelly fossils remained cosmopolitan even as faunas became moderately diverse, suggesting that there were no major barriers to dispersal at the time. Evidence has been presented for continental rifting around the edges of North America close to the Precambrian–Cambrian boundary (Bond *et al.* 1985). Continental dispersion, perhaps involving the break-up of a Precambrian supercontinent, seems to have occurred during the Precambrian–Cambrian transition. The proximity of newly separated continents during the early stages of continental drift, plus the equatorial positions of many continents, allowed early animals to spread widely. Wide oceans and latitudinal temperature gradients were not present to impede the migration of Precambrian–Cambrian transition faunas between continental shelves.

Provinciality began to develop throughout the world after archaeocyathans and trilobites were established world-wide. The growth of progresively larger oceans between continents throughout the Cambrian undoubtedly promoted provinciality. Archaeocyathans and calcareous algae contributed to the formation at this time of carbonate banks along the margins of several continents. Carbonate banks form by the accumulation of calcium carbonate shells and, since they restrict access to the open sea, they can allow animals living in the lagoonal area between the bank and the shore-line to evolve in isolation from related animals on other continents. This effect is particularly evident in archaeocyathans (Rozanov 1980) and trilobites (Palmer 1973).

AUTOTROPHY VERSUS HETEROTROPHY

As noted above, the body shapes of many of the soft-bodied Ediacaran organisms, and by inference their biology, are very unlike those of any living organisms (Seilacher 1984). The maximum thickness of *Dickinsonia* (assigned to the annelid worms by some palaeontologists) has been estimated to have been no more than 6 mm, although its pancake-shaped body can reach more than 1 m in diameter (Runnegar 1982). An unusual food-gathering strategy may explain the unusual shapes of *Dickinsonia* and other flattened members of the Ediacaran biota. The flat and thin (or in some cases sac-shaped) body plans characteristic of Ediacaran organisms would have maximised their surface area to volume ratio and been well suited for chemo- or photoautotrophic feeding.

Photoautotrophy in animals involves a symbiotic relationship between photosynthetic algae (or their chloroplasts) and animals. The photosymbiont algae receive protection from the host's tissues and release nutrients to and remove waste products from the host. For such photosymbiosis to work, substantial parts of the host's body must be exposed to sunlight so that the algae can photosynthesise. Many modern reef corals and a few tropical bivalve molluscs have photosymbiotic algae in their tissues. Chemoautotrophy, the direct uptake of energy-supplying nutrients from seawater, can also involve internal symbiosis (the harbouring of chemosynthetic bacteria is a common strategy for modern deep-sea hydrothermal vent animals), but some animals seem to absorb dissolved nutrients directly without the aid of bacteria. The high surface area of the flattened or sac-shaped bodies of many members of the Ediacaran fauna would have allowed efficient uptake of seawater nutrients, or alternatively, greater light absorption for photosynthesising endosymbiotic algae.

The bioenergetics of the host–photosymbiont association are favoured by low-nutrient waters (Hallock and Schlager 1988). Members of the

Ediacaran fauna may have been well adapted to marine conditions prevailing during the late Precambrian, which are thought to have been oligotrophic (nutrient-deprived) compared to the Precambrian–Cambrian boundary interval which followed immediately afterwards (Cook and Shergold 1984; McMenamin 1986).

The ingestion of other organisms to obtain nutrients (heterotrophy) became increasingly important near the end of the Precambrian, as evidenced by the fossil record of stromatolites. These dome- or column-shaped organosedimentary structures built by filamentous algae reached their peak in diversity and abundance at about 900 Myr ago. After this time, stromatolites underwent a marked decline in diversity which has been attributed to the appearance of algae-eating metazoans. Microbial sheaths known from microfossiliferous cherts also became more robust at this time, possibly in response to metazoan grazing activity (Awramik *et al.* 1985). Trace fossils of scavenging and deposit feeding animals also appear about 700 Myr ago, and these early trace-makers are probably ancestral to the Cambrian shelly organisms which followed. Stromatolites exist today, but most modern examples are restricted to environments where grazing animals are rare or absent.

The Cambrian evolutionary radiation initiated the first complex heterotrophic metazoan communities, as well as rapidly establishing new phyla and classes. Much attention has been focused on the origin of higher taxa near the Precambrian–Cambrian transition, but of even greater importance for the marine biota were the ecological changes that occurred during this interval. Earlier assumptions that predators were unimportant in Cambrian communities have been challenged by new evidence. Skeletons probably first appeared in a number of higher animal taxa as a direct response to increased predation pressure (Hutchison 1961), and the appearance of predators may have resulted in the demise of the Ediacaran fauna. Not all the new skeletons were formed by biochemical secretion of calcium carbonate and phosphate. The Early Cambrian worm tube *Onuphionella* (Figure 4.9) formed its shell by gluing together flat mica flakes. The flakes were attached together in an imbricate fashion like the shingles on a slate roof, and allowed the tube to flex without kinking (Signor and McMenamin 1988). The enigmatic Cambrian fossil *Xenusion* (Figure 4.10) may be a relict member of the Ediacaran fauna which developed spines (of unknown composition) to protect itself from the onslaught of Cambrian predators (McMenamin 1986).

Three lines of evidence suggest that predation pressure was intense during the Early Cambrian: specimens of damaged and sometimes healed prey; fossils of predators; and anti-predatory adaptations. Many small shelly fossils have boreholes in them which resemble the holes made by the activity of Recent shell-boring predators. Numerous trilobite specimens are known with bites taken out of the carapace. The wounds to these trilobites have healed, indicating that the scar does not represent damage

Figure 4.9. *Onuphionella durhami*, an agglutinated Early Cambrian worm tube. The tube is composed of imbricate mica flakes 0.5 mm in greatest length. The imbrication of the mica flakes may have given the tube more flexibility than rounder or less well-orientated grains. Total length of tube 9.8 cm.

to a shed moult. Healed bites such as these represent unsuccessful attacks—successful attacks would generally have left no trace!

In addition to the grasping spines of *Protohertzina* discussed earlier, there is another convincing example of an Early Cambrian predator. *Anomalocaris* (Figure 4.2) has been reconstructed on the basis of nearly complete specimens from the Middle Cambrian Burgess Shale, as a gigantic (by Cambrian standards) and bizarre predatory metazoan (Whittington and Briggs 1985). Damage inflicted on trilobites from the Lower and Middle Cambrian of North America has been attributed to the activities of *Anomalocaris*. It does not closely resemble any known living animal and is probably an example of a short-lived, experimental phylum.

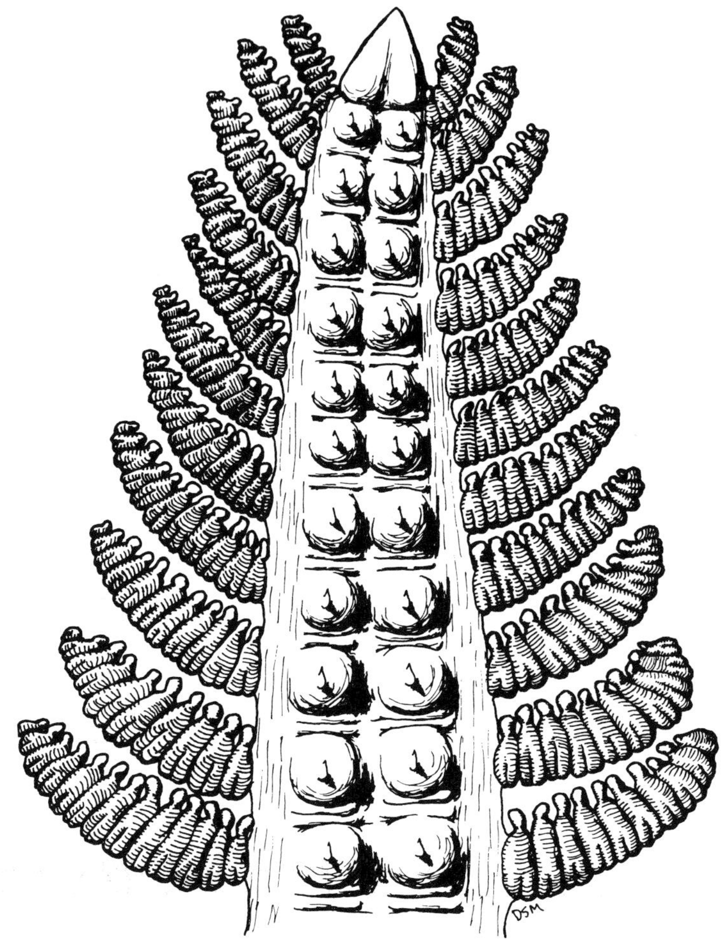

Figure 4.10. *Xenusion auerswaldae* is an Early Cambrian soft-body fossil which may be a relict of the Ediacaran fauna. This species, known from only three isolated specimens in Germany, bears a close resemblance to frond fossils from the Mistaken Point (Vendian) soft-bodied fossil locality in southeastern Newfoundland. Spines project from the top of each of the paired humps along the midline of the frond. Length of frond approximately 8 cm.

Many features of Early Cambrian animals can be interpreted as antipredatory adaptations. Deep burrows provided protection from predators which could not move through sediment. Spines on trilobites such as *Laudonia* (Figure 4.11) made them less susceptible to attack by contemporaneous predators such as *Anomalocaris*. Some Early Cambrian brachiopods and helicoplacoids also sported spines, and the 'helen' appendages on some Cambrian hyolithids projected outward from the shell in a spike-

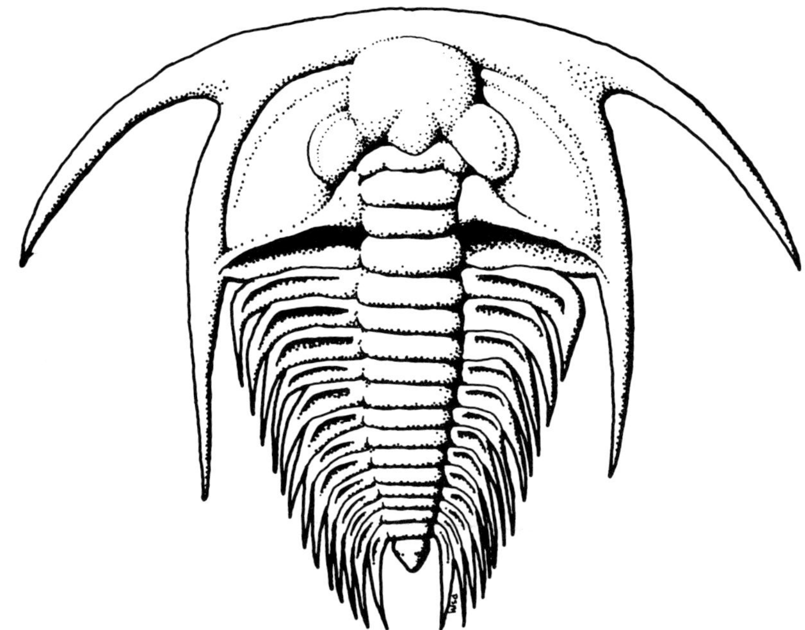

Figure 4.11. *Laudonia*, an Early Cambrian trilobite with pronounced genal and metagenal spines projecting from the cephalon. This genus is an index fossil for the base of the *Bonnia–Olenellus* Zone in western Northern America. Greatest dimension approximately 4 cm.

like fashion. It has been convincingly shown that many Early Cambrian small shelly fossils are actually sclerites, disarticulated parts of an originally multi-element spiny mail-coat. This armour probably protected the upper surface of slow-crawling metazoans (Figure 4.12) so that in life, these animals were heavily protected by spines.

Spines are easily formed by many shell-bearing lineages, and are a relatively unsophisticated defence against predators. More advanced defences are seen in other Cambrian animals that apparently utilised chemotoxins to discourage predation. Early Cambrian mickwitziid brachiopods have numerous punctae running through their shelly walls. These punctae may have delivered chemical deterrents to the shell surface, discouraging predatory attacks and parasitic borrers (McMenamin 1986; 1987).

We now know that predators and evasive prey are characteristic of Cambrian marine environments, and that most major types of animal first appear at this time. This is in stark contrast to the complete lack of evidence for predators and predation for the relatively low-diversity Ediacaran fauna. Hard parts necessary for protection were co-opted for other functions such as support and feeding, and allowed the evolution of

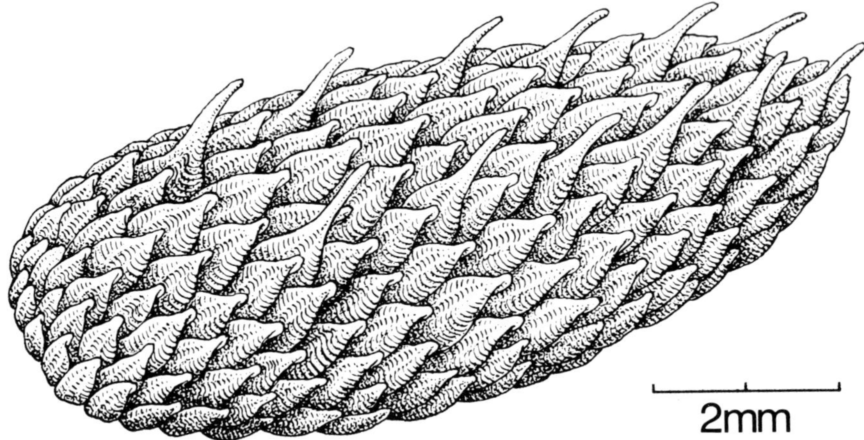

Figure 4.12. *Lapworthella*, a sclerite-bearing Early Cambrian animal, reconstructed with the sclerites forming an interlocking armour over the dorsal surface. (From 'The Emergence of Animals' by M.A.S. McMenamin. Copyright © 1987 by Scientific American, Inc. All rights reserved.)

strikingly new body plans which were dependent on skeletons in order to function.

Brachiopods, for example, depend on their skeleton for feeding. They are lophophorates, and, like phoronids (a tube-dwelling lophophorate phylum), brachiopods filter feed with the aid of a ciliated feeding structure called a lophophore. However, despite their similarity, the filter feeding systems of phoronids and brachiopods are not comparable in their functional details. In a phoronid, the feeding lophophore is extended into the water column, whereas in a brachiopod the lophophore is encased in a bivalved shell within which the brachiopod directs a precise laminar flow. This laminar stream flows through the protected lophophore, thus greatly increasing filter feeding efficiency. Brachiopods cannot operate their efficient internal filter feeding currents without the aid of a bivalved shell.

MARINE NUTRIENTS AND TIMING OF THE CAMBRIAN EXPLOSION

The new information about Cambrian predators and prey helps explain the palaeoecology of Cambrian communities. Why the Cambrian revolution happened when it did, and not tens or hundreds of millions of years earlier, remains a question. This is especially puzzling since active, peristaltically burrowing animals predate the Cambrian boundary by as much as 150 Myr (Awramik *et al.* 1985).

Ocean chemistry perturbations have been invoked as a trigger for the appearance of skeletised animals. The marine concentrations of phosphate, as well as carbon, sulphur and strontium isotopes, underwent dramatic but poorly understood fluctuations during the Precambrian–Cambrian transition (Brasier 1986; Lambert *et al.* 1987). One promising line of investigation concerns economic phosphate deposits that are abundant in Precambrian–Cambrian sediments in many parts of the world. It has been suggested that these deposits represent an episode of global phosphogenesis (Cook and Shergold 1984), when, for reasons of favourable palaeoceanographic conditions, abundant shallow-marine nutrients allowed the evolution of early phosphatic skeletons. Since calcareous shells are probably more prevalent than phosphatic ones at the Precambrian–Cambrian boundary, it may be more accurate to view this episode of phosphate deposition as a palaeoceanographic eutrophication event that discouraged autotrophic animals and increased the number of heterotrophic feeders. Thus the connection of eutrophication to skeletalisation may have been less direct than envisioned by Cook and Shergold (1984). These ocean chemistry changes may have favoured heterotrophic animals to such an extent that macrophagous predators evolved where none had been present before. The appearance of large top carnivores or keystone predators put an end to an early, low-competition episode of animal history, and protective skeletons can be thought of as a byproduct of this event. The late Precambrian interval of limited ecological interaction between animals, characterised by simple food webs and low biotic competition, was brought to an end by Cambrian predators.

This low-competition pattern of ecological interaction, which has been called the 'Garden of Ediacara' (McMenamin 1986), seems to be characterised by a marked physical two-dimensionality. All benthic marine life of the time, from the microbial mat-forming monerans to large, flattened or bag-like Ediacaran creatures, were restricted to a vertically narrow zone at or near the marine sediment–water interface. Although some of the frond fossils may have projected into the water column, at least one species of the elongate frond-fossil *Pteridinium* seems to have lain prostrate on the sea-floor in life. The reasons for this two-dimensionality of the Precambrian biota are not yet clear (perhaps it had to do with maximisation of light-gathering ability), but it seems certain that the modernisation of the burrowing fauna and the appearance of deep burrowers added a previously unavailable, third dimension to marine habitats. Diversity is strongly correlated to the dimensionality of a given environment (Briand and Cohen 1987), and the new ecological dimension opened up by the deep burrowing niche must have contributed significantly to the 'explosion' in animal diversity at the beginning of the Cambrian (McMenamin 1988).

This increase in environmental dimensionality was accompanied by two related changes in the marine biosphere. The first was the nearly complete destruction of the autotrophic, moneran/microbial biocoenose that had dominated the sea-floor, and which had formed stromatolites and other microbial mat-stabilised sediments for hundreds of millions of years (Awramik *et al.* 1985). This destruction was partly accomplished by the burrowing and overgrazing activities of the newly emergent animals, but was also a direct consequence of the rise of shell-bearing organisms. Moneran mats, thought to have been responsible for most Precambrian stromatolites, produce essentially fine-grained stromatolites as they trap and bind sediment. Monerans have difficulty trapping coarse bioclastic sediment (Awramik and Riding 1988), and the change in the texture of sea-floor sediment when shells became abundant must have had a detrimental affect on the ability of monerans to form microbial mats and to stabilise marine sediments. Nutrients that during the Precambrian had been sequestered beneath moneran mats were suddenly free to be stirred back into the water column, and this further fuelled the marine eutrophication that led to the establishment of heterotrophic communities.

The second factor related to environmental dimensionality involves the structure and abilities of the animals themselves. The first fossil evidence for advanced visual systems comes from Early Cambrian trilobites, and it is reasonable to expect that many (but not necessarily all) of the predators on these trilobites had visual acuity and a level of encephalisation (brain mass in excess of the minimum required to control a body of a given size; Russell 1981) that exceeded that of the trilobites themselves. Once predator–prey relationships were established during the Cambrian in a heterotrophic ecosystem, an escalating system of positive feedback between predators and prey was initiated which resulted in (among many other things) the evolution of life forms with increasingly complex behavioural capabilities.

Interestingly, Russell (1981) suggested that there is a positive correlation between the rate of encephalisation increase in marine environments during the Phanerozoic and the amount of nutrients in ocean waters. This is in accordance with evidence presented above suggesting that high marine nutrient flux at the Precambrian–Cambrian transition led to the establishment of the first metazoan-dominated heterotrophic communities. The presence of higher trophic levels, as well as increased dimensional complexity of the environment, are cited by Russell (1981) as factors contributing to the evolution of animals with large brains and complex behaviour.

A WHIMPER OR A BANG?

The earliest Cambrian animal communities brought with them innovations (shells, predators with relatively large brains, deep burrowers) that rapidly spread throughout the world's oceans and changed for ever the ecological and sedimentological characteristics of the sea-floor. Burrow-forming metazoans resembling those living today are present but not very conspicuous during the Precambrian, but during the Precambrian–Cambrian transition, their descendants quickly radiated to establish diverse heterotrophic communities with keystone predators. New ecological niches and new body plans emerged in concert with a new kind of community, one characterised by food webs more complex than those present during the Precambrian.

The Cambrian radiation of animals is often referred to as an 'explosion', a term which implies that the appearance of these animals was a geologically sudden event. Not all palaeontologists agree with this characterisation of the Cambrian evolutionary radiation, and some prefer to see this radiation as marking the moment in geologic time when animals first appear as shelly *fossils*, not necessarily when these types of animal first evolved. Therefore, there are two ways of viewing the Cambrian explosion: either it was the nearly simultaneous evolution of a number of animal phyla (the 'bang' hypothesis); or Cambrian animal phyla had long Precambrian histories that were not recorded as fossils (the 'whimper' hypothesis). In other words, did the Cambrian commence with a whimper or a bang?

The whimper hypothesis has one major point in its favour. There are animal trace fossils occurring well before the Cambrian boundary, and arthropods (Jenkins 1988) and perhaps even echinoderms (Gehling 1987) are known from Vendian strata. One line of argument, however, acts as a fatal blow to the whimper hypothesis. With only four or five exceptions, all of the well-skeletalised animal phyla living today first appear as fossils in the Cambrian. Furthermore, many phyla that are now extinct first appeared in the Cambrian. As many as 100 phyla may have existed during Cambrian, and no more than five of these phyla show any evidence of a Precambrian ancestry. Thus, the genesis of most animal phyla must date back to the Cambrian and not before then.

In the late 1960s, Valentine (1969) realised that large numbers of phyla burst upon the scene during the Cambrian radiation, and that virtually no new phyla appeared during the many subsequent evolutionary radiations of animals. This effect also held true for other animal higher taxa (classes and orders). Valentine (1969) noted not only that the Cambrian contained most of the originations of animal phyla, but also that it had a disproportionate share of class-level originations. For example, in Phylum Mollusca, gastropods, bivalves, monoplacophorans, and rostroconches *all* first appear in the Early Cambrian.

COMPETING HYPOTHESES

Two principal hypotheses have been offered over the years to explain the emergence of all these higher taxa during the Cambrian radiation: the conventional ecological hypothesis and the genomic hypothesis.

According to the conventional ecological hypothesis, major evolutionary innovations could occur during the Cambrian because marine diversity was low and there existed a limited level of competition. Such an ecological setting offered greater opportunity for the establishment of innovations that might not have survived in a more crowded ocean.

The genomic hypothesis holds that the genetic programs of animals were more easily changed early in metazoan history (Valentine and Erwin 1987). As the generations passed, significant changes in the animals' genetic codes became increasingly unlikely. Particularly successful sequences of development became 'locked in' and less forgiving of major restructuring. This 'locking in' is often called canalisation, by analogy with a stream cutting its channel so deeply that it becomes increasingly difficult to switch the stream flow to a new direction. The genomic hypothesis implies that major mutational change in animals could only occur when genes were young and unsophisticated, so I like to call it the 'green genes' hypothesis.

Both the conventional ecological and the 'green genes' hypotheses attempt to explain why the Cambrian explosion could not happen again. After the severe Permo-Triassic mass extinction about 200 Myr ago, as many as 96% of all marine animal species became extinct. Erwin *et al.* (1987) calculated that after this mass extinction, animal lower taxa were reduced to Vendian levels of diversity. In spite of this, there was no repeat of the Cambrian explosion. No new phyla appeared during the Mesozoic.

The 'green genes' hypothesis predicts a lack of new Mesozoic phyla because animal genetic codes were neither as flexible nor as able to undergo radical change as they had been in the Cambrian. The conventional ecological hypothesis, on the other hand, favoured by Erwin *et al.* (1987), explains that no new phyla appeared after the Permo-Triassic mass extinction because nearly all of the Palaeozoic phyla survived. The Mesozoic marine world was already 'seeded' with representatives of all the basic types of animal, which quickly radiated to repopulate the extinction-ravaged marine world of the early Mesozoic.

The 'green genes' and conventional ecological hypotheses both have serious drawbacks that render them inadequate to explain the Cambrian explosion. The main problem with the conventional ecological hypothesis concerns early land animals (see Chapter 6). Animal life on land probably dates back to the Ordovician (Retallack and Feakes 1987). Before this time, terrestrial environments were presumably empty of animals, and offered a very open and uncrowded ecological setting. However, no new animal phyla appeared on land, during a time early in animal history when animal genes might be expected to retain much of their Early Palaeozoic 'green genes' flexibility, and despite the fact that relatively few animal phyla were ever able to flourish in niches on land or in freshwater environments. This failure to replay the Cambrian explosion, under extremely uncrowded ecological conditions and without previous 'seeding' of the land environment with animal taxa, is the downfall of the conventional ecological hypothesis.

The land environment is more physically challenging for animals than the marine environment (See Chapter 6), and this might be argued as the reason why no animal phyla emerged during land colonisation. Temperature fluctuations are greater, and animals require more structural support and adaptations to prevent dehydration. However, these difficulties are offset by the fact that terrestrial environments offer more in the way of primary food production. Land plants, even the earliest ones, had access to more energy than their aquatic predecessors because sunlight is more intense in air than under water. Further, the animals that did solve the problems of life at the bottom of the air (as opposed to bottom of the sea) were able to diversify with great success. The physical rigours of the terrestrial environment should not have prevented the evolution of new phyla on land. Under the conventional ecological hypothesis, in fact, one would have *expected* the appearance of completely new types of animal to cope with land's physiological challenges and to fill vacant niches.

The 'green genes' hypothesis also fails because of the animal trace fossils known from strata deposited perhaps as long ago as 50–150 Myr before the Cambrian boundary. The genomic hypothesis is at a loss to explain why Vendian metazoans, with the greenest of genes, failed to undergo a major radiation millions of years earlier than they did.

An alternative hypothesis, called the Garden of Ediacara hypothesis or the 'unconventional ecological hypothesis', avoids the problems encountered by both the conventional ecological and 'green genes' hypotheses (McMenamin, 1986; 1988; McMenamin and McMenamin, 1989). This new hypothesis shares some aspects of the ecological hypothesis. For instance, an uncrowded Cambrian environment plays a role in both, but the crucial difference lies in how each hypothesis characterises the duration of the underpopulated interval in ocean history. In the conventional hypothesis, marine environments waited passively for 3000 Myr before animals came to fill up the vacant habitats or niches. This scenario might be plausible if the Cambrian explosion was close (i.e. within a few million years) to the origin of multi-cellular organisms, but we know that it was not.

According to the Garden of Ediacara hypothesis, the Proterozoic oceans were never truly ecologically empty; the Precambrian world was *not* vacant and passively waiting for animals to appear, but was a vital world filled to the brim with successful organisms, most of which happened to participate in very short food chains. This ecosystem was both destroyed and replaced by a multi-trophic-level system dominated by heterotrophic animals. With the exception of a few burrowing and grazing heterotrophs, animals present before the Cambrian seem to have lived by Proterozoic ecological 'rules'; hence the body plans of Ediacaran animals shaped for autotrophic lifestyles appear unfamiliar.

THE FALL FROM THE GARDEN OF EDIACARA

The Garden of Ediacara was overthrown at the Cambrian boundary by heterotrophic communities in a rather short period of time. The speedy appearance of abundant shelly fossils (Landing *et al.* 1988) suggests that the turnover took only a few million years, rapidly and radically changing the basic ecology of the marine biosphere. Immediately after the breakdown of the Ediacaran ecosystem, animal diversity was still relatively low but began to build up rapidly immediately thereafter—this is what is called the Cambrian explosion. It is important to note that low taxonomic diversity does not imply low biotic competition, as required by the conventional ecological hypothesis for the explanation of the origin of new animal forms.

An important feature of the Garden of Ediacara hypothesis is that it can explain the unique taxonomic characteristics of the Cambrian explosion, or why so many higher animal taxa appeared at that time in comparison to radiations before and after. Valentine and Erwin (1987) noted that there is not nearly enough geological time available for Cambrian phyla to appear by the gradual, slow accumulation of relatively minor evolutionary changes. Whatever led the phyla to appear caused them to appear quickly.

McMenamin (1988) suggested that the predators which came as part of the Cambrian heterotrophic revolution played a crucial role in creation of Cambrian higher taxa such as new phyla. Many of the animals that were able to defend themselves against predators (by burrowing deeply, forming shells, and other strategies) *incidentally* found themselves in a position to capitalise on food resources that had been weakly exploited or ignored by animals during the Precambrian. Shells, for instance, greatly aided both filter feeders (e.g. brachiopods, eocrinoids) and deposit feeders (e.g. trilobites, which used their shells as a sediment-processing factory). Deep burrowers excavated food resources unavailable before (perhaps trapped below a moneran carpet).

The explosive potential for behavioural innovation inherent in animals began to be realised during the Cambrian. Cambrian predators forced other animals to make the jump to successful new lifestyles, changes which might never have been made in a carnivore-free ocean. Probably the only jumps that were successful were those that were made quickly, while ecological conditions on the sea-floor were in a state of turmoil.

The Garden of Ediacara hypothesis also explains why no new animal phyla appeared with the animal invasion of land. Land colonisation by animals occurred some time 50–100 Myr after the Cambrian radiation, and the first successful colonisers brought with them a multitrophic level ecosystem inherited from the Cambrian explosion. Early land animal fossil communities are rich with arthropod predators; so rich that in one well-preserved early terrestrial fauna, all but one of the identifiable components of the fauna are predators (Shear *et al.* 1984). Opportunities for major evolutionary innovation and the evolution of new animal phyla were present on land before colonisation, but these opportunities were decisively cut off when the Cambrian ecosystem was transposed onto land and freshwater environments. Most importantly, this migration of an ecosystem was accomplished with only a very few phyla (initially perhaps only two: arthropods and annelid worms). According to the tenets of the conventional ecological hypothesis, the presence of these two phyla should not have prevented evolutionary innovation of animals in land environments. After all, arthropods and worms were presumably around during the Vendian, but they did not halt the Cambrian explosion.

Perhaps if land colonisation by animals had occurred earlier, say before the Cambrian explosion rather than during the Late Ordovician or Silurian, there might have been a better chance of evolving distinctive new body plans and radically new types of living on land. The body plans of land animals are relatively minor modifications of body plans that evolved in the sea during the Cambrian explosion. The bodies of the land fauna remain well suited for marine life; indeed, many terrestrial vertebrates (cetaceans, several reptiles, penguins) have returned to the sea over geologic time. The simplest way to explain the lack of new phyla on land is that early land faunas were initiated with a predator-rich heterotrophic ecosystem, which eliminated any possibility of a truly new ('new' in the sense of unconventional body plans and food-gathering capabilities) fauna evolving on land. This would have been the case regardless of which animal phyla first colonised land after the Cambrian explosion. Whereas marine Cambrian predators encouraged the origin of new phyla by selecting for novel body plans during an interval of ecological confusion, Ordovician and later terrestrial predators prevented the evolution of new body plans by perpetuating and extending an established, multi-level trophic hierarchy as land colonisation proceeded.

The Cambrian explosion is a geologically abrupt event of great biotic significance, and represents the collapse of an old marine ecosystem dominated by autotrophs and filter feeders and its replacement by a more competitive ecology characterised by multi-level trophic pyramids. The apparent suddenness of the event (at most a few million years) is real, and new feeding types and anatomical designs of animals rapidly evolved during this interval. The old order vanished or was driven to marginal environmental situations, and a new order was quickly established. Many of the groups which have been most successful in Phanerozoic oceans achieved their dominance as a consequence of good fortune during a time of profound ecological change. Skeletalisation proved to be a successful strategy for coping with the ecological changes occurring during the transition, but was by no means the only way. Vendian anatomies and behaviours that were able to capitalise on or fruitfully to coexist with emergent heterotrophic feeding styles and changes in the sea-floor substrate rapidly proved to be the most successful. The characteristics of oceanic ecosystems have not changed much since, and the younger land ecosystems are also much the same.

The arguments presented above imply that most animal phyla came into being during a brief episode of ecological chaos, during the transition from the Garden of Ediacara to the modern-style biosphere ecological system. Opportunities appear to have opened up during this time of ecological turmoil which remained unavailable during less chaotic times. Evolution appears to have operated under conditions of 'chaos with feedback' (Joseph Ford *in* Gleick 1987, p. 314) at least during the Vendian–Cambrian boundary; in fact, the biotic feedback processes that result in escalation (Vermeij 1987) may operate fully *only* during times of ecological chaos. The barriers to innovation, such as exist in the structural checks and balances of an ordered ecosystem, are temporarily opened during the formative stages of a new ecosystem. After the new system is established, the doors to major innovation may be closed until the next episode of traumatic ecological disruption. Feedback such as predator–prey escalation occurred after the Cambrian explosion (Vermeij 1987), but was comparatively weak and ineffectual at bringing about major new innovations. The greatest innovations can only occur during the most pronounced episodes of ecological change. Metazoans surely originated well before the Cambrian boundary, but their potential for evolutionary radiation and their ecological impact were not fully realised until the Precambrian–Cambrian boundary.

REFERENCES

Awramik, S.M., McMenamin, D.S., Yin, C. Zhao, Z. and Zhang, S., 1985, Prokaryotic and eukaryotic microfossils from a Proterozoic/Phanerozoic transition in China, *Nature*, **315**: 655–8.

Awramik, S.M. and Riding, R., 1988, Role of algal eukaryotes in a subtidal columnar stromatolite formation, *Proceedings of the National Academy of Sciences (USA)*, **85**: 1327–9.

Bengtson, S., 1986, Introduction: The problem of problematica. In A. Hoffman and M.H. Nitecki (eds), *Problematic Fossil Taxa*, Oxford University Press, New York, pp. 3–11.

Bond, G.C., Christie-Blick, N., Kominz, M.A. and Devlin, W.J., 1985, An Early Cambrian rift to post-rift transition in the Cordillera of North America, *Nature*, **315**: 742–6.

Brasier, M., 1986, Why do lower plants and animals biomineralize?, *Paleobiology*, **12**: 241–50.

Briand, F. and Cohen, J.E., 1987, Environmental correlates of food chain length, *Science*, **238**: 956–60.

Cloud, P. and Glaessner, M.F., 1982, The Ediacarian Period and System: Metazoa inherit the Earth, *Science*, **217**: 783–92.

Cook, P.J. and Shergold, J.H., 1984, Phosphorus, phosphorites and skeletal evolution at the Precambrian–Cambrian boundary, *Nature*, **308**: 231–6.

Darwin, C., 1859, *On the origin of species by means of natural selection*, John Murray, London.

Donovan, S., 1987, The fit of the continents of the late Precambrian, *Nature*, **327**: 139–41.

Durham, J.W. and Caster, K.E., 1963, Helicoplacoidea: a new class of echinoderms, *Science*, **140**: 820–2.

Erwin, D.H., Valentine, J.W. and Sepkoski, J.J., Jr, 1987, A comparative study of diversification events; the early Paleozoic versus the Mesozoic, *Evolution*, **41**: 1177–86.

Fedonkin, M.A., 1987, The non-skeletan fauna of the Vendian and its place in the evolution of metazoans, *Akademiya Nauk SSSR, Trudy Paleontologicheskogo Instituta, Tom*, **226**. (In Russian.)

Gehling, J.G., 1987, Earliest known echinoderm—a new Ediacaran fossil from the Pound Subgroup of South Australia, *Alcheringa*, **11**: 337–45.
Glaessner, M.F., 1985, *The dawn of animal life, a biohistorical study*, Cambridge University Press, Cambridge.
Gleick, J., 1987, *Chaos: Making a New Science*, Viking Penguin Inc., New York.
Hallock, P. and Schlager, W., 1988, Nutrient excess and the demise of coral reefs and carbonate platforms, *Palaios*, **1**: 389–98.
Hutchison, G.E., 1961, The biologist poses some problems, *American Association for the Advancement of Science Publications*, **67**: 85–94.
Jenkins, R.J.F., 1988, Functional and ecological aspects of Ediacaran assemblages. In B. Klein-Helmuth and D. Savold (eds), *Abstracts of Meetings of the 154th National Meeting, Boston, American Association for the Advancement of Science Publication*, **87–30**: 14.
LaBarbera, M., 1981, Water flow patterns in and around three species of articulate brachiopods, *Journal of Experimental Marine Biology and Ecology*, **55**: 185–206.
Lambert, I.B., Walter, M.R., Zang, W., Lu, S. and Ma, G., 1987, Palaeoenvironment and carbon isotope stratigraphy of Upper Proterozoic carbonates of the Yangtze Platform, *Nature*, **325**: 140–2.
Landing, E., Narbonne, G.M. and Myrow, P. (eds), 1988, *Trace Fossils, Small Shelly Fossils and the Precambrian–Cambrian Boundary*, New York State Museum Bulletin 463.
McMenamin, M.A.S., 1982, A case for two late Proterozoic–earliest Cambrian faunal province loci, *Geology*, **10**: 290–2.
McMenamin, M.A.S., 1986, The Garden of Ediacara, *Palaios*, **1**: 178–82.
McMenamin, M.A.S., 1987, The emergence of animals, *Scientific American*, **255**: 94–102.
McMenamin, M.A.S., 1988, Paleoecological feedback and the Vendian–Cambrian transition, *Trends in Ecology and Evolution*, **3**: 205–8.
McMenamin, M.A.S. and McMenamin, D.L.S., 1989, *The Emergence of Animals*, Columbia University Press, New York.
Narbonne, G.M., 1987a, Trace fossils, small shelly fossils and the Precambrian–Cambrian boundary, *Episodes*, **10**: 339–40.
Narbonne, G.M., Myrow, P.M., Landing, E. and Anderson, M.M., 1987b, A candidate stratotype for the Precambrian–Cambrian boundary, Fortune Head, Burin Peninsula, southeastern Newfoundland, *Canadian Journal of Earth Sciences*, **24**: 1277–93.
Palmer, A.R., 1973, Cambrian trilobites. In A. Hallam (ed.), *Atlas of Palaeobiogeography*, Elsevier, Amsterdam, pp. 3–11.
Pearse, V., Pearse, J., Buchsbaum, M. and Buchsbaum, R., 1987, *Living invertebrates*, Blackwell, Palo Alto, CA.
Retallack, G.J. and Feakes, C.R., 1987, Trace fossil evidence for Late Ordovician animals on land. *Science*, **235**: 61–3.
Rozanov, A. Yu., 1980, Tsentry proiskhozhdeniya kembriiskikh faun. In B.S. Sokolov and A.I. Zhamoida (eds), *Mashdunarodnyi Geologicheskii Kongress*, Izdatel'stvo Nauka, Moscow, pp. 30–4.
Runnegar, B., 1982, Oxygen requirements, biology and phylogenetic significance of the late Precambrian worm *Dickinsonia*, and the evolution of the burrowing habit, *Alcheringa*, **6**: 223–39.
Seilacher, A., 1984, Late Precambrian and Early Cambrian Metazoa: preservational or real extinctions? In H.D. Holland, and A.F. Trendall (eds), *Patterns of Change in Earth Evolution*, Springer-Verlag, Berlin, pp. 159–68.

Seilacher, A., 1985, Discussion of Precambrian metazoans, *Philosophical Transactions of the Royal Society of London,* **B311**: 47–8.

Shear, W.A., Bonamo, P.M., Grierson, J.D., Rolfe, W.D.I., Smith, E.L. and Norton, R.A., 1984, Early land animals in North America: evidence from Devonian age arthropods from Gilboa, New York, *Science*, **224**: 492–4.

Signor, P.W. and McMenamin, M.A.S., 1988, The Early Cambrian worm tube *Onuphionella* from California and Nevada, *Journal of Paleontology*, **62**: 233–40.

Szaniawski, H., 1982, Chaetognath grasping spines recognized among Cambrian protoconodonts, *Journal of Paleontology*, **56**: 806–10.

Valentine, J.W., 1969, Patterns of taxonomic and ecological structure of the shelf benthos during Phanerozoic time. *Palaeontology*, **12**: 684–709.

Valentine, J.W. and Erwin, D.H., 1987, Interpreting great developmental experiments: the fossil record. In R.A. Raff and E.C. Raff (eds), *Development as an evolutionary process*, Alan R. Liss, Inc., New York, pp. 71–107.

Vermeij, G.J., 1987, *Evolution and Escalation*, Princeton University Press, Princeton, NJ.

Whittington, H.B. and Briggs, D.E.G., 1985, The largest Cambrian animal, *Anomalocaris*, Burgess Shale, British Columbia, *Philosophical Transactions of the Royal Society of London*, **B309**: 569–609.

Chapter 5

PATTERNS OF EVOLUTION AND EXTINCTION IN INVERTEBRATES

Christopher R.C. Paul

Interest in patterns of evolution and extinction, particularly in marine invertebrates, has been stimulated since the early 1970s by three ideas: first, Eldredge and Gould (1972) argued that evolution was punctuational not gradual; second, Alvarez *et al.* (1980) suggested that the mass extinction at the Cretaceous–Tertiary boundary was due to the impact of a major extraterrestrial object (a bolide); and third, Raup and Sepkoski (1984) argued that there has been a periodicity of 26 Myr in mass extinctions over the last 250 Myr. All three ideas have precipitated a great deal of debate (and some additional data-gathering) and are still regarded by many as controversial or unproven. Others doubt whether the fossil record preserves adequate data to test such ideas.

RELIABILITY OF PATTERNS FROM THE FOSSIL RECORD

Everyone knows that the fossil record is incomplete. This fact can be conveniently exploited as an excuse to ignore any evidence from the fossil record that does not agree with currently accepted theories. The reasons for the incompleteness are obvious. The chances against an individual being preserved are astronomically large; soft-bodied organisms are only very rarely preserved, as are those, like land snails, which live in areas of active erosion; many organisms that become preserved are subsequently destroyed by erosion, diagenesis and metamorphism; and the resulting remnant is poorly known anyway. New species of fossils are still being described at an increasing rate and, even if some will eventually prove to be synonymous with previously named taxa, there is no sign of an end to new discoveries. The incompleteness of the fossil record is an established fact, but what really matters is its significance. No science is based on complete knowledge, so why should the fossil record be singled out because it, too, is incomplete and incompletely known? For example, there have recently been three or four discoveries in astronomy that have revolutionised our view of the Universe, yet no one regards astronomical information as unreliable or astronomy as anything other than a serious scientific discipline. If the Universe started with a 'Big Bang' as most current theories hold, it is likely that many relatively short-lived phenomena have left no trace; others may lie beyond the reach of our most powerful telescopes. Astronomers have yet to document, let alone study, even a majority of visible stellar objects, only one of which can be examined close up (if 150 million kilometres away can be regarded as close up) and until very recently all astronomical information was filtered through the Earth's atmosphere and distorted to a greater or lesser extent. No planetary system other than our own is known for certain. If the arguments used to judge the fossil record were applied to astronomy, few scientists would take it seriously. However, this is not an attack on astronomy, quite the reverse. If our poor knowledge of the Universe is a stimulus to discover more, the same should be true of the fossil record. If recent theories make astronomy seem an exciting and rapidly advancing science, the same should be true of palaeontology. Evidence from the fossil record is comparable with that available in any other science.

The great strength of the fossil record lies in its long time-span. It is the only objective record of the actual course of evolution and extinction on Earth and, despite its incompleteness, it preserves the sequence of events faithfully and their duration reasonably accurately. Figure 5.1A depicts the total periods of existence of four hypothetical species of fossil (not their known stratigraphic ranges). Consider the first appearances of species A and B, which never coexisted. To emphasise the point, imagine that they are a Cambrian trilobite and a Pleistocene snail, respectively. There is clearly no chance that any fossils of these two species, let alone the first known examples, could be preserved in the opposite order to that in which they evolved, because they never coexisted. It is true that derived fossils are preserved out of sequence, but for derived fossils to mislead two conditions must be met: first, they must not be recognised as derived fossils from their state of preservation or occurrence; and second, no specimens from the deposit in which they were originally preserved must be known. Otherwise the upper limits of their ranges would simply be extended. The coincidence of both requirements is unlikely, so it follows that when species did not originally coexist their first appearances in the fossil record cannot be preserved in the wrong order.

On the other hand, species C and D in Figure 5.1A did coexist and therefore it is possible that their apparent order of evolution or extinction could be preserved the wrong way round, that is, that the first known specimen of species D could be stratigraphically below that of species

C, and the same applies to the last known specimens of both species. Although this is possible, in fact even when only one specimen of each species is known there is nearly always a higher probability that they will occur in the correct order, and this probability increases very rapidly as more specimens of each species are discovered (Paul, 1982, p. 105). More importantly, as soon as a single specimen of species C is discovered from level 1, first appearances of these two species must be in the correct sequence. Equally, as soon as a specimen of species D is found from level 3, last occurrences must be in the correct order (unless derived examples of species C exist in a still younger horizon and are not recognised as being derived). Since the total lifetimes of species are unknown, there is no way to tell when this has happened; nevertheless, it must happen quite commonly.

Finally, to estimate the absolute maximum likelihood that the fossil record preserves events in the wrong order, suppose that all pairs of species which did coexist give unreliable data. The basic question then becomes: 'when comparing the ranges of fossil species, on average, how often will a randomly chosen pair of species have coexisted?' The answer depends on two things: how long the average species survived and how overall species diversity has changed throughout the Phanerozoic. Taking 6 Myr as the mean lifetime of a species, 600 Myr as the duration of the Phanerozoic, and accepting a total diversity curve like that shown in Figure 5.2A, on average

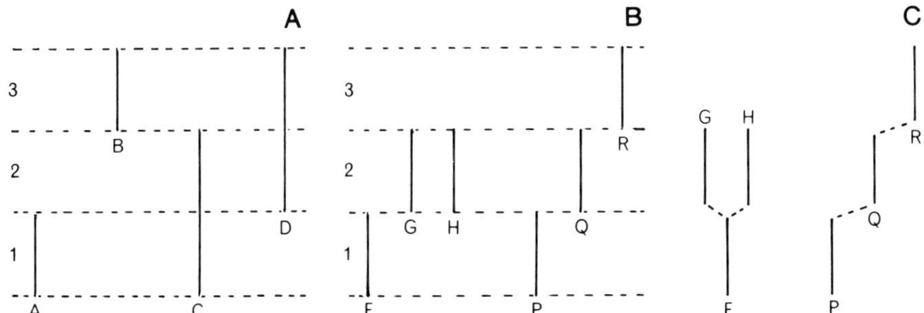

Figure 5.1. Total periods of existence, known stratigraphic ranges and patterns of evolution of different species. A: Total periods of existence of four species. A and B did not coexist and therefore cannot be preserved in the fossil record in the wrong order. C and D did coexist (at time 2) and therefore their first and last known specimens could be in the wrong order. Nevertheless, as soon as a single example of species C is found from level 1, and an example of species D from level 3, both their first and last known appearances will be in the correct sequence. B: Known stratigraphic ranges of two sets of closely related species. Since G and H have coincident ranges, true speciation (i.e. bifurcation of a lineage) has occurred. However, P, Q and R have sequential ranges and only transformation occurred. C: Patterns of evolution (phylogenetic trees) inferred from both morphological and stratigraphic information. Note that in lineage P, Q, R neither speciation nor extinction occurred.

approximately 3% of the species that existed in the entire Phanerozoic coexisted at any one time. In other words, only in 3% of random comparisons of known stratigraphic ranges of fossil species is there any possibility whatsoever that first or last occurrences might be in the wrong order with respect to the true original sequence of evolution or extinction. Put another way, the usual 95% confidence limits would occur if the average lifetime of a species had been about 10 Myr throughout the Phanerozoic, and this is assuming that all species which coexisted give unreliable data, which is a manifestly false assumption. In reality only a small proportion may do so.

This theoretical argument confirms that the fossil record preserves the sequence of evolutionary events reliably. If it did not, correlation using fossils would not work. Note, too, that this result is not affected by the completeness or otherwise of the fossil record. Indeed if the fossil record consisted of just two fossils, the chances are overwhelming that they would be preserved in the correct order!

STRATIGRAPHIC AND EVOLUTIONARY PATTERNS

If the sequence of events in the fossil record is reliable, then two things follow. First, it is worth documenting the sequence as accurately as possible. Second, other relative data are probably equally reliable so patterns observed in the fossil record probably mean something and are not just artefacts of its incompleteness. Shaw (1971) argued that the only adequate stratigraphic data for fossils consist of measured positions relative to an objective lithological marker horizon in a logged section. He and Paul (1985) documented how far short of this ideal much of current palaeontological literature falls. It seems that many practising palaeontologists do not record adequate stratigraphic data because they do not think that such information is useful. This is a self-fulfilling prophecy: if the data are not available they cannot possibly be of use. Even among biostratigraphers current stratigraphic practice leaves much to be desired. Citing fossil ranges in terms of zones automatically increases apparent stratigraphic ranges, even for long-ranging taxa, because fossils rarely appear or disappear exactly at zonal boundaries. It also destroys the sequence of appearances and disappearances, resulting in those all too common range charts in which all terminations of ranges coincide with zonal boundaries. Worst of all, it defines the level of stratigraphic precision (to the nearest zone) before research commences. No matter how many new samples are collected or specimens discovered, nothing new will be revealed about the sequence of events (see Paul 1985, Figure 5).

More important in the context of this chapter is what sorts of pattern occur and how they relate to evolutionary processes. Consider patterns for individual species first. Figure 5.1B shows patterns of stratigraphic distribution for two sets of three species which on morphological grounds are thought to belong in two monophyletic clades (e.g. two genera). Species G and H have coincident ranges so that true speciation (i.e. two daughter species arising from a single parent) must have occurred, and diversity has increased with time. However, species P, Q and R have sequential stratigraphic ranges, there has been no true speciation, and diversity has not increased at all. They represent sequential transformations of species in a single lineage. Figure 5.1C illustrates the evolutionary patterns that can be deduced from both the morphological and stratigraphic information. It is important to distinguish between these patterns of evolution, because in the latter case not only is there no true speciation, but neither is there any real extinction, only a 'pseudoextinction'. When species P transformed into species Q nothing became extinct. Condensed sedimentary sequences may contain species which actually lived at separate times. However, in normal sedimentary sequences evidence of the coexistence of closely related species implies a branching pattern of evolution (i.e. true speciation), whereas evidence of sequential occurrence implies neither speciation nor extinction.

Now consider the general pattern of evolution. Individual species are adapted to their unique ecological requirements. Species are intimately linked in food webs—therefore the evolution of a new species, or the extinction of an already existing species, should have little effect on the general biota. Even when specialised organisms have a single food source which becomes extinct, only these two species should become extinct simultaneously. Thus theoretical considerations argue, first, that extinction probably results from a combination of causes, and second, that biological causes should normally result in random patterns of evolution and extinction. If, however, there are external (i.e. non-biological) causes such as major climatic changes, substantial fluctuations in sea level, intense volcanic activity or impacts of large extraterrestrial objects, then numerous extinction events might be expected to coincide, i.e. a mass extinction (see Chapter 2). Put the other way round, the fossil record undoubtedly shows periods of apparently coincident mass extinction. It is likely, therefore, that these were due to non-biological causes. Furthermore, unless there is a major radiation which follows a mass extinction and apparently fills vacated ecospace, radiations are unlikely to coincide in the way that mass extinctions affect a wide variety of taxonomic groups simultaneously. Even when new ecospace becomes available, it may be exploited by different organisms at widely different times. Birds, bats, pterosaurs and several groups of insects became adapted for flight at various times since the Carboniferous.

Finally, in considering patterns, available evidence indicates that overall diversity (i.e. total number of species) has increased through time (see Sepkoski *et al.* 1981). This means that, with a few notable exceptions, origination rates generally exceeded extinction rates. Thus branching patterns (i.e. true speciation) must have played a significant (though not necessarily dominant) role in the history of life. Transformations are neutral in terms of diversity because, at least until the lineage finally becomes extinct, originations equal pseudoextinctions. It is important to know the relative frequency of speciation versus transformation if we are to understand the history of life on Earth. Radiations arise when speciation becomes significant and greatly exceeds extinction rates; standing diversity will remain constant either when speciation equals extinction or when only transformations occur. One of the most fundamental questions of evolutionary theory is how new barriers to cross-fertilisation arise (i.e. a general theory of speciation). The allopatric speciation model, which relies on the geographic isolation of small populations, seems inadequate as a general explanation for speciation, especially in the marine environment. Geographic isolation is only one of the means by which genetic isolation of related populations is maintained. It seems inherently unlikely that only one of the possible mechanisms causes true speciation, especially when that mechanism is difficult to envisage in a continuous marine environment.

To summarise: patterns of stratigraphic distribution reflect underlying evolutionary processes; evidence of coexistence in related species is evidence for branching of lineages, i.e. true speciation; coincidence of extinctions is likely to have non-biological causes, and finally, the overall increase in diversity through the Phanerozoic means that true speciation played a significant role in the history of life.

PATTERNS OF EVOLUTION

Punctuation versus gradualism

In writing *The Origin of Species*, Darwin's task was, in essence, to convince his readers that things which they had hitherto taken to be immutable had in fact changed through time. It is not surprising, therefore, that he stressed how gradual the changes had been. Indeed, in his efforts to do so he produced one of the very rare overestimates of the length of geologic time, arguing that the evolution of life from the middle of the Mesozoic (i.e. Middle Jurassic) to the present day would require about 300 Myr, whereas current radiometric estimates indicate 150 Myr. Gradualism was not essential to Darwin's arguments; nevertheless it became entrenched in our view of evolution, at least until Eldredge and Gould (1972) questioned it. Eldredge set about documenting in detail the evolution of phacopid trilobites in the Middle Devonian, only to find that through increasingly long periods of the Devonian they just did not change (see Eldredge 1972). He had expected to see gradual morphological change with time, instead he found long periods of no change, punctuated by short intervals of considerable morphological change. Unlike many scientists before him, and much to his credit, he did not dismiss these observations as artefacts of the incompleteness of the fossil record, but questioned whether this pattern indeed reflected what had actually happened. It was not long before he and Gould came up with their punctuational model and a theoretical justification for it. This was the allopatric speciation model which holds that new species arise in small isolated populations confined to small geographic areas. It follows that the evolution of a new species would be most unlikely to be preserved in the fossil record, but once that species became successfully established, its subsequent migration into new areas would be preserved. Thus Eldredge and Gould (1972) argued that well-adapted species should show no significant change for considerable periods of time and then suddenly be replaced, or accompanied, by new species which are morphologically distinct. They suggested that the general pattern of evolution in the fossil record was that depicted by Eldredge's (1972) trilobites, one of stasis (i.e. no morphological change) punctuated by short periods of evolutionary innovation. In other words the 'sudden' appearance of fossils was precisely what evolutionary theory would predict and was not an artefact of the incompleteness of the fossil record.

The most significant contributions of Eldredge and Gould's theory are the acceptance of patterns as preserved in the fossil record, and the recognition of stasis (Lewin 1986). Hitherto, no morphological change had been equated with no data, and just ignored. With the benefit of hindsight, it is amazing that palaeontologists could have accepted gradual evolution as a universal pattern on the basis of a handful of supposedly well-documented lineages (e.g. *Gryphaea*, *Micraster*, *Zaphrentis*) none of which actually withstands close scrutiny. (For example *Micraster* shows sudden appearances of new taxa (Stokes 1977, Figure 2) and relatively sudden changes in morphological features (Drummond 1983, Figure 1).) The evidence that the vast majority of species appeared suddenly, had well-defined periods of existence, and then disappeared equally suddenly, was just ignored. Furthermore, because evolution was known to be gradual, very few palaeontologists documented actual patterns preserved in the fossil record. Eldredge and Gould (1972) did a great service in prompting a re-examination of the evidence. Darwin himself was aware of stasis. In discussing the reasons why the fossil record did not preserve much evidence of gradual changes, he wrote in *The Origin of Species*:

> although each species must have passed through numerous transitional stages, it is probable that the periods, during which they underwent modification, though many and long as measured by years, have been short in comparison with the periods during which each remained in an unchanged condition. (Darwin 1872, pp. 488–9.)

That is a fair summary of punctuated equilibria made more than a century earlier. Darwin undoubtedly recognised stasis: indeed the only diagram in the book (between pp. 140 and 141) shows one species (F) remaining unchanged and not giving rise to any descendants through the entire period of time represented.

The punctuation versus gradualism debate has become sterile partly because some protagonists view the two alternatives as mutually exclusive rather than as end members of a continuum (but see Fortey 1985), partly because 'gradualism' and 'punctuation' are relative, and partly because both punctuation and stasis can only be recognised in comparison with each other. Thus it is not simply that a rate of change which is gradual to a population ecologist may be extremely rapid to a palaeontologist, there is also the fact that the same amount of change over say 200 000 years is gradual for a species that only existed for those 200 000 years, but rapid for another which had remained unchanged for the previous 20 Myr. Documentation of patterns of evolutionary change in morphological characters requires an objective measure of rates of change.

General patterns in the Phanerozoic

The basic pattern of evolution has been one of overall increasing diversity, interrupted by a few periods of mass extinction, notably at the ends of the Palaeozoic and Mesozoic eras (see Sepkoski *et al.* 1981). Within this overall pattern, Sepkoski (1979; 1981) argued that there were three main contributary radiations: Cambrian; Early Palaeozoic; and, following the Permo-Triassic life crisis, an Early Mesozoic radiation. He identified the composition of these three faunas which he named the Cambrian, the Palaeozoic and the Recent. These faunas were not totally exclusive in time. The Cambrian fauna dominated the Cambrian Period and declined thereafter. The first progenitors of the Palaeozoic fauna were present in the Cambrian, but rose to predominance in the Ordovician and dominated throughout the rest of the Palaeozoic. Similarly the first members of the Recent fauna arose well back in the Palaeozoic, but only became the dominant after the end-Permian mass extinction. At the other end of the time-scale, some members of Sepkoski's Cambrian fauna are still extant. Thus all three of his faunas occur throughout the Phanerozoic (Figure 5.2A), and it is shifts in their relative importance which define them.

Sepkoski's (1979; 1981) idea of three faunas has been criticised (see, for instance, Smith 1988) on three main grounds: first, that data on families are at too crude a taxonomic level; second, that the data do not distinguish between true extinction and pseudoextinction; and third, that insufficient account has been taken of so-called Lazarus taxa, which are families that apparently disappear from the fossil record only to reappear higher up the geologic column. Sepkoski is currently compiling a generic data base and indications so far are that it shows a very similar pattern to the family-level data set (see below). Furthermore, since Sepkoski used total ranges, the difference between 'normal' and 'Lazarus' taxa should not bias the data. The distinction between true and pseudoextinction is more important and, at least as far as echinoderms are concerned (Smith, 1988), can lead to dramatically different interpretations.

In discussing patterns of extinction and survival in echinoderms Paul (1988) argued that a thorough treatment requires: phylogenetic analysis of

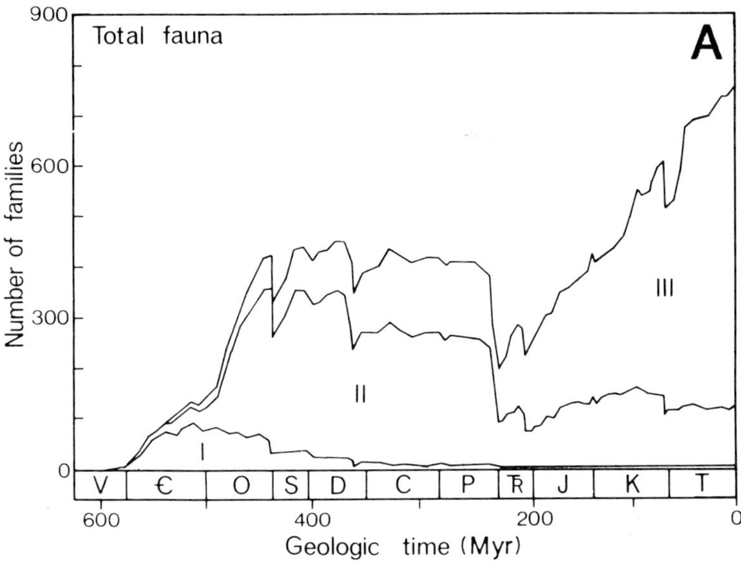

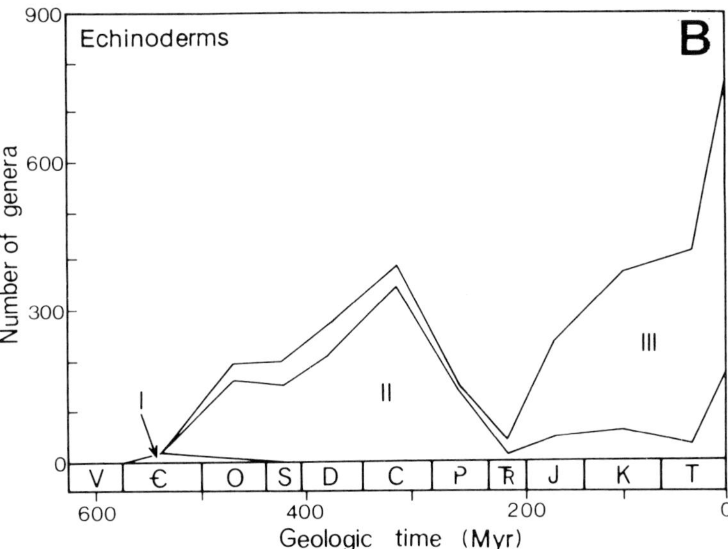

Figure 5.2. Patterns of taxonomic diversity in the Phanerozoic among A: marine families and B: echinoderm genera. Roman numerals indicate Sepkoski's three faunas. (Redrawn from Sepkoski 1981, Figure 5, and Paul 1988, Figure 8.5.)

evolution within a group to distinguish pseudoextinction from true extinction; analysis of the relative completeness of the fossil record throughout the time interval concerned; and treatment at specific level. No available data meet all these requirements. Paul (1988) used genera of fossil echinoderms because they were the only available data. His analysis involved the same kind of data as that of Sepkoski (1979; 1981) and showed that the echinoderm fossil record apparently also falls into three faunas (Figure 5.2B), a Cambrian fauna dominated by 'eocrinoids', a Palaeozoic fauna dominated by crinoids and a post-Palaeozoic fauna dominated by eleutherozoans (starfish, sea urchins and brittlestars). He concluded that the echinoderm fossil record is representative of the fossil record of marine invertebrates as a whole. Smith (1988) has since undertaken a much more penetrating analysis of the early echinoderm record which goes a long way towards meeting the three requirements cited by Paul (1988) and mentioned above, and allows assessment of the effects of pseudoextinction and Lazarus taxa.

Smith analysed cladistically all well-characterised species of echinoderm known from the Lower Cambrian to the Arenig (Lower Ordovician), in order to assess the validity of the so-called Cambrian fauna. He concluded that the Cambrian fauna only exists because of the 'mystique' of the Cambrian and is almost entirely an artefact of taxonomic practice. The 'eocrinoids' are a primitive group largely distinguished by the absence of the characters which are used to define higher taxa among their descendants. They lack, for example, the diplopores of diploporites, the rhombs of rhombiferans and the hydrospires of blastoids. Nevertheless, they were ancestral to all these groups and hence the clades initiated within the 'Eocrinoidea' did not become extinct with the last accepted 'eocrinoid', they were simply renamed. Once the basic data are subjected to a thorough phylogenetic analysis, so that we can record only true extinctions, a completely different pattern emerges—one of continued success without a late Cambrian decline in diversity and certainly without a Palaeozoic fauna replacing a Cambrian one.

Since Paul's (1988) analysis of all echinoderm genera produced a similar pattern to Sepkoski's (1979; 1981) for all families of marine animals, and both were based on the same type of data, it is reasonable to ask if a similar reappraisal of patterns would result from phylogenetic analysis of other phyla. Pseudoextinction is very common and it distorts results. Patterns of evolution and extinction should be analysed, not patterns of (sometimes arbitrary) taxonomic practice, particularly if causes are sought for the patterns detected. Even then patterns need careful assessment. The patterns of evolution and extinction of blastoid genera, which until Smith's (1988) analysis were the best available data for echinoderms, mirror each other. Maxima for appearances and disappearances coincide with maxima for blastoid diversity. They do not reflect real radiations and extinctions, merely the very patchy fossil record of blastoids (Paul 1988, Figure 8.4).

Clade shapes

In extinct groups the entire pattern of evolution and extinction is preserved, whereas extant groups reveal an incomplete history. In groups that have apparently peaked, the timing and number of peaks in diversity vary. It is also possible to distinguish groups with an apparently 'smooth' evolutionary history from those whose evolution was punctuated by significant and sometimes repeated extinctions—a 'boom and bust' evolutionary pattern.

Clades with an early peak in diversity

The cystoids *sensu lato* (blastozoans of Sprinkle 1973) typify a pattern of rapid early diversification followed by a relatively gradual and protracted decline (Figure 5.3A). Cystoids first appeared in the late Early Cambrian, reached maximum diversity in the Middle Ordovician, but survived to the end of the Palaeozoic. After the earliest Late Devonian they were represented by a single class, the blastoids. 'Living fossils' such as the monoplacophoran *Neopilina* or the coelacanth *Latimeria*, belong to extant groups with a similar pattern—at least as regards the long period of relatively low diversity towards the end of their known range. Genera of nautiloids (Figure 5.3B) exhibit a similar pattern with an early peak and gradual decline, but punctuated by repeated extinctions. After the first extinction, subsequent peaks reached progressively lower levels of diversity.

Clades with subcentral peaks in diversity

Families of Palaeozoic bryozoans (Figure 5.4A) present a pattern with a rapid rise, a plateau, and a rapid decline in diversity. They arose in the Early Ordovician and declined sharply in the Late Permian. There is a minor decline in total diversity through the Late Silurian and Early Devonian, but overall diversity remained remarkably consistent from the Late Ordovician to the Middle Permian. Rugose corals (Figure 5.4B) show a similar pattern except that it is punctuated by a major extinction in the middle. They originated in the Early Ordovician and survived to the end of the Permian. Peaks of diversity in the Middle Devonian and Early Carboniferous are separated by a major extinction at the Frasnian–Famennian boundary, which reduced generic diversity to about a third of the Middle Devonian level.

Clades with late peaks in diversity

Ammonoid genera present a pattern of overall increase in diversity punctuated by repeated extinctions and radiations (Figure 5.5A). They arose in the Devonian and became extinct at the end of the Cretaceous. In between were repeated wide fluctuations in diversity, but unlike nautiloids, successive

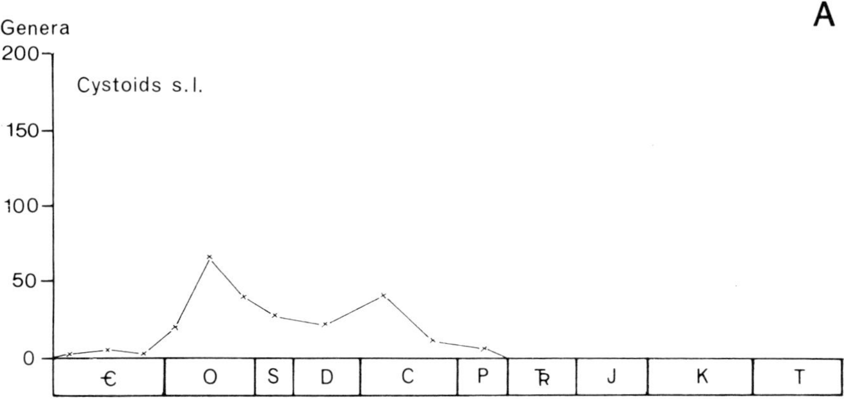

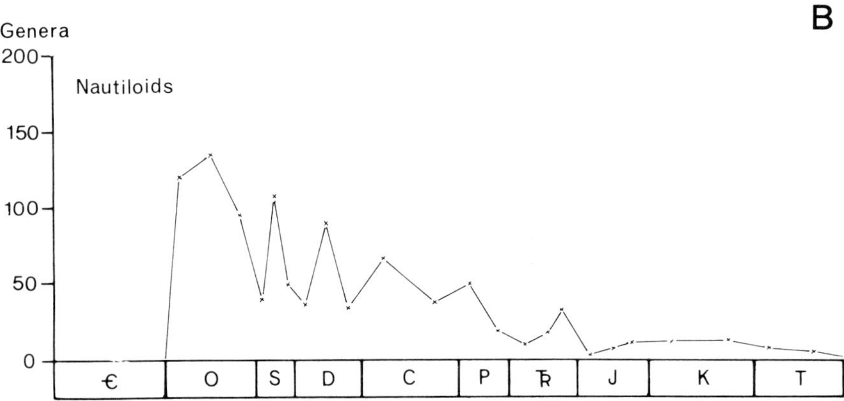

Figure 5.3. Clades with early peaks in diversity. A: Cystoids with a relatively smooth curve (data from Paul and Smith 1984, Figure 14). B: Nautiloids with repeated fluctuations but overall decline in diversity (redrawn from House 1988, Figure 7.2).

peaks show overall increase. Ammonoids have perhaps the clearest 'boom and bust' pattern of evolution among major marine clades.

Among groups that have apparently not yet reached peak diversity, the record of the bivalves (Figure 5.5B), which shows a continuous, virtually uninterrupted increase in diversity from their first appearance in the Early Ordovician until the present, can be contrasted with, for example, that of the echinoids (Figure 5.5C) which also appeared in the Ordovician, but did not begin to increase in diversity significantly until the Early to Middle Jurassic. Many other groups that survived the end-Permian or end-Cretaceous mass extinctions show a pattern of more or less continuous increase since. Such groups include post-Palaeozoic bryozoans and echinoderms as

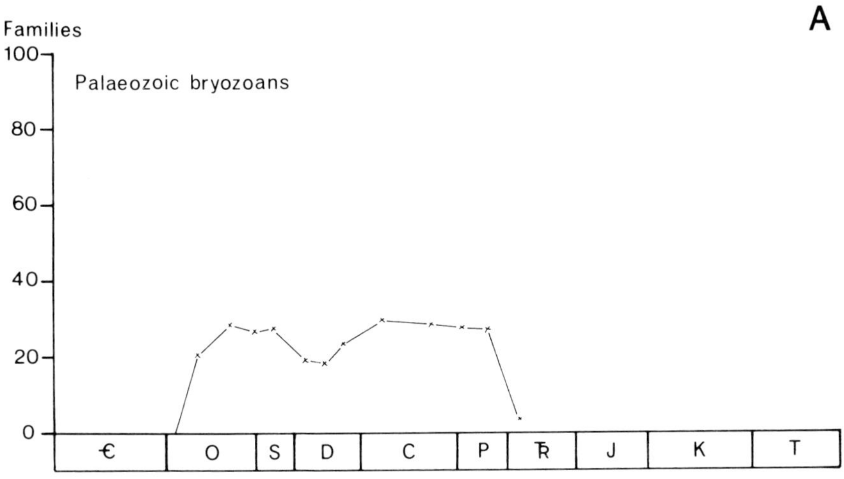

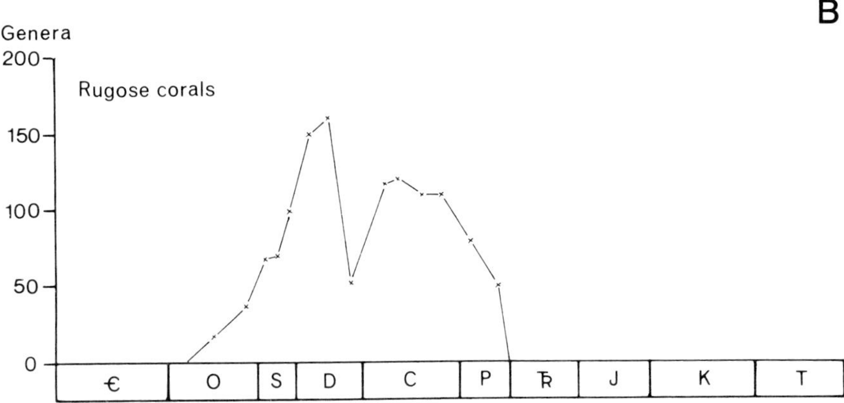

Figure 5.4. Clades with subcentral peaks in diversity. A: Palaeozoic bryozoans with a broad plateau. (Redrawn from Taylor and Larwood 1988, Figure 5.1.) B: Rugose corals with the overall pattern interrupted by a single major extinction. (Redrawn from Scrutton 1988, Figure 3.3.)

well as post-Mesozoic reptiles, birds and mammals (Benton 1988, Figure 13.3).

Thus it seems that, within major taxa, patterns and timing of diversification vary widely. Within extinct taxa where the complete history is known, some groups exhibit smooth diversity curves, others jagged patterns reflecting 'boom and bust' evolutionary cycles. Peaks in diversity may be rapidly or gradually achieved and occur early, subcentrally, or late in the history of the group. This variation in clade shape is evident at all taxonomic levels. Curves presented by Sepkoski and Hulver (1985) for

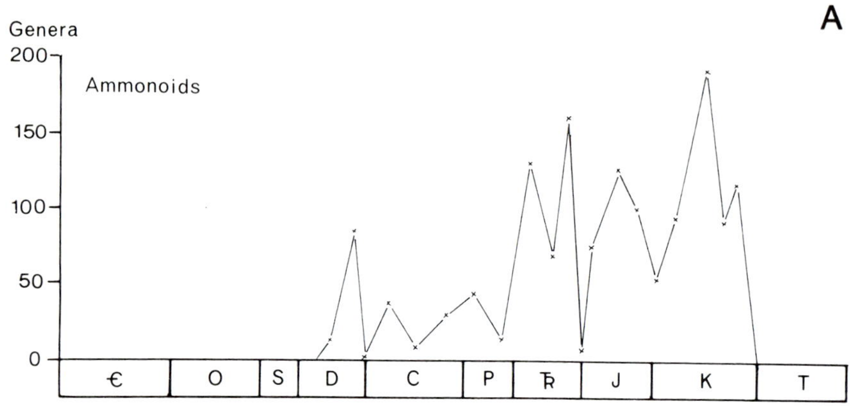
A
Genera
200
150
100
50
0
Ammonoids
€
O
S
D
C
P
Ŧ̵R
J
K
T

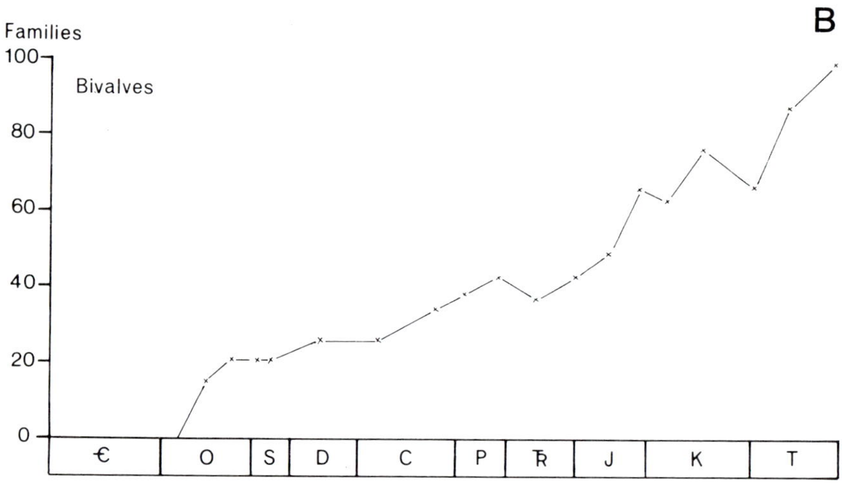
B
Families
100
80
60
40
20
0
Bivalves
€
O
S
D
C
P
Ŧ̵R
J
K
T

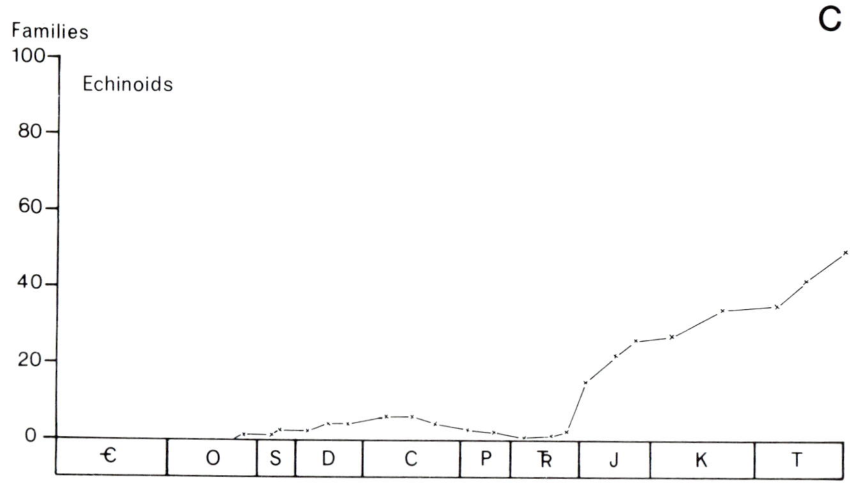
C
Families
100
80
60
40
20
0
Echinoids
€
O
S
D
C
P
Ŧ̵R
J
K
T

familial diversity within phyla, classes and orders of marine and continental organisms vary widely. Ward and Signor (1985) showed similar variation in clade shapes for ammonite genera within families. There seems to be no general rule governing radiations except perhaps that overall diversity is probably greater now than at any time in the past due to a common increase through the Tertiary for many extant groups. Certainly there is little evidence that ecospace was filled early in the history of a group and maximum diversity was maintained thereafter. Only Palaeozoic bryozoans appear to fit the pattern predicted by this idea, at least until the mass extinction at the end of the Permian.

PATTERNS OF EXTINCTION

Effects on individual clades

Much of the pattern of extinction within major taxa is as random as the pattern of radiations (see below). Nevertheless, at certain times all or a majority of groups seem to have suffered significant extinction simultaneously, and total diversity declined markedly. Such events are known as mass extinctions and, as pointed out earlier, they are inherently likely to be caused by external, non-biological causes. Some mass extinctions coincided with the total demise of major taxa, the extinction of the dinosaurs at the end of the Cretaceous being the best-known example. Even when total extinction of a clade did not occur, reductions in diversity were often dramatic. *Miocidaris* is the only genus of echinoid known to have survived from the Permian to the Triassic (Smith 1984); only two genera of ammonites are believed to have crossed the Triassic–Jurassic boundary. Even groups with a record of apparently continuous diversification, such as bivalves and marine gastropods, suffered slight declines at the most conspicuous mass extinctions. Note, however, that Lazarus taxa may grossly distort patterns if their total stratigraphic ranges are not taken into account. Nakazawa and Runnegar (1973) demonstrated this effect for bivalves across the Permo-Triassic boundary, as did Paul (1980) for the cystoids across the Ordovician–Silurian boundary. Mass extinctions are generally accepted as real events and require explanation. There is, however, continued debate about how mass extinctions can be recognised, how intense they were, and how many occurred.

Figure 5.5. Clades with late peaks in diversity. A: Ammonoids with repeated fluctuations superimposed on an overall increase, followed by terminal extinction. (Redrawn from House 1988, Figure 7.4). B: Bivalves with virtually continuous increase in diversity. (Redrawn from Hallam and Miller 1988, Figure 6.1.) C: Echinoids with prolonged low diversity in the Palaeozoic and Early Mesozoic, followed by continuous increase to the present. (Data compiled from Smith 1984, Figure A.1.)

In considering patterns of extinction in marine invertebrates, it is instructive to note when individual clades suffered maximum decline in diversity. Although both rugose and tabulate corals, for example, became extinct at the end of the Permian, the sharpest decline in generic diversity occurred in the Late Devonian at the Frasnian–Famennian boundary. Tabulates also show a minor extinction at the Ordovician–Silurian boundary (Scrutton 1988). Palaeozoic byrozoan families show only the terminal Permian mass extinction and post-Palaeozoic forms only minor extinctions at the end of the Triassic and Cretaceous (Taylor and Larwood 1988). Bivalve families show only minor reductions in diversity at the end of the Triassic, Early Jurassic, Early, and end-Cretaceous. Rudists (Hippuritoida) are the only order of bivalves ever to become extinct (Hallam and Miller 1988). Brachiopods suffered maximum decline at the Permo-Triassic boundary (Sepkoski and Hulver 1985), while both nautiloids and ammonoids show classic 'boom and bust' patterns with repeated significant declines in diversity (House 1988). Echinoderms show steep declines at the end of the Carboniferous and Permian, the latter being a real life crisis, but were virtually unaffected by the terminal Cretaceous mass extinction (Paul 1988). Thus although terminal extinctions of major clades are obviously important events in the history of life, by no means all of them are accompanied by a significant decline in diversity. Periods of major extinction, both in absolute numbers of taxa and in the proportion of extant fauna, may precede terminal extinction by a significant period of time.

This brief review of some major taxa indicates that periods of significant decline in diversity differed for different clades. What makes mass extinctions conspicuous is the coincidence of significant levels of extinction in unrelated taxa at the same time. There was, for example, almost no taxon unaffected by the Permo-Triassic life crisis and Raup (1978, Figure 4) estimated that it was equivalent to 85 Myr of background extinction at Palaeozoic rates. This coincidence of extinctions suggests external causes and debate has recently been stimulated by the discovery of an iridium anomaly at the Cretaceous–Tertiary boundary (see Chapter 2).

The Cretaceous–Tertiary boundary and bolides

Alvarez *et al.* (1980) discovered a peak of iridium concentration at the end of the Cretaceous in three principal sections. They argued that the iridium derived from planetary interiors, not the crust, and that the anomaly resulted from the impact of a large extraterrestrial object, a bolide. They suggested further that the impact caused the mass extinction at the Cretaceous–Tertiary boundary. From the iridium concentrations they calculated that the bolide was of the order of 10 km in diameter, and since the oceans are on average 4 km deep, its impact would have had similarly catastrophic effects wherever it landed. The impact scenario stimulated considerable debate and led to a re-examination of many other sections which cover the relevant interval. It also led to the idea that impacts of extraterrestrial bodies may provide a general explanation of mass extinctions, and has promoted a search for iridium anomalies at other stratigraphic levels associated with mass extinctions. Several such discoveries have been claimed, although documented detail is still scarce. Clearly there is a need for knowledge of background frequencies of significant peaks in the concentration of iridium and other trace elements. If iridium anomalies are as common as magnetic reversals (for example), it is highly likely that several will coincide with mass extinctions merely by chance. Furthermore, if only levels associated with mass extinctions are searched, since some are bound to coincide with iridium anomalies, spurious confirmation of this idea will result. If, at the other extreme, there were only one iridium anomaly and it coincided precisely with mass extinctions at the end of the Cretaceous, this would argue very strongly for a link between them, perhaps cause and effect. No doubt the truth lies somewhere in between these extremes (see Chapter 2).

The ideas of Alvarez *et al.* (1980) are of considerable importance in themselves, but in the present context their significance is that they led indirectly to a new general theory that mass extinctions result from extraterrestrial causes.

Periodicity

Further analysis of Sepkoski's (1982) data on marine fossil families led Raup and Sepkoski (1982) to suggest that mass extinctions lay well above the average level of background extinction rates, and later (Raup and Sepkoski 1984) that there was a periodicity of 26 Myr between mass extinctions over the last 250 Myr (see Chapter 2). They offered the fluctuations of the solar system through the spiral arms of the galaxy as a possible cause for this periodicity. Reactions to this idea have been varied and fall into three main groups: criticisms of the data as inadequate thus producing false patterns (e.g. Patterson and Smith 1987); arguments that the periodicity is an artefact of the statistical methods (e.g. Quinn 1987) or time-scale used (Hoffman 1985); acceptance of the data as a basis for alternative analyses (e.g. Fox 1987).

Raup and Sepkoski (1984) analysed a subset of Sepkoski's data consisting of families with the most reliable ranges and excluding all extant families. They showed that certain peak extinction rates significantly exceed the background rate, defined as the proportion of families which did not survive from one period to the next. No allowance for the differing lengths of periods was made because, they argued, durations of the same stage in different time-scales varied by as much as the durations of different stages, so accurate estimates of percentage extinction per Myr are impossible. Peaks in extinctions (defined as having troughs on either side) revealed a 26 Myr periodicity through the Mesozoic and Tertiary. The criticisms levelled at the Sepkoski data base (p. 106) have been invoked for this analysis, too. Some of them, however, seem to be more an attempt to explain away, than to explain, the pattern. Patterson and Smith (1987) argued that the periodicity is largely generated by noise, not true extinction rates. They recognised seven categories of data on extinction of families of fish and echinoderms. Of these categories only one (monophyletic families, the extinction of which was assigned to the correct stage) provided a true signal. The other six (monophyletic families, the extinction of which was assigned to the wrong stage; paraphyletic, polyphyletic, and non-monophyletic families, which probably record pseudoextinction rather than true extinction; monotypic and non-marine families: see Chapter 9, Figure 9.2 for illustrations of some of these terms) were regarded as generating noise. Noise represented 75% of the total extinctions of fish and echinoderm families used by Raup and Sepkoski (1984). Patterson and Smith (1987, Figure 2b) demonstrated a remarkable correlation between peaks in extinctions of the noise component and Raup and Sepkoski's peaks. While this implies that a large proportion of Raup and Sepkoski's data were inaccurate or unreliable, it does not explain why the periodicity occurs. Noise should randomise the data, not produce spurious periodicity.

Hofmann and Ghiold (1984) postulated that rates of radiation and extinction behaved like a random walk, with equal probability of increasing or decreasing at any given time. They were unable to falsify this hypothesis, except that they could demonstrate too many coincident peaks of both radiation and extinction. They argued, however, that these were due to fossil Lagerstätten (deposits with exceptional preservation of soft-bodied faunas) which inevitably give the impression of a radiation within the Lagerstätten followed by significant extinction in the next normal deposit above.

Sepkoski (1986) subsequently used a data set of about 25 000 marine invertebrate genera, analysed at substage level, which gave a total of 161 intervals. Of the genera, roughly a third (33.5%) had known ranges that could be resolved to substage level and another third (36%) to stage level. The rest had coarser stratigraphic resolution and their extinctions were assigned within series or systems according to the distributions of extinctions of the more reliable genera. Thus they affected the absolute, but not the relative, values for extinction rates. Sepkoski produced two figures (1986, Figures 2a,b; see Figure 2.4 in Chapter 2) which depicted percentage extinction and total extinctions per ma million years. Both show a pattern of wide fluctuations in extinction rate, and a periodicity of mass extinctions. Sepkoski also argued that background extinction rates declined from a high of about 50% in the Cambrian. How much of this is a distortion due to pseudoextinction and whether the decline, if real, was gradual or stepped, are still unknown.

Fox (1987) applied Fourier analysis to a larger generic data set (28 800 genera) but only over the last 260 Myr. He concluded that the first ten harmonics coincided very closely to a periodicity of 26 Myr (Fox 1987,

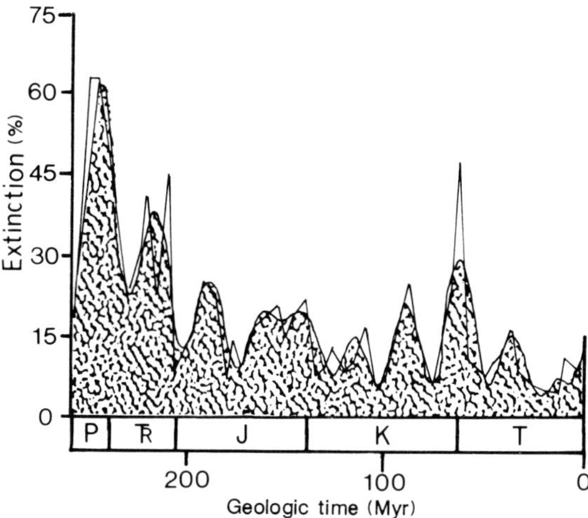

Figure 5.6. Correlation between mass extinctions of marine genera over the last 250 Myr and the first ten harmonics of a Fourier analysis (stippled). (Redrawn from Fox 1987, Figure 3.)

Figure 3, reproduced here as Figure 5.6) and that this coincidence was significant. This analysis confirmed that generic data mirror the pattern for families first detected by Raup and Sepkoski (1984). However, Quinn (1987) concluded that there was no evidence for periodicity in either familial or generic extinction using lag time, i.e. the intervals between extinction peaks. He argued that if there were a rigorous periodicity, then all intervals would be 26 Myr. If one or two peaks were missing or undetected, there would be a second group of intervals of 52 Myr. Hence a frequency plot of lag times should confirm periodicity. He found no such confirmation using several definitions of mass extinction, and analysing Sepkoski's data in three ways: for all families; for those families with a robust skeleton which were likely to have reliable stratigraphic ranges; and for all genera (Figure 5.7).

Other criticisms of Raup and Sepkoski's original analysis (e.g. Hoffman, 1985) suggest: that the definition of mass extinctions is artificial; that the omission of extant families increases the significance of minor extinctions in the most recent past, and that the analysis is on a crude time-scale which inevitably tends to produce a periodicity. There is an element of truth in these criticisms. Identifying peaks merely by the presence of adjacent troughs is hardly rigorous. However, if one takes an arbitrary cut-off such as 10%, only two of Raup and Sepkoski's (1984) original peaks are removed, and only three troughs lie above this arbitrary level. Hoffman (1985) pointed out that because extant families were excluded from the analysis, the extinction of just five families in the Middle Miocene becomes a possible mass extinction, whereas the extinction of 19 families in the

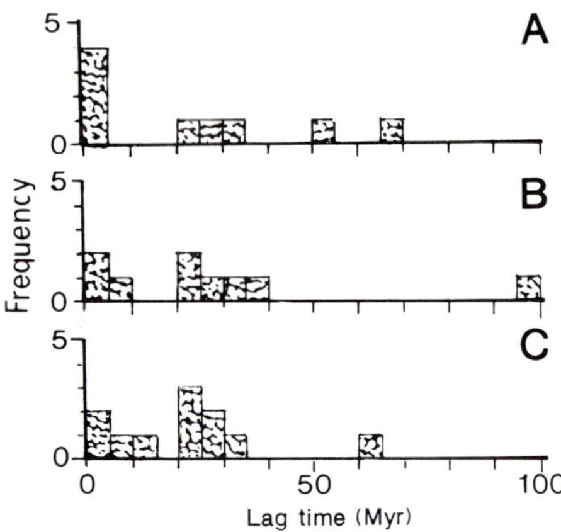

Figure 5.7. Frequency histograms of lag times (intervals between extinction peaks) over the last 260 Myr, for A: all marine families, B: selected marine families, and C: all marine genera. (Redrawn from Quinn 1987, Figure 3.)

Campanian represents a period of low extinction. As to time-scale, periodicities which are approximate multiples of the average interval are clearly more likely to occur than other values. If all intervals were 5 Myr, a periodicity of 22 Myr would be impossible, but one of 20 or 25 Myr highly probable. Raup and Sepkoski's (1984) intervals averaged 6.2 Myr but some of their analyses seem to show that only the 26 Myr interval fits sufficiently well to confirm a periodicity in their data. Equally, Sepkoski's (1986) and Fox's (1987) analyses of generic data are at a finer time-scale (an average interval of 3.5 Myr), yet produced a similar pattern, and still revealed a 26 Myr periodicity.

Shaw (1987) suggested that non-linear dynamics tends to generate periodicities. In linear dynamics changes of any action produce commensurate (and therefore predictable) changes in the reaction. Non-linear dynamics involves systems where changes in an action produce quite disparate (and therefore less easily predictable) reactions. The two are analogous to the difference between the bounce of a spherical soccer ball and that of an ellipsoidal rugby ball. Non-linear dynamics attempts to describe phenomena such as the patterns of bounce of rugby balls, and Shaw (1987) argued that these patterns are sometimes periodic. If true, the apparent periodicity of mass extinctions may not reflect any particular cause. Finally, familial or generic originations have yet to be subjected to a similar analysis. Is there periodicity in radiations?

Different authors using the same basic data but different analytical methods have clearly come to widely different conclusions about periodicity in extinction peaks. Note, however, that all the objections are potentially self-fulfilling prophecies. If we accept that periodicity in extinction rates is an artefact of the data, the method of analysis, or the perversity of nature, then it is not worthy of further investigation. Yet if we do not investigate it the true cause will never be known. Thus while doubt still justifiably prevails about periodicity in extinction rates, it is essential that the idea be pursued. Indeed, the real merit of all three major ideas discussed in this chapter (see p. 99) has been their stimulus to detailed collecting and documentation of the patterns preserved in the fossil record. Even if all three should eventually be rejected, they will have advanced the state of knowledge of the fossil record and rendered invaluable service to palaeontology and evolutionary science in general.

REFERENCES

Alvarez, L.W., Alvarez, W., Asaro, F. and Michel, H.V., 1980, Extraterrestrial cause for the Cretaceous–Tertiary extinction, *Science*, **208**: 1095–1108.

Benton, M.J., 1988, Mass extinctions in the fossil record of reptiles: paraphyly, patchiness, and periodicity (?) In G.P. Larwood (ed.), *Extinction and survival in the fossil record*, Systematics Association Special Volume **34**, Oxford University Press, Oxford, pp. 269–94.

Darwin, C. 1872, *The origin of species by means of natural selection* (6th edn), John Murray, London.

Drummond, P.V.O., 1983, The *Micraster* biostratigraphy of the Senonian White Chalk of Sussex, southern England, *Géologie Méditerranéenne*, **10**: 177–82.

Eldredge, N., 1972, Systematics and evolution of *Phacops rana* (Green, 1832) and *Phacops iowensis* (Delo, 1935) (Trilobita) in the Middle Devonian of North America, *Bulletin of the American Museum of Natural History*, **147**: 45–114.

Eldredge, N. and Gould, S.J., 1972. Punctuated equilibria, an alternative to phyletic gradualism. In T.J.M. Schopf (ed.), *Models in Paleobiology*, Wiley, New York, pp. 82–115.

Fortey, R.A., 1985, Gradualism and punctuated equilibria as competing and complementary theories, *Special Papers in Palaeontology*, **33**: 17–28.

Fox, W.T. 1987, Harmonic analysis of periodic extinctions. *Paleobiology*, **13**: 257–71.

Hallam, A. and Miller, A.I., 1988, Extinction and survival in the Bivalvia. In G.P. Larwood (ed.), *Extinction and survival in the fossil record*, Systematics Association Special Volume **34**, Oxford University Press, Oxford, pp. 121–138.

Hoffman, A. 1985, Patterns of family extinction depend upon the definition and geological timescale, *Nature*, **315**: 659–62.

Hoffman, A. and Ghiold, J., 1984, Randomness in the pattern of 'mass extinctions' and 'waves of origination', *Geological Magazine*, **122**: 1–4.

House, M.R., 1988, Extinction and survival in the Cephalopoda. In G.P. Larwood (ed.), *Extinction and survival in the fossil record*, Systematics Association Special Volume **34**, Oxford University Press, Oxford, pp. 139–54.

Lewin, R., 1986, Punctuated equilibrium is now old hat, *Science*, **231**: 672–3.

Nakazawa, K. and Runnegar, B., 1973, The Permian–Triassic Boundary: a crisis for bivalves? In A. Logan and L.V. Hills, (eds), *The Permian and Triassic Systems and their mutual boundary*, Canadian Society of Petroleum Geologists, Memoir 2, pp. 608–21.

Patterson, C. and Smith, A.B., 1987, Is the periodicity of extinctions a taxonomic artefact?, *Nature*, **330**: 248–52.

Paul, C.R.C., 1980, *The natural history of fossils*, Weidenfeld and Nicolson, London.

Paul, C.R.C., 1982, The adequacy of the fossil record. In K.A. Joysey and A.E. Friday (eds), *Problems of phylogenetic reconstruction*, Systematics Association Special Volume **21**, Academic Press, London, pp. 75–117.

Paul, C.R.C., 1985, The adequacy of the fossil record reconsidered, *Special Papers in Palaeontology*, **33**, 7–15.

Paul, C.R.C., 1988, Extinction and survival in the echinoderms. In G.P. Larwood (ed.), *Extinction and survival in the fossil record*, Systematics Association Special Volume **34**, Oxford University Press, Oxford, pp. 155–70.

Paul, C.R.C. and Smith, A.B., 1984, The early radiation and phylogeny of the echinoderms, *Biological Reviews*, **59**: 443–81.

Quinn, J.F., 1987, On the statistical detection of cycles in extinction in the marine fossil record, *Paleobiology*, **13**: 465–78.

Raup, D.M., 1978, Cohort analysis of generic survivorship, *Paleobiology*, **4**: 1–15.

Raup, D.M. and Sepkoski, J.J. Jr, 1982, Mass extinctions in the marine fossil record, *Science*, **215**: 1501–3.

Raup, D.M. and Sepkoski, J.J. Jr, 1984, Periodicity of extinctions in the geologic past, *Proceedings of the National Academy of Sciences of the USA*, **81**: 801–5.

Scrutton, C.T., 1988, Patterns of extinction and survival in Palaeozoic corals. In G.P. Larwood (ed.), *Extinction and survival in the fossil record*, Systematics Association Special Volume **34**, Oxford University Press, Oxford, pp. 65–88.

Sepkoski, J.J. Jr, 1979, A kinetic model of phanerozoic diversity II. Early Phanerozoic families and multiple equilibria, *Paleobiology*, **5**: 222–51.

Sepkoski, J.J. Jr, 1981, A factor analytic description of the Phanerozoic marine fossil record, *Paleobiology*, **7**: 36–53.

Sepkoski, J.J. Jr, 1982, A compendium of marine fossil families, *Contributions in Biology and Geology*, Milwaukee Public Museum, **51**: 1–125.

Sepkoski, J.J. Jr, 1986, Phanerozoic overview of mass extinctions. In D.M. Raup and D. Jablonski (eds), *Patterns and processes in the history of life*, Dahlem Konferenzen: Life Science Report **36**, Springer-Verlag, Berlin, pp. 277–95.

Sepkoski, J.J. Jr, Bambach, R.K., Raup, D.M. and Valentine, J.W., 1981, Phanerozoic marine diversity and the fossil record, *Nature*, **293**: 435–7.

Sepkoski, J.J. Jr and Hulver, M.L., 1985, An atlas of Phanerozoic clade diversity diagrams. In J.W. Valentine (ed.), *Phanerozoic diversity patterns—Profiles in Macroevolution*, Princeton University Press, Princeton, NJ, pp. 11–39.

Shaw, A.B., 1971, The butterfingered handmaiden, *Journal of Paleontology*, **45**: 1–5.

Shaw, H.R., 1987, The periodic structure of the natural record and non-linear dynamics, *Eos*, **68**: 1651–65.

Smith, A.B., 1984, *Echinoid palaeobiology*, George Allen & Unwin, London.

Smith, A.B., 1988, Patterns of diversification and extinction in early Palaeozoic echinoderms, *Palaeontology*, **31**: 799–828.

Sprinkle, J., 1973, *Morphology and evolution of blastozoan echinoderms*, Special Publication, Museum of Comparative Zoology, Harvard, Cambridge, MA.

Stokes, R.B., 1977, The echinoids *Micraster* and *Epiaster* from the Turonian and Senonian of England, *Palaeontology*, **20**: 805–21.

Taylor, P.D. and Larwood, G.P., 1988, Mass extinctions and the pattern of bryozoan evolution. In G.P. Larwood (ed.), *Extinction and survival in the fossil record*, Systematics Association Special Volume **34**, Oxford University Press, Oxford, pp. 99–119.

Ward, P.D. and Signor, P.W., 1985, Evolutionary patterns of Jurassic and Cretaceous ammonites. An analysis of clade shapes. In J.W. Valentine (ed.), *Phanerozoic diversity patterns—Profiles in Macroevolution*, Princeton University Press, Princeton, NJ, pp. 399–418.

Chapter 6

COLONISATION OF THE LAND

Paul A. Selden
and
Dianne Edwards

THE PHYSICAL ENVIRONMENT IN THE EARLY PALAEOZOIC

The image of the colonisation of the land by plants as a spectacular event, when barren wastes suddenly became verdant triggered by a change in some extrinsic environmental factor, is fading. It derived from the appearance and rapid diversification of vascular plants, unequivocal land colonisers, and, almost simultaneously, the earliest terrestrial arthropods, in the late Silurian and early Devonian. The environmental stimulus usually cited hitherto was a reduction in ultraviolet (UV) radiation due to ozone production when oxygen levels increased. A growing amount of evidence suggests that terrestrialisation was a gradual process, occurring throughout the earliest Palaeozoic, possibly with origins in the Precambrian. As the atmosphere evolved, particularly with respect to partial pressure of oxygen (P_{O_2}), so land surfaces were modified as mats of microbes and thallophytes produced substrates more convivial for colonisation by larger plants (macrophytes) and animals (Holland 1984; Retallack 1985).

Thus to define the physical parameters of atmosphere and hydrosphere associated with the colonisation of land is to trace their history throughout the early Palaeozoic. Direct evidence is lacking. Considering the hydrosphere, analysis of fluid inclusions in evaporites suggests that the major constituent composition of seawater, including Na^+ and Cl^-, has changed little over the last 900 Myr. Values from rocks as old as Permian confirm this general observation, but Silurian results are less secure (Holland *et al.* 1986). Freshwater systems may have been depleted in minerals prior to the establishment of extensive terrestrial vegetation and associated chemical weathering, even allowing for retention of some minerals within organic material in land plants (Holland 1984). Partial pressure of atmospheric CO_2 (P_{CO_2}) may have been higher than at present, while P_{O_2} would have been lower but may have increased markedly on the advent of terrestrial higher plants, a new source of photosynthetic oxygen production. The estimated minimum P_{O_2} value of about 10% of present atmospheric level (PAL) for the early Cambrian is based on the respiratory requirements of its diverse communities of aquatic metazoans, and it may well have reached 50% PAL towards the end of the Silurian. Both P_{O_2} and P_{CO_2} could have varied considerably, but Holland *et al.* (1986) argued that the 'continuity of animal life precludes wild fluctuations'. Based on putative P_{O_2} values, the density of ozone is unlikely to have increased significantly through the Palaeozoic; reduction in UV radiation was not a trigger for the appearance of the terrestrial higher plants.

Without extensive macrophyte cover, chemical and physical properties of terestrial substrates would have been very different (Beerbower 1985). Carbon dioxide generated in decaying vegetable matter promotes the release of major mono- and divalent cations from silicates and carbonates, thereby enhancing chemical weathering; thus early substrates would have been nutrient-poor. Low availability of potassium is indicated by the preponderance of potassium feldspars in pre-Silurian sediments. Without organic matter in the soil, there would have been no bacterial nitrogen fixation. Combined nitrogen levels would have been low, and input limited to nitrogen dissolved in rainwater. Restricted chemical weathering would have resulted in low production of clay minerals with silts predominating.

In the absence of rooted plants as windbreaks and sediment binders, erosion would have been greater, and sandy, rocky soils with high permeability and low water retention developed. The absence of abundant soil, organic matter and clays would also have limited retention of essential ions. Lack of sediment binding would have affected the nature of stream systems: braided streams would have predominated over sinuous ones, resulting in a shortage of stable moist environments such as point bars. Early terrestrial environments would consequently have been nutrient-poor, physically unstable, and subject to violent fluctuations in temperature and water content (Beerbower 1985). Exceptions may have been lowland alluvial or lacustrine areas with fine-grained water- and nutrient-retentive soils. It is in such environments that the first land plants of any size may have become established and created suitable habitats for terrestrial animals.

PHYSIOLOGICAL CONSIDERATIONS

Water

Water is essential for life: virtually all biochemical reactions work in an aqueous medium, and water is necessary for the transport of solutes.

Water is also necessary for the maintenance of cell volumes which are vital for the internal cell architecture, ensuring correct functioning of biochemical pathways, and for the turgor which provides support in plants. On land, water loss is a particular problem because water-vapour pressure is much lower in air than in the aquatic medium. Since availability of water is highly variable on land, adaptations are required to maintain a continuous supply or to reduce loss during drought periods. Water must be available in some form in every terrestrial environment for life to exist there at all, although supply may vary from the highly unpredictable flash flood in the otherwise arid hot desert to the constant humidity of the tropical forest. Indeed, if terrestrial and aquatic habitats did not interdigitate then it seems unlikely that life could ever have made the sudden leap from sea to land (Little 1983, p. 2).

All organisms can be placed into two broad categories depending on their responses to fluctuations in available water. *Homoiohydric* forms maintain an internally hydrated environment and are thus immune from external changes. *Poikilohydric* forms have no such control: intolerant examples are killed on desiccation, but tolerant forms possess ecophysiological adaptations to withstand it. Most homoiohydric plants are desiccation-tolerant.

Living terrestrial organisms can be classified into four main groups on the basis of their water relations (Little 1983, p. 3):

1. *Aquatic organisms*. Microscopic soil animals (e.g. nematodes, ostracodes, protozoans) and plants (e.g. unicellular algae) which can only survive immersed in a film of water are, strictly speaking, aquatic rather than terrestrial, and are poikilohydric and desiccation-intolerant.
2. *Cryptic forms*. These inhabit soil, leaf litter and tropical forests where a constant high humidity is assured. They differ from the first group in that the majority are macroscopic. They are intolerant of desiccation and may use behavioural adaptations to prevent it. Among animals, examples are earthworms, leeches, flatworms, slugs, woodlice, insect larvae, burrowing amphibians, and some myriapods. Plants include algae, some bryophytes and, most importantly, the gametophytes of homosporous pteridophytes. Although some of the latter may be subterranean, most require light for photosynthesis and thus live on the surface.
3. *Poikilohydric organisms with desiccation tolerance*. These, too, require high humidity for activity, but withstand desiccation by ecophysiological adaptations. The cytoplasm of some has the capacity to dry out and then rehydrate to function normally (cryptobiosis: e.g. cyanobacteria, algae, mosses and liverworts, rotifers, mites); others aestivate during dry periods (e.g. land snails, lungfish). Plants in this category are common primary colonisers of unstable environments. In some extreme instances, desiccation tolerance is exhibited by resting stages such as the seeds of angiosperms or the eggs of fairy shrimps, the organisms themselves being opportunists, flourishing and completing their life histories under ephemeral favourable conditions.

4. *Homoiohydric organisms*. These are the true conquerors of the land. As well as occupying the environments already mentioned, they have invaded a great diversity of arid niches, and created new habitats themselves. They maintain an internally hydrated environment, and include most tracheophytes, reptiles, birds, mammals, insects, arachnids, and some isopods and snails.

Homoiohydry has been achieved in plants and animals by the use of waterproof cuticle covering all aerially exposed parts. In both plants and arthropods, long-chain lipids (waxes) in the cuticle form the main impermeable barrier (for review, see Hadley 1981), and the lowest H_2O permeability coefficients recorded are of the order of 10^{-5} m s^{-1} for plant cuticles and 10^{-6} m s^{-1} for arthropod cuticles (Raven 1985). Lipids also occur in the epidermis of terrestrial vertebrates, but their waterproofing effectiveness is probably low compared to the keratinised layers in these animals. The waxes may have evolved initially as a network of patches on the cuticle to repel external water in small arthropods which obtain oxygen from air by diffusion through the cuticle. A water-impermeable cuticle may also help in maintaining internal osmotic concentration where that of the animal differs from that of the environment, thus it may have already evolved in aquatic ancestors. In plants, too, patchy wax would favour CO_2 diffusion, while its water-repellant properties would prevent the development of a water-film barrier to diffusion. Cuticles may also have conferred some protection from harmful UV radiation and from attack by fungi (and arthropods?).

Arthropod epicuticle, which contains water-repellant lipids, is rarely preserved in fossils, but in scorpions the waterproofing seems to extend into the upper part of the exocuticle, termed the hyaline exocuticle. Evidence for this comes partly from *in vivo* experiments on scorpions which indicate that water loss is not greatly increased by removal of the epicuticle (Hadley and Quinlan 1987). Hyaline exocuticle is known only from scorpions, and possibly xiphosurans and eurypterids. The presumed lipids appear to enhance the preservation potential of scorpion exocuticle when that of other arthropods oxidizes, since fossil scorpion cuticle is abundant in non-marine horizons in the Carboniferous to the virtual exclusion of all other arthropod remains (Bartram *et al.* 1987).

Water balance and transport

Primary aquatic plants may have specialised organs for attachment, but not for water absorption. The latter must have evolved in response to terrestrialisation, and the earliest were probably hair-like unicellular outgrowths (rhizoids) on horizontal underground stems (rhizomes). Compared with an extensive root system, the volume of substrate exploited by rhizomes is low and such a system is relatively inefficient. Before development of extensive plant cover, the substrate itself would have been deficient in mineral nutrients, e.g. phosphorus, and so it has been argued that the transmigrant algae, or earliest vascular plants, could have survived only in the presence of a symbiotic fungus in the absorbing system. Today, the roots of plants colonising nutrient-poor soils commonly possess such a fungus, termed a mycorrhiza. Hair-like fungal hyphae emanating from the roots greatly increase the volume of substrate exploited for minerals, while fungal metabolism produces phosphorus- and probably nitrogen-containing compounds which are subsequently absorbed by the higher plant. In addition, organic acids derived from the fungus enhance chemical weathering of the substrate.

Water-transport systems are known only in land plants and are best developed in homoiohydric forms. Flow of water is maintained by transpiration (evaporation from aerial surfaces), and the conducting cells, initially tracheids, have implosion-resistant modifications in the form of rings or helices formed from lignin. This polymer is unknown in algae, and may also have been involved in support in the early land colonisers. Only homoiohydric plants can achieve a height greater than 2 m in air, and this is made possible by additional structural tissues, a gas-exchange regulation system, and xylem, which is dead tissue with a specific conductance to water a million times greater than that of living parenchyma (Raven 1985). This greater burden of non-photosynthetic tissues (e.g. for absorption, conduction, support, and reproduction) in larger plants would have been supported by the greater phosynthetic yields possible on land. Higher O_2 concentrations would allow their growth and maintenance.

Water loss

Water loss in animals has been studied particularly closely in some arachnids, and especially desert-dwelling forms (for a review, see Little 1983, p. 109). It is not easy to obtain data recorded under comparable conditions for all arachnids, since water loss varies according to various factors, including air humidity, temperature, wind speed, and the activity and size of animal. The lowest rates of water loss, expressed as percentage of body weight per hour at close to 0% relative humidity, are found in desert living scorpions (0.032–0.091), ticks (0.004–0.033), and Solifugae (0.090) (Little 1983, Table 6.3). Because of the waterproofing of the cuticle in terrestrial arthropods, gas exchange occurs through specialised respiratory surfaces which are internalised to reduce water loss.

The major route of water loss in plants is via open stomata (see p. 134). Opening and closure is usually related to internal concentrations of CO_2, but under extreme water stress this control is overridden and the stomata close.

Water loss during excretion is solved in terrestrial animals by the production of insoluble waste. In general, relatively insoluble and non-toxic guanine or uric acid is excreted in preference to the widespread production of highly toxic and soluble ammonia or less toxic but soluble urea of aquatic ancestors.

Osmoregulation

Life originated in saltwater, and the body fluids of organisms contain solutes in aqueous solution. Osmosis is the movement of solvent from a dilute to a more concentrated solution across a semipermeable membrane (i.e. one that is impermeable to the solute). Osmosis is passive, and can be prevented by the application of hydrostatic pressure. Thus the difference in solution concentration can be expressed as the pressure required to prevent osmosis: osmotic pressure. Alternatively, the osmotic gradient which exists across a membrane can be visualised as a gradient in the potential for chemical activity. Water will pass passively from a dilute solution with high potential to a more concentrated solution with lower potential. Plant cell walls act against the resulting increase in volume so it is useful in botany to distinguish this counteracting pressure (wall or turgor potential) from the concentration gradient (solute or osmotic potential). Animal cell membranes offer little resistance to osmosis so the osmotic pressure referred to in this context is theoretical, and generally expressed as a concentration; osmotic concentration is a better description.

Osmotic concentration can be expressed in moles per litre (molarity), moles per kilogram (molality), osmoles (Osm) per litre (osmolarity), or osmoles (Osm) per kilogram (osmolality) (for details see Rankin and Davenport 1981), and osmotic potential in bars or pascals. The concern here is the difference in concentration between the environment (salt- or freshwater) and the body fluids. Solutions of the same osmotic concentration are iso-osmotic; if one solution has a greater osmotic concentration than another then the former is hyperosmotic and the latter hypo-osmotic. Isotonic, hypertonic, and hypotonic are almost equivalent terms used in botany.

Animals

The observation that animal body fluids resemble seawater in ionic composition, together with the fact that all animal phyla occur in the sea, yet several do not occur on land or in freshwater, provides strong evidence that all are primitively marine. The osmolarity of seawater is about 1000 mOsm l^{-1}. Radiation of animals into environments of greater osmolarity (hypersaline: 1200–2300 mOsm l^{-1}; brines: above 2300 mOsm l^{-1}) or lower osmolarity (brackish: 15–900 mOsm l^{-1}; fresh below 15 mOsm l^{-1}) required active mechanisms for the regulation of the osmotic concentration of their body fluids (osmoregulation), or changes in metabolism to tolerate low internal osmolarity. Accordingly, the present osmotic concentration of terrestrial animals can give clues to the routes taken onto land. Crustaceans have been most intensively studied because they occur in a wide range of salinities and habitats (for a review, see Little 1983).

Most marine crustaceans are osmoconformers, that is, their haemolymph osmotic concentration is essentially the same as that of seawater. Among the vertebrates, only the hagfishes are osmoconformers. The remaining marine fishes, and a few branchiopod crustaceans, are osmoregulators, maintaining their body fluids at concentrations well below that of seawater. Decapod crustaceans living in brackish estuaries hyporegulate or hyperregulate weakly, according to the external concentration. Shore-living crustaceans face particular problems: salinities can vary from strongly hypersaline in evaporating pools to fresh during rainstorms. Therefore, like estuarine forms, littoral crustaceans are usually weak osmoregulators but are euryhaline, i.e. tolerators of a wide range of salinities.

On the other hand, freshwater crustaceans hyperregulate, some strongly, maintaining osmotic concentrations close to that of seawater even in freshwater. Others have very weak haemolymph concentrations and their regulatory ability breaks down in brackish water; these can be termed stenohaline freshwater forms (as opposed to stenohaline marine osmoconformers), and include freshwater crayfish, amphipods, and most freshwater fish.

To summarise, freshwater-adapted animals are mainly regulators, whereas littoral and estuarine forms are tolerators of a wide range of environmental osmotic concentrations; in this respect, the tolerators are better adapted to the variable terrestrial environment.

Plants

Cell water relations in plants are fundamentally different to those in animals due to the presence of an enclosing, relatively inextensible, cellulose cell wall and an intracellular vacuole. The latter provides the opportunity for using toxic substances, such as Na^+ and Cl^-, as osmoregulatory material, isolating them from sites of general metabolic activity. When such a cell is placed in water or a hypotonic medium, water will tend to enter in response to the concentration of organic and inorganic solutes in the vacuole and cytoplasm (the osmotic potential of the cell), but as the cell increases in volume there is an opposing pressure from the cell wall (turgor potential). The resultant of these two forces is called the water potential of the cell, and at equilibrium is equivalent to the potential of the bathing medium. Such cells are said to be turgid, and this condition is essential for both metabolism and growth, and incidentally provides hydrostatic support. For most vascular plants the osmotic potential of cell contents (-7 to -20 bars), and hence the water potential, is lower than that of the soil solution, and water is taken in. Halophytes take up water because they have much lower cell osmotic potentials (see Table 6.1). Vascular plants, marine and some freshwater algae respond to changes in environmental water potential by maintaining turgor (turgor regulation) through osmotic adjustment, i.e. by removal or addition of solutes (Turner and Jones 1980). However, certain freshwater characean green algae, believed to be related to the group from which vascular plants evolved, maintain constant osmotic potential (osmolarity regulation), but not turgor. More needs to be known of the distribution of turgor and osmolarity regulation mechanisms in

Table 6.1. Ranges in approximate values of osmotic potential in ecologically different plants (from various sources).

Plant group	*Habitat*	*Osmotic potential (bars)*
angiosperms	freshwater	−2 to −16
	mesic meadows (herbs)	−5.5 to −20
	dry woodland (herbs)	−8 to −32
	deciduous woodland (trees)	−10 to −27
	desert	−15.5 to −40
	mangroves (halophytes)	−30 to −40
	saltmarshes (halophytes)	−25 to −50
thalloid liverworts	moist	−5 to −15
mosses	mesic to dry	−15 to −25
	very dry	−30 to −60
algae brown	marine	−24 to −32
green	marine	−40
green	freshwater	−1.5 to −8
diatoms	freshwater	−4 to −8
cyanobacteria	freshwater	−5

pteridophytes and charophytes before hypotheses relating osmotic properties and routes on to land can be applied to plants (Raven 1985).

Temperature

Temperature, like many of the other physical properties discussed here, is far more variable and unpredictable on land than in water; in part this is because the rate of transfer of heat in air is much faster than in water. Cold temperatures tend to slow down biochemical reactions, and problems are caused by the correlation of temperature with other factors, such as water loss. Animal life in hot deserts is made possible through numerous behavioural modifications such as burrowing, nocturnal activity, and 'stilting' gaits over hot surfaces. Temperature control is also achieved in cold climates by the antifreeze properties of the haemolymph of some arthropods. Homoiothermy was made possible through the transition to land life and the greater availability of oxygen (see below) for metabolism.

Upper temperature survival limits (45–50°C) are similar in plants and animals but there are no active mechanisms for temperature control in plants. Transpiration can reduce overheating but is of little use in xerophytes (plants adapted for life in very dry environments) which often have to endure high temperatures accompanied by water stress. Indeed, most individuals occasionally sun-scald or photosynthesise suboptimally in elevated temperatures. In contrast, certain plants can become physiologically acclimated to survive low temperature, the most important environmental factor which controls plant distribution. In some cases the cytoplasm itself may not freeze, even in temperatures as low as −40°C, but ice crystals form extracellularly. Most information comes from seed plants, particularly crops, and little is known of low-temperature responses in pteridophytes, particularly of the free-living sexual phase (gametophyte) in their life cycle. More data are available for mosses, which in the dehydrated state can survive major extremes of temperature (−100°C to +100°C). They are not so tolerant when moist, except for certain boreal and arctic mosses, some of which can survive wetting by snow melt during the day and encasing in ice at −20°C overnight (Richardson 1981).

Gas exchange

There is more oxygen in air (8.65 mol m^{-3}) than in water (0.262 mol m^{-3}) and the potential flux of O_2 across a respiratory membrane is 0.17 mol m^{-2} s^{-1} in air and 4.2×10^{-5} mol m^{-2} s^{-1} in water (Raven 1985). However, its availability to organisms depends on a variety of physical and biological factors; for example, since diffusion to one side of a membrane causes depletion at the other, there must be constant replenishment. For this reason, aquatic animals normally irrigate their gills mechanically by pumping water over them, by vibrating them, or by the use of ciliary currents. In air, simple diffusion is adequate for many invertebrates, but ventilation lungs occur particularly in the more active or homoiothermic animals. Other factors which govern the rate of diffusion of O_2 at the respiratory membrane are: the efficiency of oxygen-binding molecules; the circulatory system; and diffusion rates within the tissues. The effects of these factors can be measured as the percentage difference in oxygen content between the inhalent and exhalent respiratory currents, which is termed utilisation. Figures for utilisation exceed 50% (and may reach 80%) in fish and cephalopods, vary from around 30% to 75% in polychaetes and crustaceans, but are generally well below 10% in bivalves (Jones 1972). In higher vertebrates utilisation figures lie in the middle range, being around 22% in man and 31% in the pigeon. Finally, the metabolic costs of ventilation and irrigation must be considered. It is costly in energy to move large amounts of water, in contrast to the far less dense and less viscous air, yet irrigation is far more important for the supply of oxygen in aquatic animals than is ventilation in air-breathers. Particularly in small animals, the cost of supplying oxygen to muscles to improve irrigation must be carefully balanced against the improved oxygen uptake produced.

The advantages to be gained by breathing air are offset to some extent by the problem of water leakage across respiratory membranes, since the H_2O molecule is smaller than the O_2 molecule. To reduce water loss, air-breathing organs (lungs) are internalised, in contrast to gills, which are evaginations for gas exchange in water. The methods by which animals have made the transition from obtaining oxygen in water to air-breathing is the main concern here, a subject complicated by the great variety of oxygen-uptake methods employed by animals living in transitional environments.

Cutaneous gas exchange

Cutaneous respiration alone is sufficient for small, aquatic animals, but this method is also used in many other groups in conjunction with respiratory organs. Cutaneous respiration in air occurs in virtually all terrestrial animals to some degree, but is particularly well developed in the amphibians. Air-breathing amphibians utilise the skin as well as the lungs and mouth cavity for gas exchange. In some frogs, for example, the skin accounts for 30–50% of oxygen uptake, but 75–86% of carbon dioxide loss (Jones 1972). In this, as in other amphibians studied, temperature, and hence seasonality, affects the relative percentages of oxygen obtained by pulmonary, buccal, and cutaneous gas exchange. The importance of cutaneous respiration in amphibians is shown by observations reported in Jones (1972, p. 45) that some newt species utilise the skin for all oxygen uptake in air, but need to use lungs in addition when in water!

Gills

Gills are used for breathing air in many Crustacea and some amphibious fish. Terrestrial amphipods and isopods respire partly through the skin but mainly through the gills, and in isopods an additional pseudotracheal system has developed on the pleopods. The gills of air-breathing crabs remain moist in the branchial chamber, and in many species accessory lungs are also developed (see below). The gills of Palaeozoic aquatic scorpions may have been used for air-breathing since it is likely that the book-lungs of these arachnids have evolved directly from book-gills. Air-breathing by arthropod gills can be achieved simply by selective thickening of the cuticle to strengthen gill lamellae supports and the development of struts to keep the lamellae apart to allow air circulation. In fish such adaptations are not possible and additional mechanisms for oxygen uptake, or tolerance of anaerobic metabolism, are used by amphibious fish.

Lungs—the gastropods

True lungs occur in the gastropod molluscs, arthropods and vertebrates. In the gastropods, six families of prosobranchs and 30 of pulmonates have terrestrial representatives (see Little 1983, p. 34). In all these, the lung is formed from a highly vascularised part of the mantle cavity wall, and gills are reduced or lost. Secondary gills are developed in freshwater snails, and are used together with cutaneous and pulmonary respiration in relative proportions dependent on habitat. In the terrestrial pulmonates these lungs are ventilated by air inhaled by muscular action of the mantle cavity, in contrast to the lungs of prosobranchs which rely on diffusion alone.

Crustacean lungs

The lungs of decapod crustaceans are, like those of the pulmonate molluscs, highly vascularised areas of the branchial chamber wall. There is a general trend among crabs to reduction in gill area and increase in lung area with increasing terrestrial habitat, but gills are never lost. Three respiratory strategies occur in air-breathing crabs (Greenaway 1984). In the first, appendage activity causes inhalation of air at the bases of the legs. This passes through the gills, which are stiffened to prevent collapse and facilitate drainage after emersion, and then into the lungs. In the second strategy, aerial respiration via lungs merely supplements the gills. In these crabs, water is held in the branchial chamber, and circulated by the scaphognathites (flattened branch of the maxilla) over hair patches where the water is reoxygenated before being passed back over the gills. The most highly terrestrialised 'land' crabs (e.g. *Holthuisiana*) adopt a third strategy. These crabs utilise a perfusion lung, inhaling and exhaling air by means of thoracic musculature; the gills are not used in aerial respiration.

Chelicerate respiration

Microscopic palpigrades and some mites, among the arachnids, have no specialised respiratory organs. Extant scorpions, uropygids, amblypygids, schizomids, and some spiders have only book-lungs. Most spiders, however, use book-lungs and tracheae (Little 1983, p. 114). Book-lungs are internal, lamellate structures in the opisthosoma, opening by a small pore, the stigma, and are homologous with the book-gills of aquatic chelicerates: *Limulus* and the aquatic Palaeozoic scorpions. It would appear that scorpions colonised land without the use of secondary lungs. The so-called 'gill' of the extinct eurypterids closely resembles crab lungs or isopod pseudotracheae and could not have functioned as a gill; true gills, presumed to have been present, are as yet undiscovered (Selden 1985). Sieve tracheae are developed from book-lungs and are similar to them with the exception that cylinders replace the lamellae; they occur most commonly in small spiders or those living in xeric habitats, as well as in ricinuleids and pseudoscorpions. Tube tracheae, on the other hand, are a secondary development of ramifying diverticula arising from the atrium (entrance chamber) of the book-lung; the lung itself may be lost. The fast and ferocious desert-dwelling solifugids utilise an extensive, ventilated tracheal system (tube tracheae also occur in opilionids, and some spiders and mites). Because of the morphological similarity between book-gills and book-lungs, it is not easy to distinguish them in fossils unless stigmata or similar restricted openings are present.

Uniramian respiration

Two small groups of myriapods, Pauropoda and symphyla, are small, primitive, and cannot withstand desiccation; they obtain oxygen either cutaneously or by the use of tracheae. Chilopoda and Diplopoda breathe by means of tracheae also, as do the Onychophora in spite of their flexible cuticle which lacks a wax layer. The hexapods apparently evolved on land from myriapod-like uniramians. They have a well-developed tracheal system which in pterygotes, unlike that of most myriapods and apterygotes, includes a mechanism for closure of the spiracles to reduce water loss.

Vertebrate lungs

Vertebrate lungs are generally thought to have arisen in fishes living in those freshwater environments where oxygen levels are very low. However, there is no reason why lungs could not have developed in marine, perhaps intertidal, conditions, since air-breathing occurs in gobies and blennies, and most Devonian lobe-finned fishes were marine (Thompson 1980). Many modern fishes from numerous families breathe air to supplement the poor oxygen supply in swamps and seasonally anaerobic waters. Commonly a highly vascularised part of the buccal and pharyngeal walls is used, but in some species more distal parts of the alimentary canal serve for oxygen uptake, and expired air is regurgitated or passed out through the anus. Comparison with the secondary lungs of pulmonate gastropods and decapod crustaceans is obvious. However, it is the separate pharyngeal gas bladder, which has arisen numerous times among the fishes, which eventually gave rise to the tetrapod lung. Though modern teleosts use it as a hydrostatic organ, it is quite probable that the swim bladder arose for air-breathing. In lungfish, the lung arises primitively from the ventral side of the pharynx (the teleost swim bladder is dorsal in position), and coupled with this are a series of circulatory and physiological modifications (for a review, see Little 1983, p. 183). *Neoceratodus* is essentially a gill-breather and uses its lungs only as accessory air-breathing organs. However, *Lepidosiren* and *Protopterus* are obligate air-breathers and use the much reduced gills primarily to eliminate CO_2.

The bimodal system of gas exchange of the lungfish, in which O_2 is taken up primarily by the lungs but CO_2 is expelled mainly through the gills, also occurs in amphibians. The earliest tetrapods had scaly skin, so the respiratory system of modern amphibians, utilising cutaneous, buccal, and pulmonary gas-exchange, may therefore be a secondary adaptation (Gans 1970). However, rich faunas of Lower Carboniferous tetrapods from Scotland (Milner *et al.* 1986) suggest a broad diversity of forms which probably originated in the Devonian. Modern anurans inhale air by means of a buccal pump (their dorsoventrally flattened heads may be an adaptation to increase the efficiency of this), but exhale using flank muscles; higher tetrapods have well-developed ribs (except chelonians, which use hyoid cartilages) and thoracic musculature for inhalation and exhalation (aspiration ventilation). As soon as the tetrapod gait developed, and the body could be supported off the ground, it is likely that aspiration ventilation developed and the gills became reduced. Carbon dioxide could be lost through the skin, but there would nevertheless be a rise in P_{CO_2} in the blood (acidosis). Internal CO_2 receptors developed, which, coupled with the improvement in pulmonary breathing, would have come to control the ventilation rate, and short-term acid-base regulation to prevent harmful acidosis would thus have passed from the gills to the lungs. With the need for longer-term acid-base regulation out of water, this role passed to the kidneys, and the skin could become waterproof and scaly; this stage was apparently reached in the labyrinthodonts (Little 1983, p. 198).

Plants

For photosynthesising land plants, maintenance of a sufficient supply of CO_2 to green cells would have been the major problem (Raven 1985), there being no mechanical ventilation. Increase in area for uptake is achieved by the development of an intercellular space system with maximum exposure of the surfaces of mesophyll (photosynthetic) cells to this internal atmosphere. Such surfaces must be moist and there results a problem of excessive water loss. The latter is reduced by the internalisation of the surfaces for gas exchange and cuticularisation of the non-photosynthetic epidermis. Contact with the atmosphere is via pores with variable aperture, the stomata.

Conclusions

The major limiting factor in the change from the aquatic life to air-breathing in animals is not the supply of oxygen but the problem of CO_2 excretion in air. In tetrapods a secondary perfusion lung and bimodal gas exchange developed; a strikingly similar system occurs in land crabs, which may be considered to be at the 'lungfish level' of evolution in this comparison (Innes and Taylor 1986). Apparently, secondary lungs developed in eurypterids but this group never became fully terrestrialised (Selden 1985). In contrast, the closely related arachnids developed a diffusion lung and became terrestrial, but, as Little (1983, p. 117) pointed out, the arachnid book-lung is an inefficient organ, existing only because it evolved directly from the book-gill, and it is most improbable that it would have evolved directly in any group which originated on land. Tracheae have been developed by isopod Crustacea, arachnids, and uniramians, apparently after they emerged onto land. Because rates of water loss are high in tracheate arthropods without closeable spiracles, it is likely that emergence from the cryptozoic niche only occurred after the development of spiracular closing mechanisms; a striking parallel therefore exists between tracheates and land plants, with their stomata.

Support and locomotion

Smaller land plants, such as mosses, rely on a hydrostatic skeleton, and turgid tissues are also important for support in some vascular plants. However, most homoiohydric vascular plants also possess thick-walled structural tissues, where cell walls are composed of extra cellulose or cellulose impregnated with lignin. The latter polymer probably evolved as the rigid wall component in xylem, but its use in extraxylary structural tissues allowed the achievement of the increased height permitted by homoiohydry. In addition, there is considerable evidence in early vascular plants that support was achieved by virtue of growth in dense, monospecific stands.

Small animals such as slugs and worms can use hydrostatic skeletons on land. Arthropods and tetrapods moving onto land evolved the hanging stance for stability, and both groups developed some form of ankle joint to prevent the newly acquired plantigrade foot from twisting on the ground, and thereby being abraded and losing grip. Such skeletal adaptations are among the most useful indicators of terrestriality in fossils. Aquatic scorpions retain the digitigrade foot, for example. Arthropods become vulnerable to predation and desiccation during moulting, and it is possible that pioneer terrestrial forms returned to the water for ecdysis (Dalingwater 1985).

Sense organs

Eyes used in air differ from aquatic visual organs because of the differences in refractive index of the two media, and organs of hearing used in air are capable of perceiving higher-frequency sounds than in water. The stridulatory organs found on fossil trigonotarbid arachnids from the Devonian of Aberdeenshire and New York suggest their terrestrial mode of life, and their trichobothria (fine hairs which respond to air vibrations) prove it (Shear *et al.* 1987).

Digestion

Størmer (1976) emphasised the development of a preoral cavity for external digestion in terrestrial arachnids, myriapods and insects, a feature which is discernible in fossils.

Reproduction and dispersal

Organs for internal fertilisation and elaborate courtship devices are indicative of, but not exclusive to, terrestrial forms. Extant decapod Crustacea and amphibians return to the water to breed, but the amniote egg of higher tetrapods has removed the dependence on aquatic environments in all stages in their life cycles. Insect eggs have a complex coat to prevent both drowning and water loss. In extant bryophytes and pteridophytes the male gametes are motile and require a film of water for external fertilisation. Internal fertilisation is seen in seed plants, which appeared in the Late Devonian. Dispersal in early land plants was achieved by asexual meiospores produced on the sporophyte; meiospore walls are impregnated with sporopollenin which may originally have evolved as a chemical defence against infection in algae. These protective walls also confer resistance to desiccation and possibly UV radiation in both poikilohydric bryophytes and homoiohydric pteridophytes.

THE NATURE OF THE FOSSIL EVIDENCE

Palaeontological evidence for colonisation of the land comes from two sources: external evidence from the sedimentary environment, including palaeosols and trace fossils; and morphological and anatomical evidence from the fossils themselves.

External evidence

The Rhynie Chert in Aberdeenshire (Table 6.2) is the only example of an early, autochthonous, petrified, terrestrial ecosystem, providing a unique glimpse of a wetland community in Devonian times. Its exact age remains uncertain, but is probably Pragian/Emsian. The chert comprises thin accumulations of peat showing clear evidence of fungal decomposition, separated by plants in growth position. Parts of the peat surface were covered by cyanobacterial mats, and algae also flourished in surface pools. Animal remains in the Rhynie Chert include freshwater crustaceans, predatory trigonotarbid arachnids, and rare mites and collembolans. Petrological evidence suggests that there were several periods of flooding by siliceous water, each separated by a significant time-lag. The source of the silica may have been from the weathering of volcanigenic sediments or from hot springs.

There are occasional records of plants preserved in growth position in clastic sediments. Schweitzer (1983) recorded *Sciadophyton laxum*, thought to be the gametophyte of *Zosterophyllum rhenanum*, in environments interpreted as intertidal or lower saltmarsh, in the Rhenish Lower Devonian. The associated fauna includes eurypterids and the bivalve *Modiolopsis*. Rayner (1984) illustrated Scottish Emsian *Drepanophycus spinaeformis* rooted in fluviatile sediments.

Fossils are recovered from sediments deposited in continental, brackish-water and marine environments. As all these sediments are water-lain, the

Table 6.2. The three major Devonian terrestrial faunas compared. Key: Flora (not exhaustive): *L*. lycopsids, including prelycopods; *R*. rhyniophytes; *P*. progymnospermopsids; *A*. Algae; *F*. Fungi. Aquatic/amphibious fauna: *E*. eurypterids; *S*. scorpions; *X*. xiphosurids; *C*. freshwater crustaceans; *Other* includes molluscs and fish. Terrestrial fauna: *T*. trigonotarbids; *S*. spiders; *M*. mites; *P*. pseudoscorpions; *A*. arthropleurids; *C*. centipedes; *c*. collembolans. Abundance: √ = presence, c = common, r = rare, ? = questionable record.

				Flora					*Aquatic/amphibious fauna*					*Terrestrial fauna*						
Locality	*Age*	*Lithology*	*Habitat*	*L*	*R*	*P*	*A*	*F*	*E*	*S*	*X*	*C*	*Other*	*T*	*S*	*M*	*P*	*A*	*C*	*c*
Gilboa, New York	Givetian 380 Myr	Grey shale	Freshwater delta swamp	√		√			r	?				c	?	r	r	√	√	
Alken-an-der-Mosel, Germany	Emsian 390 Myr	Black shale	Brackish lagoon	√	√		√		c		√	√	c	r	?			r		
Rhynie, Scotland	Pragian–Emsian 400 Myr	Chert	Terrestrial bog near hot spring	√	√		√	√				c		c	?	r				r

original habitats of the entombed organisms remain obscure and recognition of land colonisers is critical.

The earliest Precambrian palaeosols merely indicate emergent subaerial surfaces and are termed extinct or abiotic (Holland 1984). Later Precambrian representatives (from 2400 Myr ago onwards), with abundant disseminated carbon, may indicate the activities of microbial mats. Retallack (1985) described in detail early Palaeozoic examples and found indirect evidence of a soil fauna from bioturbation structures.

Trace fossil evidence of land life is equivocal because the crawling traces tend to be preserved in wet, cohesive sediment, and could have been made by amphibious animals, or subaqueously (see reviews by Pollard 1985; Rolfe 1980; and Selden 1984).

Morphology and anatomy

Direct comparison with living relatives can be used as a criterion of terrestriality in some fossil groups; this is only possible if extant relatives exist and the preservation of the fossils is sufficiently good to be certain of affinity. Thus Silurian *Baragwanathia*, preserved as impressions, is considered a terrestrial vascular plant because it shares sufficient morphological characters with later Lower Devonian forms which on grounds of morphology and anatomy are closely related to lycopods. Silurian *Cooksonia* on the other hand, which is known only from marine sediments, lacks cuticle, stomata, and vascular tissues. Its claim to land-plant status rests on its axial architecture and hence presumed self-supporting erect growth habit and *in situ* spores with sporopollenin. It shows little morphological similarity with extant pteridophytes. The discovery (Edwards *et al.* 1986) of support tissues (sterome) and stomata in Lower Devonian representatives establishes the land-plant status of *Cooksonia* by this time, but does nothing to confirm the affinity or habitat of Silurian examples. In the case of the earliest land plants which are morphologically completely different from later and extant groups (e.g. Rhyniophytina, Trimerophytina, Zosterophyllophytina), anatomical evidence can be used to demonstrate land status. Thus in the common coalified compression fossils of vascular plants, anatomical and biochemical adaptations associated with homoiohydry (e.g. cuticle with cutin, xylem with lignin) have high fossilisation potential and persist when all other tissues are converted to coal. Indeed, because abundant strengthening tissues may have slowed down decay as well as maintaining the integrity of the whole organ, they increase fossilisation potential. Such plants are probably preferentially represented in the fossil record, while vascular plants with hydrostatic skeletons (e.g. *Rhynia gwynne-vaughanii* from the Rhynie Chert) do not occur as compression fossils in more typical Old Red Sandstone assemblages (Edwards 1979). Similarly, poikilohydric plants, e.g. bryophytes, will also be under-represented.

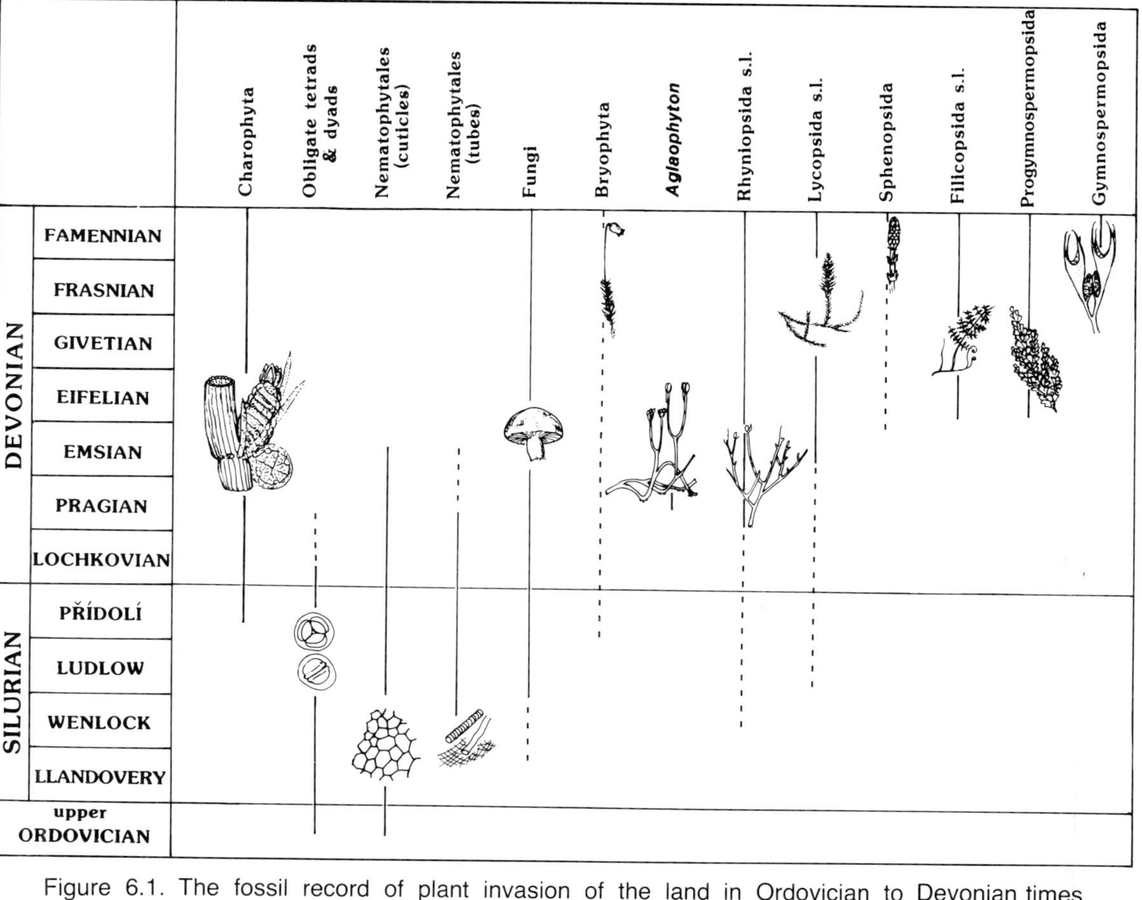

Figure 6.1. The fossil record of plant invasion of the land in Ordovician to Devonian times.

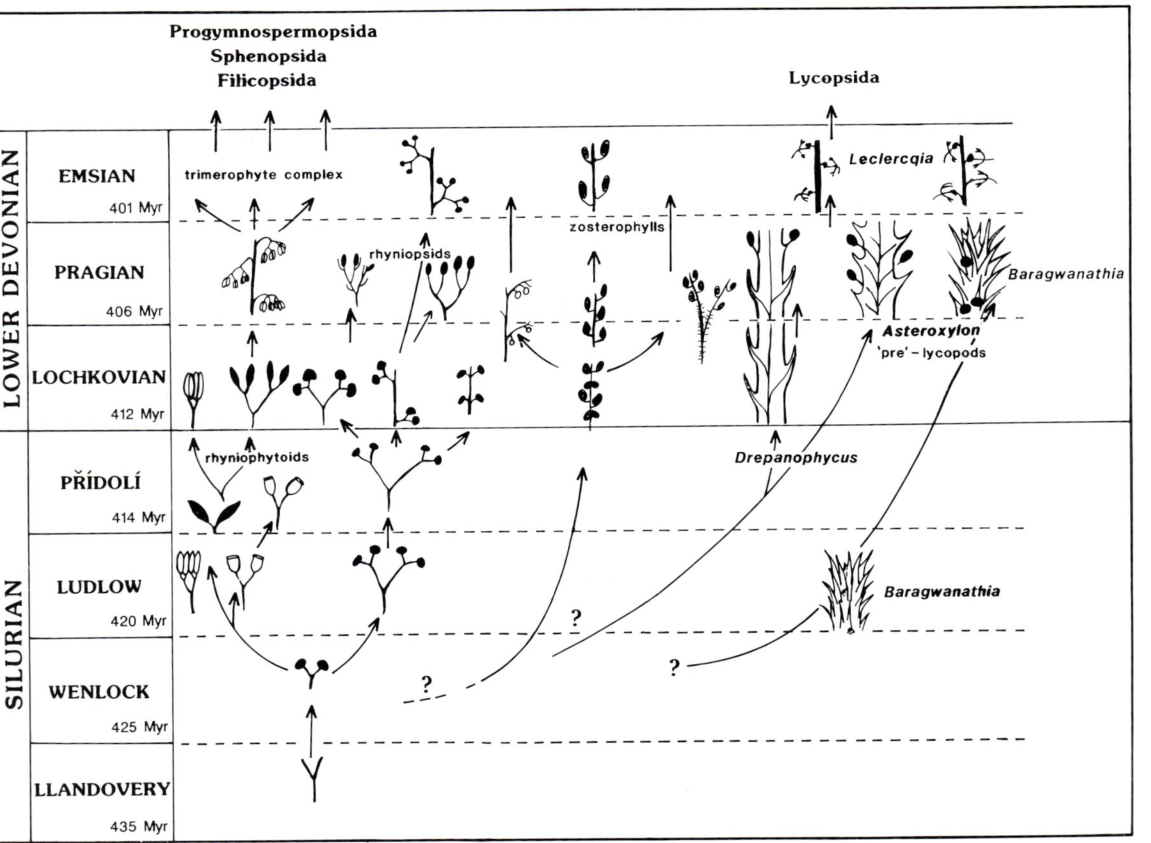

Figure 6.2. Possible phylogeny of early vascular plants during the Silurian and Devonian, showing increasing complexity of organisation.

The best three-dimensional anatomy is found in permineralisations, where cells are preserved in pyrite, calcite, or silica (Edwards 1986). In the Rhynie Chert, *Rhynia gwynne-vaughanii* has all the attributes of a vascular land plant, including an intercellular space system and rhizomes with rhizoids. The associated *Aglaophyton (Rhynia) major*, which is not very dissimilar and would undoubtedly also be placed in the Rhyniophytina were it preserved as a compression fossil, is shown to have a moss-like conducting strand and not tracheids. It thus possesses a new combination of moss and homoiohydric vascular plant characters (D.S. Edwards 1986). Similarly, while all extant scorpions are terrestrial, the presence of external gills and a digitigrade tarsus, as well as sedimentological evidence, confirms the view that early scorpions were aquatic.

The potential of microfossils in the recognition of terrestrial organisms has only recently been exploited. Spores and fragments of cuticle extend the record of land plants into the Ordovician (Figures 6.1 and 6.2), although the types of plant that produced them remain highly conjectural (Gray 1985). Micropalaeontological techniques are similarly proving extremely useful in the study of the earliest terrestrial animals (Shear *et al.* 1987).

THE FOSSIL RECORD

Plants

Colonisation of the land by animals was achieved by a number of groups at different times. In contrast, the plant story appears very simple. This may merely mask our ignorance, as the record is predominantly of vascular plants. Thus algae may have invaded the land on a number of occasions, but such attempts went unrecorded because of their low fossilisation potential. Similarities between bryophytes and tracheophytes, particularly the egg-containing apparatus (archegonium), indicate a common ancestor. Whether the presumed archegoniate intermediate migrated onto land just once, or whether it evolved on land, perhaps as a drought evader or with ecophysiological tolerance, or indeed whether there were several independent migrations from semi-aquatic forms, cannot be determined. Whatever their origins, poikilohydric bryophytes and homoiohydric tracheophytes as we know them evolved on land, and probably more or less at the same time (although the earliest moss, *Muscites*, is recorded in the Carboniferous and the liverwort *Palaviciniites* in the Upper Devonian). Possible earlier examples are the Lower Devonian *Sporogonites* with some moss characters (unbranched fertile axes, stomata immediately below the sporangium), and Přídolí *Tortilicaulis* with an unbranched twisted axis characteristic of liverworts. Evidence for diversity among the early colonisers comes from anatomically preserved plants with anomalous conducting tissues formerly considered tracheophytes (Edwards and Edwards 1986). These include

Taeniocrada dubia and *Aglaophyton major*. However, in the absence of gametophytes, relationships to archegoniates cannot be determined.

As to the tracheophytes themselves, the earliest examples with *in situ* tracheids occur in Ludlow sediments. *Cooksonia* is usually considered the earliest fertile member, but Silurian representatives, including the oldest Wenlock ones, lack cuticle, stomata, and vascular tissues (Edwards *et al.* 1983). Their claim to land-plant status rests on their axial architecture and hence presumed self-supporting erect growth habit, and *in situ* spores with sporopollenin. Support tissues and stomata in Lower Devonian representatives confirm the land-plant status, but no relationship with the tracheophytes. The earliest fertile rhyniophytes with *in situ* tracheids are Pragian, e.g. *Uskiella spargens*. They persisted in small numbers into the Emsian, but because of their predominantly determinate mode of growth, limited sporing capacity, and small size they were outcompeted by the Trimerophytina from which all major lines of vascular plants, except the club mosses, evolved.

Also present in marine Ludlow sediments from Australia is *Baragwanathia longifolia*. Anatomical and morphological features in Lower Devonian examples demonstrate lycopod affinities, although the diagnostic relationship between sporophyll and single axillary sporangium has not been demonstrated. Based on records from Old Red Sandstone localities in the present northern hemisphere (Edwards and Fanning 1985), lycopods are traditionally thought to have evolved from the Zosterophyllophytina, fertile examples of which do not appear until the Lochkovian. Thus it is possible that plants with tracheids evolved more than once (i.e. the zosterophyll and rhyniophyte lines), but it has to be emphasised that this would have occurred on land.

Spores

The earliest spores thought to derive from vascular plants or their immediate ancestors occur in Llandovery sediments. Older microfossils suggest the existence of a completely different kind of land vegetation (Gray 1985). Both bryophytes and pteridophytes possess spores with triradiate marks, and there is currently no way of distinguishing between them in the dispersed spore record. Gray argued most persuasively that the dispersed permanent (obligate) tetrads (Figures 6.1 and 6.2) found in palynomorph assemblages from late Ordovician to late Silurian sediments derive from land plants at the liverwort grade of organisation, with either ecophysiological adaptations to drought or ephemeral vegetative stages. The affinities of the tetrad producers, e.g. as bryophyte ancestors, are highly conjectural.

Nematophytales

Associated with the earliest tetrads in Ordovician sediments, and present throughout the Silurian and Lower Devonian, are fragments of cuticle with a reticulate pattern quite unlike that produced by the epidermis of vascular plants. Assuming that the functions of the cuticles are similar, these microfossils provide the most compelling evidence for terrestrial vegetation in Ordovician times, although the nature of the organism below the cuticle still eludes us. Later in the Silurian and Devonian such cuticles are found associated with wefts of large and small tubes, some of the former possessing thickenings reminiscent of tracheids. Lang (1937), calling this thalloid organism *Nematothallus*, believed that its unique organisation indicated a new and now extinct phylum of land plants that were neither algal nor tracheophytes. Certainly the abundance of such cuticles in Lower Devonian sediments indicates that *Nematothallus* was an important coloniser of the land.

Fungi

Abundant vegetation on land would eventually have resulted in accumulation of plant matter forming substrates for exploitation by terrestrial fungi. The Rhynie Chert provides excellent examples of such saprotrophs (Stubblefield and Taylor 1988). The evolution of cutin, lignin, and sporopollenin would have presented new sources of energy for fungi. Lignin, for example, can be metabolised only by basidiomycetes and a few ascomycetes. Hyphae and spores of terrestrial Ascomycetes are recorded in the Silurian (Sherwood-Pyke and Gray 1985) and thyriothecia typical of extant Hemisphaeriaceae in the Lower Devonian (Stubblefield and Taylor 1988). Hyphae present in an Upper Devonian wood, which shows a pattern of decay similar to that in white rot, may belong to ascomycetes but are more probably basidiomycetes. Mycorrhizal associations between Zygomycetes and higher plants are suggested by the presence of resting spores in axes of Rhynie Chert plants and *Psilophyton* (Stubblefield and Taylor 1988).

Animals

Figure 6.3 shows the fossil record of terrestrial animals, which is sparse (Rolfe 1980; 1985). There were many animal invasions of the land, at least one for each class, and numerous separate invasions within arthropod groups such as the Chelicerata and the Crustacea. The record suggests that the primary period for these terrestrialisations was the Silurian, and most terrestrial animal groups invaded the land alongside the plants. The major exception to this is the Crustacea, groups of which are attempting colonisation at the present time.

The oldest terrestrial animals may be the Siluro-Devonian kampecarid myriapods (Almond 1985), but in spite of their diplopod features, there is doubt regarding both this affinity and their terrestriality (cf. scorpions). Recent records of myriapod-like forms from undoubted marine sediments of Silurian age, Wisconsin (Mikulic *et al.* 1985), and the Cambrian of Utah (Robison 1987), suggest a long aquatic ancestry for this group. Records of other myriapod-like forms from the Siluro-Devonian generally prove, on reinvestigation, to be misidentified (e.g. *Necrogammarus*; see Selden 1986).

The oldest undoubted terrestrial animal body fossils are from the Lower Devonian Pragian–Emsian (400-Myr-old) Rhynie Chert of Aberdeen, from the Emsian (390-Myr-old) of Alken-an-der-Mosel, Germany, and the Givetian (380-Myr-old) of Gilboa, New York, all yielding comparable material (Table 6.2). Among the terrestrial invertebrates in these faunas are the earliest records of many arachnids: trigonotarbids (extinct spider

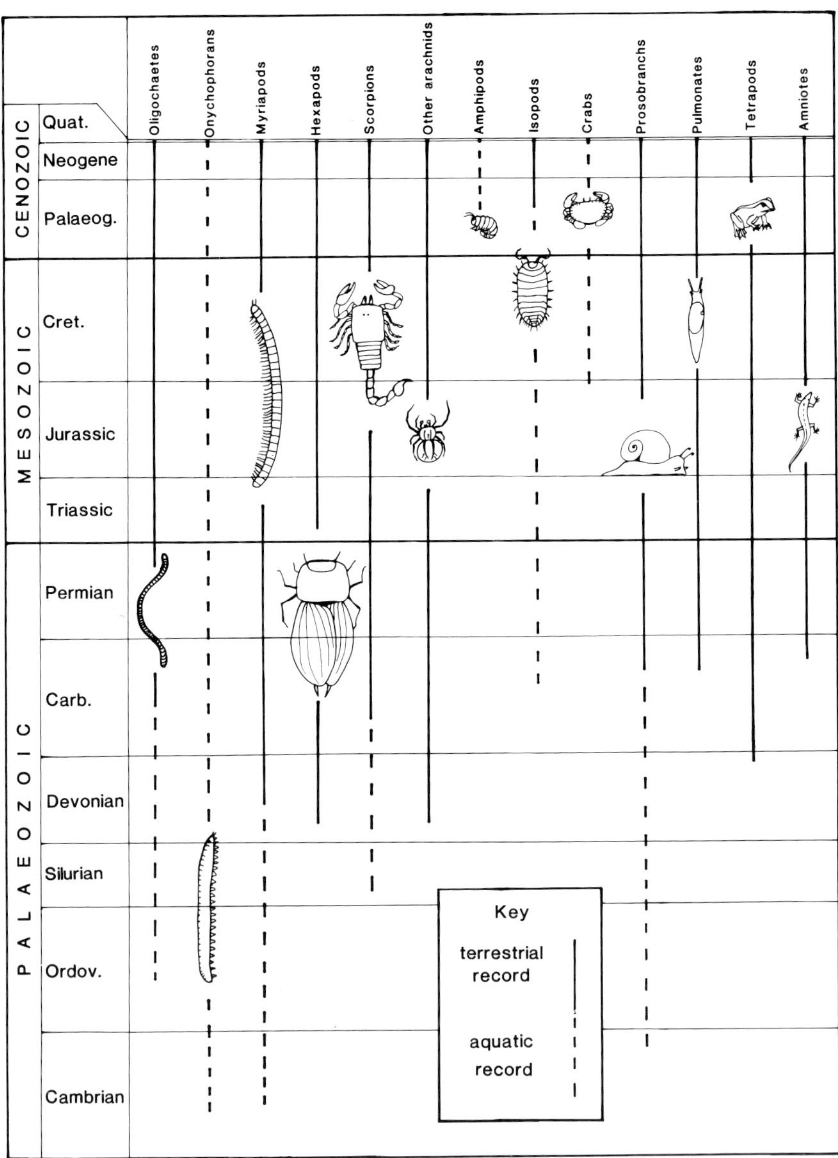

Figure 6.3. The fossil record of colonisation of the land by animals.

relatives), a possible spider, mites, a pseudoscorpion, arthropleurids, a new order of centipedes (Devonobiomorpha) among the myriapods, and collembolans. The fully terrestrialised features of these animals suggest origins in the Silurian or earlier. Thus these three faunas present glorious vistas on early land life at the start of the history of terrestrial ecosystems.

There are no records of terrestrial planarians, nemerteans or nematodes, although fossil examples of parasitic and aquatic nematodes are known (Conway Morris, 1981). Oligochaete annelids are known from the Carboniferous, and their traces (burrows and faecal pellets) occur in palaeosols from the Carboniferous onwards. Land snails, both pulmonates and helicinid prosobranchs, are recorded from the Upper Carboniferous. They were probably established members of the land fauna by that time (Solem 1985). Crustaceans have a generally good fossil record, but the ranges of terrestrial groups are short. Amphipods first appeared in the Upper Eocene, though they may have originated in the mid- to late Mesozoic; the terrestrial talitrids have no fossil record. They may have invaded the land via newly established coastal angiosperm rain forests around Cretaceous Gondwanaland (Spicer *et al.* 1987). Isopods have a fossil record from the Upper Carboniferous and supposed origins in the Devonian, but the earliest terrestrial Oniscoidea are Eocene. Crabs and crayfish first appeared in the Jurassic; the main crab radiations were Cretaceous and Eocene. Families with terrestrial representatives first appeared in the Palaeogene but true terrestrial forms not until the late Neogene.

The oldest tetrapods are from Late Devonian (Famennian) age sediments from Australia, South America and Eurasia, but especially eastern Greenland (Clack 1988). Assuming these animals had a common ancestor, their wide geographic range suggests a long pre-Famennian history. The oldest trackway assigned to tetrapod walking is from the late Silurian/early Devonian Grampians Group of Victoria, Australia (Warren *et al.* 1986). Therefore tetrapods probably share with the arthropods a Silurian origin. However, the earliest tetrapods probably laid eggs in water like modern amphibians; for the earliest amniote we must look to the Carboniferous. However, since the palaeontological criteria for recognising it as a reptile are skeletal we still cannot be sure when the amniote egg first appeared.

ROUTES ONTO LAND

Freshwater

The habitats of the transmigrant algae were probably not streams or rivers but shallow ephemeral pools, backwater lagoons, or lakes with oscillating water levels, and were probably far commoner before the advent of extensive terrestrial vegetation (Beerbower 1985). Such transitory water bodies would have provided the stimulus for the evolution of desiccation resistant structures, e.g. sporopollenin impregnated meiospores (Raven 1984b) and cuticles.

The freshwater route was used by animals as diverse as platyhelminths, amnelids, prosobranch molluscs, crayfish, some grapsid crabs, and the vertebrates (Little 1983, Table 10.1). The reasonably good correlation between osmotic concentration and routes onto land shown by crustaceans can also be demonstrated in the molluscs. The most successful are the pulmonates, which have a relatively high osmotic concentration; in contrast, the prosobranchs have a low osmotic concentration and are generally restricted to humid tropical forests where fatal desiccation is less likely (Little 1983).

The success of terrestrial vertebrates, having taken the freshwater route, contrasts with those successful invertebrate groups with largely marine ancestors. The relatively bigger tetrapods could spend longer on land without the threat of desiccation (cf. scorpions also; see Rolfe 1980). Their freshwater origins conferred the ability to osmoregulate, rather than tolerate, but their large size and waterproof skin allowed them to overcome the water problem on land. Respiratory mechanisms also give clues to the freshwater origin of vertebrates. Littoral animals need to breathe air fairly continuously for long periods while awaiting the return of the tide, so they have generally adapted pre-existing branchial structures for air-breathing. In contrast, animals in poorly oxygenated freshwater are intermittent air-breathers and so developed entirely new organs to take in large volumes of air at a time. Palaeontological evidence also points to a freshwater route for the terrestrial vertebrates.

MARINE

Estuarine, saltmarsh, and littoral environments are characterised by major fluctuations in the osmotic potential of the bathing medium. Physiology of extant plants colonising salt marshes (halophytes) is exceedingly specialised and such plants probably derive from terrestrial angiosperms. Today, pteridophytes and bryophytes (Raven 1984a) are rarely found in such stressed environments, and it seems unlikely that emergent marine or saltmarsh 'rooted' algae were the successful transmigrants. However, Schweitzer (1983) cited assemblages of early vascular plants preserved in littoral and saltmarsh environments as evidence for this route.

The marine route proved the most successful for invertebrate colonisers, and was apparently taken by nemertines, polychaetes, littorinacean prosobranchs, pulmonates, most crustaceans, chelicerates, and probably myriapods and hexapods (Little 1983, Table 10.1). Osmotic arguments favouring a marine route for Crustacea and Mollusca are particularly strong, but are less so for other groups. In particular, interstitial environments present a more gradual change in salinity from marine to terrestrial than that encountered by the epifauna. It therefore follows that among interstitial forms, such as oligochaetes and nematodes, it may be difficult to distinguish osmoregulatory ability derived from a freshwater or interstitial ancestry. Myriapods, hexapods, and chelicerates all show relatively high osmotic concentrations which, in the case of the relatively small myriapods and hexapods, can be compared with the high osmotic concentrations in small crustaceans, and a marine route is suggested. The arachnid conversion of the book-gill into the book-lung is a prime example of the use of an existing aquatic respiratory organ to breathe air for long periods in littoral animals. Interstitial animals, having emerged as tolerators, possibly later developed an osmoregulatory ability as well, since both oligochaetes and some insects can produce hypo-osmotic urine (Little 1983, p. 213). A marine-interstitial route can therefore be distinguished.

Land

Stebbins and Hill (1980) postulated that the higher green plants may have evolved from mats of filamentous algae inhabiting surfaces of moist soils. Physiologically, this would be plausible, but direct evidence is lacking, and the question of the migratory routes of the soil algae remains unanswered. Similarly, it is possible that the hexapods evolved from the myriapods on land, so that the problems of terrestrialisation had already been solved by their ancestors. Palaeontological evidence suggesting a long marine history for the myriapods, rather than their evolution from onychophoran-like forms on land, is accumulating.

EARLY LAND COMMUNITIES—AN OVERVIEW

Land surfaces in the early Palaeozoic would have been at least partially covered by microbial mats initially predominantly cyanobacterial but later with green coccoid and filamentous algae. Today such mats mainly occur on moist soils, but cyanophyte-dominated crusts also form on rock, sand and soil in dry and arid desert environments, where they accrete and bind sediments. Some of these micro-organisms would have acted as natural nitrogen fertilisers. Decaying remains would have introduced humus and promoted some chemical weathering in the substrate, and would have increased water and mineral retention, all this creating microenvironments for interstitial animals. Although such mats may have been widespread, they would have been easily disrupted and destroyed in unstable environments, and soils would have been gradually built up only in stable, perhaps low-lying areas.

Around mid-Ordovician times the microfossil record suggests a change in land vegetation, perhaps to one comprising multicellular thallophytic carpets, with nearest living examples (but not necessarily relatives) in the bryophytes (Gray 1985). From the projected size and presumed lack of rooting systems, it can be postulated that some further amelioration, and to a lesser extent stabilisation, occurred particularly in humid environments, but there was no major change in the processes operating earlier. Palaeosols are known from the late Ordovician onwards (Retallack 1985). The oldest, the Potters Mill clay palaeosol from the Ashgillian of Pennsylvania, appears to have formed in a subtropical, subhumid palaeoclimate with wet and dry seasons, and was a red-brown silt soil. Of great interest are the abundant, deep, vertical burrows, possibly made by arthropods, possibly millipedes (Retallack and Feakes 1987). Animals responsible for the palaeosol burrows may have avoided desiccation if the soil never dried out; yet the postulated wet/dry season would have encouraged cryptobiosis, and animals capable of this, such as mites and tardigrades, could have colonised the land by wind dispersal in the encysted state.

The first half of the Silurian saw the appearance of the pioneering erect pteridophyte-like plants. Initially small, of limited productivity, and lacking extensive subterranean systems, they would have made little impact on terrestrial vegetation. Suitable habitats may have been limited in number and highly stressed. This may well have depressed the vegetative and reproductive vigour of the sporophyte resulting in conservation of resources. In the surface-living gametophyte, however, in which almost all cells are photosynthetic and less energy is required for structural and conducting tissues, both vegetative and reproductive phases may have been completed quickly when moisture was available (Edwards 1979). The sporophyte then became established as a site occupier tolerant of drier conditions. The growth habit and terminal sporangia of the earliest representatives (rhyniophytes) suggest that there was a single period of spore production.

As such plants became larger and more widespread, particularly in early Devonian times, there would have been an increase in ecosystem homeostasis and a major amelioration of substrates with build-up of nutrient-rich/water-retentive humic soils. Stands of probably monotypic vegetation spreading by means of extensive rhizomatous systems would have clothed low-lying areas, with spore production occurring over much longer periods. The Rhynie Chert provides evidence of peat development. However, bearing in mind the limitations of 'external' fertilisation that characterises the pteridophytes, plants may still have been restricted to damp areas. Moderation of physical stress would have produced stable populations, however, and the potential development of food webs of increasing complexity. The oldest body fossils of land animals occur at this time, but evidence of herbivory is lacking. The development of spines on aerial surfaces of plants may have increased photosynthetic capacity, but may also have conferred protection against grazers. In the zosterophyll *Sawdonia ornata* open darkened tips to spines suggest additional chemical defence. Response to wounding, as dark-staining or necrotic areas, has been reported in Rhynie Chert plants (Kevan *et al.* 1975), but arthropod involvement is equivocal (Rolfe 1985). Regeneration of surface tissues is described in *Psilophyton dawsonii* (Banks 1981) but the cause of the wound is unknown.

In all three early terrestrial ecosystems known to date (Table 6.2), predatory arthropods predominate; the remaining arthropods are probably detritivores or fungivores. The apparent lack of herbivory has been explained in various ways—spore-feeding (Kevan *et al.* 1975; rejected by Rolfe 1980; Shear *et al.* 1987), and differential preservation (Shear *et al.* 1987)—but this lack may not be significant. The decomposer food chain is typical of the soil fauna of the present day, the only herbivores being surface-feeding macrofauna. The paucity of detritivores in the faunas of Rhynie and Gilboa can be explained by the poor preservation or recovery of the mites, collembolans, nematodes and enchytraeids, for example, because of their small size.

Suggested causes of terrestrialisation are various: escape from predators, more abundant food supply, or the filling of vacant niches, for example, but such hypotheses are not readily testable. Chance undoubtedly played a large part. In theory, autotrophs must have preceded heterotrophs onto land, and carnivores must have followed their prey. Individual taxa moved on to land separately, but the majority appear to have made the adaptive breakthrough in mid-Palaeozoic times. It is probably realistic to envisage established communities colonising the terrestrial environment at this time, rather than pioneering species.

References

Almond, J.E., 1985, The Silurian-Devonian fossil record of the Myriapoda, *Philosophical Transactions of the Royal Society, London*, **B 309**: 227–37.

Banks, H.P., 1981, Peridermal activity (wound repair) in an early Devonian (Emsian) trimerophyte from the Gaspé Peninsula, Canada, *The Palaeobotanist*, **28–29**: 20–5.

Bartram, K.M., Jeram, A.J. and Selden, P.A., 1987, Arthropod cuticles in coal, *Journal of the Geological Society of London*, **144**: 513–17.

Beerbower, R., 1985, Early development of continental ecosystems. In B.H. Tiffney (ed.), *Geological factors and the evolution of plants*, Yale University Press, New Haven, CT, pp. 47–91.

Chaloner, W.G. and Lawson, J.D. (eds), 1985, Evolution and environment in the late Silurian and early Devonian, *Philosophical Transactions of the Royal Society, London*, **B309**: 1–342.

Clack, J.A., 1988, New material of the early tetrapod *Acanthostega* from the Upper Devonian of east Greenland, *Palaeontology*, **31**: 699–724.

Conway Morris, S., 1981, Parasites and the fossil record, *Parasitology*, **82**: 489–509.

Dalingwater, J.E., 1985, Biomechanical approaches to eurypterid cuticles and chelicerate exoskeletons, *Transactions of the Royal Society of Edinburgh*, **76**: 359–64.

Edwards, D., 1979, The early history of vascular plants based on late Silurian and early Devonian floras of the British Isles. In A.L. Harris, C.H. Holland and B.E. Leake (eds), *The Caledonides of the British Isles*—reviewed, Geological Society of London, pp. 405–10.

Edwards, D., 1986, Preservation in early vascular plants, *Geology Today*, November–December: 177–81.

Edwards, D. and Edwards, D.S., 1986, A reconsideration of the Rhyniophytina Banks. In R.A. Spicer and B.A. Thomas (eds), *Systematic and taxonomic approaches in palaeobotany*, Systematics Association Special Volume, **31**, Oxford University Press, Oxford, pp. 199–220.

Edwards, D. and Fanning, U., 1985, Evolution and environment in the late Silurian–early Devonian: the rise of the pteridophytes, *Philosophical Transactions of the Royal Society, London*, **B309**: 147–65.

Edwards, D., Fanning, U. and Richardson, J.B., 1986, Stomata and sterome in early land plants, *Nature*, **323**: 438–40.

Edwards, D., Feehan, J. and Smith, D.G., 1983, A late Wenlock flora from Co. Tipperary, Ireland. *Botanical Journal of the Linnean Society*, **86**: 19–36.

Edwards, D.S., 1986, *Aglaophyton major*, a non-vascular land-plant from the Devonian Rhynie Chert, *Botanical Journal of the Linnean Society*, **93**: 173–204.

Gans, C., 1970, Respiration in early tetrapods: the frog is a red herring, *Evolution*, **24**: 723–34.

Gray, J., 1985, The microfossil record of early land plants: advances in understanding of early terrestrialization, 1970–1984, *Philosophical Transactions of the Royal Society, London*, **B309**: 167–95.

Greenaway, P., 1984, The relative importance of the gills and lungs in the gas exchange of amphibious crabs of the genus *Holthuisiana, Australian Journal of Zoology*, **32**: 1–6.

Hadley, N.F., 1981, Cuticular lipids of terrestrial plants and arthropods: a comparison of their structure, composition, and waterproofing function, *Biological Reviews*, **56**: 23–47.

Hadley, N.F. and Quinlan, M.C., 1987, Permeability of arthrodial membrane to water: a first measurement using *in vivo* techniques, *Experientia*, **43**: 164–6.

Holland, H.D., 1984, *The chemical evolution of the atmosphere and oceans*, Princeton University Press, Princeton, NJ.

Holland, H.D., Lazar, P. and McCaffrey, M., 1986, Evolution of the atmosphere and oceans, *Nature*, **320**: 27–33.

Innes, A.J. and Taylor, E.W., 1986, The evolution of air-breathing in crustaceans: a functional analysis of branchial, cutaneous and pulmonary gas exchange. *Comparative Biochemistry and Physiology*, **85A**: 621–37.

Jones, J.D., 1972, *Comparative physiology of respiration*, Edward Arnold, London.

Kevan, P.G., Chaloner, W.G. and Savile, D.B.O., 1975, Interrelationships of early terrestrial arthropods and plants, *Palaeontology*, **18**: 391–417.

Lang, W.H., 1937, On the plant remains from the Downtonian of England and Wales, *Philosophical Transactions of the Royal Society, London*, **B227**: 245–91.

Little, C., 1983, *The colonisation of land. Origins and adaptations of terrestrial animals*, Cambridge University Press, Cambridge.

Mikulic, D.G., Briggs, D.E.G. and Kluessendorf, J., 1985, A new exceptionally preserved biota from the Lower Silurian of Wisconsin, U.S.A., *Philosophical Transactions of the Royal Society, London*, **B311**: 75–85.

Milner, A.R., Smithson, T.R., Milner, A.C., Coates, M.I. and Rolfe, W.D.I., 1986, The search for early tetrapods, *Modern Geology*, **10**: 1–28.

Panchen, A.L. (Ed.), 1980, *The terrestrial environment and the origin of land vertebrates*, Systematics Association Special Volume **15**, Academic Press, London.

Pollard, J.E., 1985, Evidence from trace fossils, *Philosophical Transactions of the Royal Society, London*, **B309**: 241–2.

Rankin, J.C. and Davenport, J., 1981, *Animal osmoregulation*, Blackie, London.
Raven, J.A., 1984a, *Energetics and transport in aquatic plants*, A.R. Liss, New York.
Raven, J.A., 1984b, Physiological correlates of the morphology of early vascular plants, *Botanical Journal of the Linnean Society*, **88**: 105–26.
Raven, J.A., 1985, Comparative physiology of plant and arthropod land adaptation, *Philosophical Transactions of the Royal Society, London*, **B309**: 273–88.
Raynor, R.J., New finds of *Drepanophycus spinaeformis* Göppert from the Lower Devonian of Scotland. *Transactions of the Royal Society of Edinburgh, Earth Sciences*, **75**: 353–63.
Retallack, G.J., 1985, Fossil soils as grounds for interpreting the advent of large plants and animals on land. *Philosophical Transactions of the Royal Society, London*, **B309**: 105–42.
Retallack, G.J. and Feakes, C., 1987, Trace fossil evidence for late Ordovician animals on land, *Science*, **235**: 61–3.
Richardson, D.H.S., 1981, *The biology of mosses*, Blackwell Scientific Publications, Oxford.
Robison, R.A., 1987, A marine myriapod-like fossil from the middle Cambrian of Utah, *Geological Society of America, Abstracts with Programs*, **19**: 823.
Rolfe, W.D.I., 1980, Early invertebrate terrestrial faunas, pp. 117–57. In A.L. Panchen (ed.), *The terrestrial environment and the origin of land vertebrates*, Academic Press, London.
Rolfe, W.D.I., 1985, Early terrestrial arthropods: a fragmentary record, *Philosophical Transactions of the Royal Society, London*, **B301**: 207–18.
Schweitzer, H.J., 1983, Die Unterdevonflora des Rheinlandes. 1. Teil, *Palaeontographica*, **B189**: 1–138.
Selden, P.A., 1984, Autecology of Silurian eurypterids, *Special Papers in Palaeontology*, **32**: 39–54.
Selden, P.A., 1985, Eurypterid respiration, *Philosophical Transactions of the Royal Society, London*, **B309**: 219–26.
Selden, P.A., 1986, A new identity for the Silurian arthropod *Necrogammarus*, *Palaeontology*, **29**: 629–31.
Shear, W.A., Selden, P.A., Rolfe, W.D.I., Bonamo, P.M. and Grierson, J.D., 1987, New terrestrial arachnids from the Devonian of Gilboa, New York (Arachnida, Trigonotarbida), *American Museum Novitates*, **2901**: 1–74.
Sherwood-Pyke, M.A. and Gray, J., 1985, Silurian fungal remains; oldest records of the class Ascomycetes, *Lethaia*, **18**: 1–20.
Solem, A., 1985, Origin and diversification of land snails. In A.E. Trueman and M.R. Clarke (eds), *The Mollusca*, Vol. **10**, Academic Press, London, pp. 269–93.
Spicer, J.I., Moore, P.G. and Taylor, A.C., 1987, The physiological ecology of land invasion by the Talitridae (Crustacea: Amphipoda), *Proceedings of the Royal Society of London*, **B232**: 95–124.
Stebbins, G.L. and Hill, G.J.C., 1980, Did multicellular plants invade the land?, *American Naturalist*, **115**: 342–53.
Størmer, L. 1976, Arthropods from the Lower Devonian (Lower Emsian) of Alken-an-der-Mosel, Germany. Part 5: Myriapoda and additional forms, with general remarks regarding invasion of land by arthropods, *Senckenbergiana lethaea*, **57**: 87–183 .
Stubblefield, S. and Taylor, T.N., 1988, Recent advances in palaeomycology, *New Phytologist*, **108**: 3–25.

Thompson, K.G., 1980, The ecology of the Devonian lobe-finned fish. In A.L. Panchen (ed.), *The terrestrial environment and the origin of land vertebrates*, Academic Press, London, pp. 187–222.

Turner, N. and Jones, M.M., 1980, Turgor maintenance by osmotic adjustment: a review and evaluation. In N.C. Turner and P.J. Kramer (eds), *Adaptation of plants to water and high temperature stress*, Wiley, New York, pp. 87–103.

Warren, A., Jupp, R. and Bolton, B., 1986, Earliest tetrapod trackway, *Alcheringa*, **10**: 183–6.

Chapter 7

PATTERNS OF EVOLUTION AND EXTINCTION IN VASCULAR PLANTS

Peter R. Crane

Two kinds of pattern provide basic data for interpreting the evolutionary history of plants or animals. Patterns of character distribution define groups of organisms and thus their relationships, while patterns of systematic diversity and abundance in the palaeontological record provide information from which the dynamics of origination, extinction and perhaps other evolutionary processes, may be inferred. Integrated with other biological and geological knowledge, these patterns provide the basis for relatively testable, and comprehensive, hypotheses of the evolution of particular taxa.

PRINCIPLES AND METHODS

Cladistics

Cladistic techniques for classification provide a powerful tool for the study of evolution (Wiley 1981) but have not yet been applied comprehensively to extant and fossil plants. Discussions of groups and relationships in this chapter are presented as far as possible as cladistic hypotheses to be tested as botanical knowledge increases. Informal names are used for most groups considered to avoid implications of formal taxonomic rank.

In cladistic analysis, characters occurring in one taxon, or in all taxa under consideration, provide no information on patterns of relationships. For example, in assessing relationships between a fern, a conifer and an angiosperm, only characters which occur in two of the three taxa can provide evidence for defining a subgroup. Exactly how those characters are distributed in related taxa ('outgroup comparison') is also critical. In this case, 'bryophytes' (Figure 7.1) are an appropriate outgroup because conventional views suggest that they exhibit some but not all of the characteristics of other land plants (tracheophytes). Ferns and conifers are similar in having differentiated archegonia in their gametophytes but because this feature also occurs in the outgroup ('bryophytes') it is not a unique defining character of ferns + conifers. Conversely, angiosperms and conifers differ from ferns in having seeds, and because this feature does not occur in 'bryophytes', it is potentially a good defining character of a conifer + angiosperm group in which 'bryophytes' would not be included. Applied to more taxa and more characters this procedure generates a hierarchical pattern of internested (progressively less inclusive) characters (synapomorphies), which can be used to define a pattern of groups within groups (e.g. Figure 7.1). All characters may be useful for group definition at some level in the hierarchy, and outgroup comparison merely provides an initial hypothesis of the appropriate level at which a particular character should be applied.

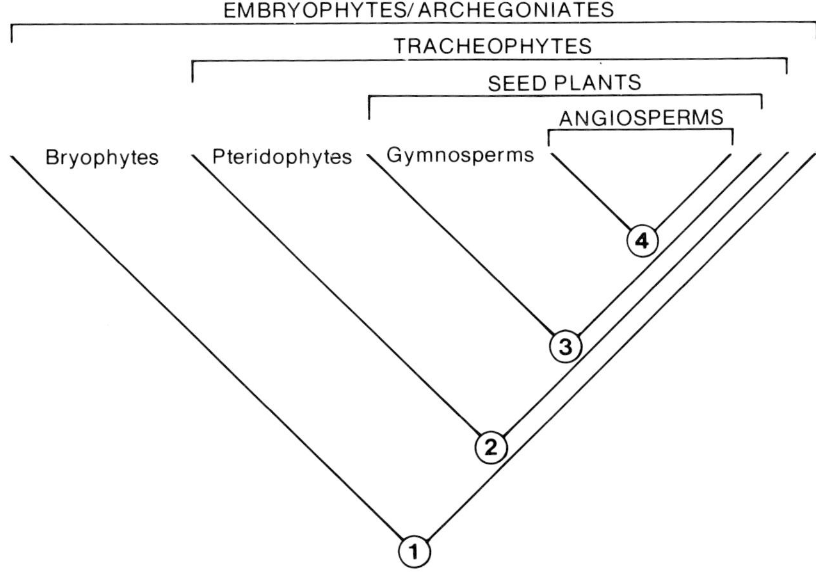

Figure 7.1. Simplified summary of the major groups of plants, excluding algae. Groups in upper case are possibly monophyletic (clades) and defined by the nested characters 1–4. Groups in lower case are paraphyletic (grades) and can be defined only by the presence and absence of characters. Numbered synapomorphies are: 1, fertilisation occurring within an archegonium and zygote undergoing mitotic divisions to produce an embryo (sporophyte). 2, true tracheids with differentially thickened secondary cell walls, branched free-living sporophyte (diploid) with multiple sporangia. 3, seeds. 4, double fertilisation, 4–16 nuclei in megaprothallus and other characters (see Figure 7.5).

Although conceptually straightforward, this procedure is often complex in practice, and the delimitation of characters and the choice of outgroup are frequently problematic. In addition, different defining characters (synapomorphies) often suggest conflicting patterns, and in such cases the principle of parsimony is applied to select the scheme which minimises such conflicts. Determining the most parsimonious ('simplest') pattern frequently requires computer analysis. In cladistic terms, characters (synapomorphies) which support the simplest pattern are interpreted as true homologous similarities. Concepts of relationship are a direct reflection of the groups recognised: e.g. conifers and angiosperms are both seed plants and are more closely related to each other than either is to ferns.

Hierarchical patterns of relationships (groups) can be summarised in a tree-like diagram (cladogram, e.g. Figure 7.1) but such patterns may be recognised independent of evolutionary preconceptions, and there is no necessary equivalence between a cladogram and a phylogenetic tree. However, cladistic diagrams are amenable to phylogenetic interpretation, with the nodes being interpreted as hypothetical common ancestors and the vertical axis being interpreted as relative time. Synapomorphies may be interpreted as evolutionary novelties and outgroup comparison, as a means of determining which characters are relatively derived (apomorphic) or primitive (plesiomorphic) at a given hierarchical level. Relationships may be defined in terms of relative recency of common ancestry: e.g. conifers and angiosperms share a more recent common ancestor than either does with ferns. Groups defined by synapomorphies may be interpreted as monophyletic (including all the descendants of a single common ancestor; see Figure 7.1). In contrast, those groups which can be defined only by a combination of primitive and advanced characters (e.g. 'gymnosperms': seeds present, angiosperm characters absent) are termed paraphyletic and constitute a phylogenetic grade rather than a clade (Figure 7.1). They are artificial in that they include some, but not all, of the descendants of a common ancestor. In a phylogenetic sense, the principle of parsimony can be viewed as a criterion for selecting the character pattern which minimises the need to invoke parallelism, convergence and reversal to account for apparent similarity.

Analyses of Diversity

Although analyses of diversity in the fossil record do not reflect ecological abundance, they can provide a link between phylogenetic patterns and ecology. In certain ecological studies it may be profitable to examine the diversity pattern exhibited by a phylogenetically unnatural group that shares common ecological characteristics but, with respect to phylogenetic diversification, examining part of a diversity pattern (paraphyletic groups) or the combined effects of unrelated diversity patterns (polyphyletic groups) is likely to confuse evolutionary interpretations (see also Chapter

9). In most cases, the use of changing patterns of biological diversity in the fossil record to infer general, or taxon specific, evolutionary patterns or properties depends on the recognition and consideration of monophyletic groups.

Direct assessment of changing palaeontological diversity confronts numerous problems, because it is clear that the fossil record preserves only a fraction of the total number of species living at a given time, and that this sample is biased in numerous and often subtle ways. Perhaps the most serious class of biases are those associated with the formation and preservation of fossils (taphonomy). These are particularly troublesome because they do not necessarily apply equally to different taxa or communities, and thus patterns which might be interpreted as of evolutionary significance may actually reflect taphonomic distortion. These problems are compounded in palaeobotany by the fragmentation of the living plant into individual organs (e.g. leaves, pollen), making the palaeobotanical record one of plant parts rather than of whole plants. Species that produce large numbers of regularly shed plant parts (e.g. leaves and pollen of deciduous wind-pollinated trees) or grow close to depositional environments (e.g. deltas, floodplains, swamps) are overrepresented in the fossil record relative, for example, to herbaceous plants which do not shed large numbers of plant parts, and to species growing in upland areas distant from potentially preservable depositional environments (see Scheihing and Pfefferkorn, 1984 and references cited therein). Other significant problems of most diversity analyses include geographic and stratigraphic sampling, stratigraphic resolution and the quality and consistency of available systematic treatments.

Notwithstanding the many problems, some patterns of changing palaeontological diversity through time are well supported, and it is possible to minimise or assess the likely effects of various biasing factors. In the palaeobotanical record one important means of testing observed patterns is by comparing different measures of the history of plant groups provided by different plant parts (e.g. pollen, leaves, fruits and seeds). Because these different parts frequently have very different taphonomic properties, congruence between patterns suggests an underlying evolutionary signal, rather than just taphonomic bias.

MAJOR PATTERNS OF LAND PLANT EVOLUTION

Phylogeny of early tracheophytes

In parallel with research on the ecology of terrestrial colonisation (Chapter 6) there has also been considerable progress with related phylogenetic issues. Ultrastructural studies of extant 'green algae' have focused on the charophytes *sensu lato* (e.g. *Coleochaetae*) as the algal group most closely related to land plants (Mattox and Stewart 1984), and preliminary phylogenetic analyses of the bryophyte grade (Figure 7.1) have resolved the mosses (Musci), hornworts (Anthocerotae) and liverworts (Hepaticae) as monophyletic groups increasingly more distantly related to the tracheophytes (Mishler and Churchill 1984). These conclusions are not unanimously supported by bryologists, but they provide a valuable starting point for integrating palaeobotanical and neobotanical data and for identifying critical phylogenetic questions. One such question is whether the tracheophytes themselves are an unquestionably natural (monophyletic) group? In an evolutionary sense, was the full suite of terrestrial 'adaptations' (Chapter 6) acquired by one, or more, lineages of early land plants?

Living tracheophytes have several characters that distinguish them from bryophytes and algae (Figure 7.1), including water-conducting cells (tracheids) with differentially thickened secondary walls. This and other features of extant taxa suggest that the tracheophytes are monophyletic, but even a cursory examination of the fossil record of early land plants reveals an almost bewildering diversity of extinct taxa, many with unusual characters or combinations of characters, which obviously need to be accounted for as part of any general understanding of tracheophyte evolution.

Palaeobotanical studies have revealed the existence of two major groups in the Early Devonian which seem to represent the primary dichotomy of the putative tracheophyte clade (Banks 1975; Chaloner and Sheerin 1979). The lycophytes *sensu lato* (zosterophylls + lycopods, Figure 7.3) comprise the first group, and include taxa such as *Asteroxylon* and *Zosterophyllum* which are apparently closely related to living clubmosses (*Lycopodium*, *Selaginella*) and quillworts (*Isoetes*) as well as to many extinct taxa (Figure 7.3). Lycophytes are characterised by laterally borne, globose to kidney-shaped sporangia with the line of dehiscence extending over the apex of the sporangium, usually in a transverse position. The ontogeny of vascular tissues in the aerial axes is also characteristic; maturation of the water-conducting elements proceeding from the outside to the inside (exarch primary xylem). The second group (rhyniophytes) includes plants such as *Rhynia* and *Horneophyton*, some of which are thought to be more closely related to ferns, sphenopsids and lignophytes, than to lycophytes (Figure 7.3). This group includes Devonian taxa with terminally borne, elongated sporangia which have unspecialised or longitudinal dehiscence. Primary xylem development in these plants proceeded from the inside toward the outside (endarch) and this is perhaps the primitive (plesiomorphic) condition in tracheophytes as a whole, although the pattern of maturation of 'bryophyte' conducting elements is unknown.

In cladistic terms the question of tracheophyte monophyly hinges on whether either of the two groups of primitive vascular land plant is more closely related to non-tracheophytes than they are to each other. The Early Devonian genus *Nothia* (Edwards and Edwards 1986), one of the lesser-known plants from the Rhynie Chert, appears to be particularly relevant to this question (Figure 7.2). *Nothia* shows some but not all of the defining characters of the lycophytes (lateral reniform sporangia with transverse dehiscence) but has the possible primitive pattern of differentiation of water-conducting elements (endarch) and most surprisingly lacks true

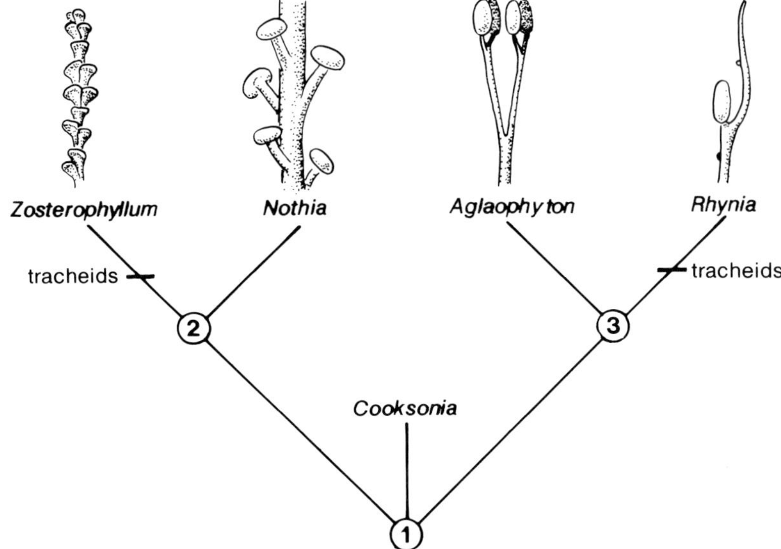

Figure 7.2. Hypothesised cladistic relationships of *Cooksonia*. *Zosterophyllum*, *Nothia*, *Aglaophyton* and *Rhynia*. This pattern of relationships implies a double origin of tracheids and assumes that the apparent absence of tracheids in *Cooksonia* is not merely a failure to detect these cells in poorly preserved material. Numbered synapomorphies are: 1, branched free-living sporophyte with multiple sporangia, sporopollenin-walled spores, stomata. 2, sporangia broader than long, borne laterally and with transverse dehiscence. 3, sporangia elongated with unspecialised or longitudinal dehiscence.

tracheids. If *Nothia* is correctly interpreted as closely related to lycophytes (Figure 7.2), then it suggests that some primitive members of the lycophyte clade did not possess true tracheids, and thus that these characteristic cells were either lost in *Nothia* or evolved within the lycophyte group.

Another Rhynie Chert plant that lacks tracheids (*Aglaophyton major*, formerly *Rhynia major*) suggests a similar pattern of tracheid evolution in the other major group of early land plants, and, taken together, *Aglaophyton* and *Nothia* suggest that the loss of tracheids (one gain, two losses) is less likely than their independent origin (two gains). Thus the likely phylogenetic placement of both *aglaophyton* and *Nothia* (Figure 7.2) suggests that the tracheids of all land plants may not be homologous and that, contrary to indications based solely on extant plants (e.g. Figure 7.1), tracheophytes (defined by the possession of tracheids) are possibly not a monophyletic group. The implication is that two groups of primitive land plants, possibly at the bryophyte grade, independently obtained the developmental capability to construct conducting cells with partial secondary cell walls (see also Chapter 6). This in turn suggests that closer examination of the structure and development of tracheids in extant lycopods and other extant 'pteridophytes' might produce further differences or similarities to refute or support the hypothesis of tracheid nonhomology. It would obviously also be of great interest to learn more about the ultrastructure of apparently anomalous conducting strands recorded among extant 'bryophytes' and fossil early land plants (see Edwards and Edwards 1986).

Although the problems of tracheid origin and the phylogenetic relationships of early land plants require much more research, these issues are very instructive of the role of palaeontological data in elucidating plant phylogeny. They show that phylogenetic analyses based only on extant taxa are inevitably incomplete and perhaps frequently misleading. The major practical problem is that relatively few of the critical early land plants are known in sufficient detail for an integrated palaeobotanical-neobotanical cladistic analysis, and there is a clear need for increased basic knowledge of critical early land plants.

Lycophytes

Despite the unanswered phylogenetic and ecological questions associated with terrestrial colonisation (Chapter 6) increases in the morphological diversity of terrestrial plant remains demonstrate unequivocally that a major evolutionary radiation took place in the Early Devonian (Lochkovian to Emsian). Through this interval the total number of both spore and macrofossil genera recognised per stage increased dramatically before apparently levelling off through the Middle and Late Devonian (Knoll *et al.* 1984). Lycophytes (Figure 7.3) are recognised very early in this diversification, the genus *Zosterophyllum* is first reported from the earliest stage of the Early Devonian (Lochkovian) and plants at the zosterophyll grade may extend back into the Silurian (Kenrick and Edwards 1988).

Taking extant and fossil lycophytes together, at least four sets of nested characters can be recognised, and these provide a useful framework (Figure 7.3) in which to examine the evolution of the group. The most inclusive clade is the lycopods *sensu stricto* defined by the possesion of simple leaves (technically microphylls) which typically have a single central vein that does not leave a leaf gap in the central vascular tissue (stele) of the stem. Within the lycopods a ligulate group can be defined by the presence of a small flap of tissue (ligule) borne on the upper (adaxial) surface of each leaf which is thought to play an important role in early leaf development. Further lycophyte groups are defined by the occurrence of distinct megaspores and microspores (heterospory), and the presence of a specialised underground stigmarian rooting organ. Although probably oversimplified, this classification based on potential synapomorphies accounts for most extant and fossil lycophytes (Figure 7.3). There are relatively few fossil taxa which are well understood and unequivocally exhibit conflicting combinations of features (e.g. heterospory and possible

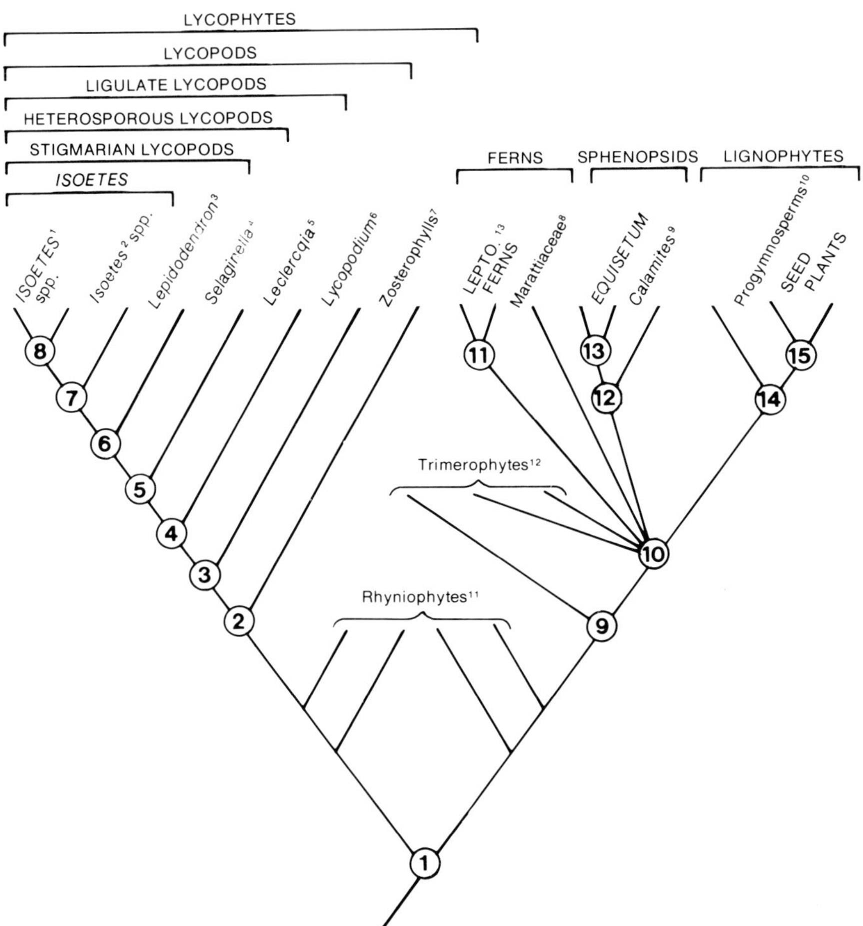

Figure 7.3. Simplified summary of the major groups of plants at the pteridophyte grade (several taxa, e.g. *Sphenophyllum*, coenopterids omitted). Groups in upper case are possibly monophyletic (clades). Taxa given in lower case are representative of other plants at that grade of evolution although in some cases the taxon listed may itself be monophyletic. Numbered synapomorphies are: 1, stomata, ?tracheids (see Figure 7.2). 2, sporangia broader than long, borne laterally and with transverse dehiscence, exarch primary xylem development. 3, microphylls. 4, ligules. 5, heterospory. 6, pseudobipolar growth and stigmarian rooting organs. 7, lacunate leaves, multiflagellate sperm. 8, reduced proximal leaf laminae (alae). 9, increased size and complex three-dimensional branching. 10, actinosteles or other advanced stelar types with mesarch primary xylem development and several protoxylem points. 11, sporangia developing from a single cell and borne on a slender stalk, sporangial wall one cell thick and producing a small number of spores. 12, whorled leaves, air (carinal) canals associated with primary xylem, spores with elaters. 13, reduced leaves, loss of secondary xylem, loss of leaves

stigmarian system but apparently no ligule in the Late Devonian lycopod *Cyclostigma*). The nested characters also appear in the fossil record in the relative chronological order predicted from the cladogram.

As judged from the only available estimates (Niklas *et al.* 1985), the diversity of lycophytes consistently increases through the Devonian to the Late Carboniferous but then declines rapidly, reaching essentially modern levels by the Middle Triassic (see below). However, measures of total diversity (total number of discrete taxa usually summed per stage or per period) do not provide any indication of either the floristic importance or ecological abundance of the lycophytes in Palaeozoic vegetation. Local floristic importance can be assessed from the number of lycopod species in individual fossil floras (usually expressed as the mean of several percentage values). In the Devonian and Early Carboniferous lycophytes typically comprise about 40% of the species present in fossil floras but gradually become much less important in the Late Carboniferous as a result of extinctions and probably also some 'dilution' by the rapidly diversifying seed plants (Knoll 1986). These estimates are largely based on impression or compression macrofloras and thus probably average a variety of different clastic, deltaic or floodplain depositional environments. However, in peat-accumulating (coal) swamps the lycophytes were much more important until close to the end of the Carboniferous.

The most primitive level in lycophyte evolution is represented by the zosterophylls, which include fossil genera such as *Zosterophyllum* and *Sawdonia*. The zosterophylls dominated the earliest phases of lycophyte history but, by the end of the Devonian, plants at this grade of organization were extinct. Zosterophylls are lycophytes characterised mainly by the absence of microphylls, although spines (without a vascular supply) are present in some taxa. The zosterophylls thus seem paraphyletic, although most species have sporangia arranged in two rows in the fertile portions of

---

between whorls of sporangia. 14, bifacial cambium producing secondary xylem and phloem. 15, seeds. [1]Includes most extant species of *Isoetes* (Hickey 1986). [2]Includes three extant species of *Isoetes* and the fossil *Isoetites* (Hickey 1986). [3]Other genera at this grade would include *Diaphorodendron*, *Lepidophloios*, *Lepidodendron*, and *Sigillaria*. [4]This grade would also include fossil genera such as *Selaginellites*. [5]*Leclercgia* is the only plant with this combination of primitive and advanced characters. [6]Other genera at this grade would include the extant taxa *Huperzia* and *Phylloglossum*, as well as fossil genera such as *Drepanophycus* and possibly *Asteroxylon*. [7]Zosterophylls include genera such as *Sawdonia* and *Zosterophyllum*. [8]Marattiaceae includes the *Psaronius* plant as well as extant genera; other eusporangiate ferns would also be included at this grade. [9]*Archaeocalamites* may also be at this grade. [10]Includes both aneurophytalean and archaeopteridalean forms. [11]Includes fossil genera such as *Renalia* and *Rhynia*. [12]Includes fossil genera such as *Pertica* and *Psilophyton*. [13]LEPTO. FERNS, leptosporangiate ferns.

the stems and more detailed analysis may show that this characteristic is homologous and defines most zosterophylls as a monophyletic group.

Within the lycopods *sensu stricto* the next grade of evolution (with microphylls but without a ligule, Figure 7.3) includes fossil genera such as *Drepanophycus*, *Baragwanathia*, and possibly also *Asteroxylon* depending on whether the leaflike appendages in that genus are interpreted as microphylls. Plants at this level are the dominant Lower and Middle Devonian lycophytes. They were typically larger than the zosterophylls and perhaps photosynthetically more efficient by virtue of the increased surface area of their aerial parts. The extant genera *Lycopodium* and *Phylloglossum* also fall within this grade and are thus cladistically the most primitive of living lycopods. In particular, the extant species *Huperzia selago* lacks any specialisation of the fertile portion of the shoot (no differentiated cones) and is very similar morphologically and anatomically to the Early Devonian Rhynie Chert fossil *Asteroxylon*. Although detailed comparisons have never been made *H. selago* seems to be a good example of evolutionary stasis.

The preligulate grade of evolution includes all of the earliest lycophytes, but the late Early–Middle Devonian genus *Leclercqia* provides the first evidence of the ligulate clade (Grierson and Bonamo 1979). All other ligulate lycopods are heterosporous and the extant genus *Selaginella*, along with fossil genera such as *Selaginellites* from the Carboniferous, are among the cladistically more primitive genera of the ligulate heterosporous clade.

From the vegetational standpoint, the stigmarian lycopods are the most important lycophyte group and include extinct arborescent taxa such as *Diaphorodendron*, *Lepidodendron*, *Lepidophloios* and *Sigillaria*. This clade underwent a major radiation in the latest Devonian and earliest Carboniferous, reaching their acme in terms of size, complexity, reproductive sophistication and ecological dominance close to the end of the Carboniferous. They are perhaps the only free-sporing heterosporous group in the history of plant life to have dominated the vegetation of large areas.

The arborescent stigmarian lycopods of the Carboniferous were some of the most extraordinary pteridophytes known, and several of their structural features show striking parallelisms to developments in other tracheophyte clades. In particular, heterospory in some stigmarian lycopods developed to an extreme level, approaching that seen in seed plants, with a single large functional megaspore retained within each megasporangium, and the megasporangium itself serving as the unit of dispersal. Similarly, these lycopods developed the arborescent habit but their architecture was totally different from that in seed plants. The primary supporting tissue in the stigmarian lycopods was a corky, barklike tissue (periderm) rather than secondary xylem, and some of the increase in girth of the stems was accounted for by a substantial expansion in the diameter of the apical meristem rather than secondary cambial activity of the kind that occurs in modern conifers and dicotyledons.

Another unusual, and important feature of the stigmarian lycopods was their anomalous 'pseudobipolar' growth, and it could be argued that this particular developmental innovation was one of the factors that facilitated their exploitation of the arborescent habit. In seed plants two distinct meristems are established in the embryo which grow to form the stem and the root of the mature plant. In contrast, most 'pteridophytes' have an embryo with only a single apical meristem, and underground portions of that plant are either produced by dichotomies of the original axis or by adventitious branching. In the arborescent lycopods a very early dichotomy of the embryonic stem apex resulted in two distinct shoot systems. One of these formed the underground stigmarian rooting organs which bore unusual deciduous rootlets that were positionally and developmentally homologous to leaves. The other shoot system developed into the aerial stem bearing true leaves (Rothwell and Erwin 1986).

Growth of both the aerial and the underground stems was strictly limited (i.e. determinate). In its most extreme and simplest form this meant that for most of their life the trees consisted of an unbranched stem clothed with long leaves, and only produced a rather sparse and limited crown towards the end of their growth. Such a crown is very different from that of a modern dicotyledonous tree, and probably did not cast much of a shadow or markedly increase the photosynthetic area over that of the young plant (DiMichele and Phillips 1985). Communities in which such trees were dominant were probably rather light and open. Other species of stigmarian lycopod, although still determinate, produced lateral deciduous branches during the growth of the main stem and this resulted in a crown much more like that of an extant dicotyledonous tree in terms of increased photosynthetic area and its presence, in changing form, for most of the plant's life (DiMichele and Phillips 1985). These variations in growth and habit also have significant effects on the duration and timing of reproduction because megasporangiate cones are often borne at, or close to, the tips of branches (DiMichele and Phillips 1985).

Arborescent stigmarian lycopods dominated the Euramerican coal swamp forests for most of the late Carboniferous but subsequently experienced a marked decline at around Westphalian–Stephanian boundary (Figure 7.4). This decrease in abundance (as judged from the percentage volume of coal accounted for by different major plant groups) has been studied in great detail and is apparently linked with the onset of drier climatic conditions which may themselves have been initiated by tectonic changes (Phillips *et al.* 1985; DiMichele *et al.* 1985). By the Permian, many of the once dominant stigmarian lycopods were extinct. During the Mesozoic, stigmarian lycopods are represented by genera such as *Pleuromeia* and *Nathorstiana*, and the large number of dispersed lycopod megaspores of this age suggest that there was perhaps a diversity of other taxa that are currently unknown as macrofossils. The only living representatives of the stigmarian lycopods (*Isoetes*) are typically 'rosette' plants of wet or submerged habitats (Figure 7.3).

Non-lycophyte 'pteridophytes'

Alongside the lycophytes the Early Devonian land plants usually considered to be more closely related to ferns, sphenopsids and seed plants are generally taken to include two groups (Figure 7.3). The simpler, and on the whole earlier, plants (e.g. *Renalia, Rhynia*) are usually included in the rhyniophytes (Edwards and Edwards 1986), whereas the generally later-occurring forms which are larger and have more complex branching (e.g. *Psilophyton, Pertica*) are usually included in the trimerophytes (Banks 1975; Chaloner and Sheerin 1979). Rhyniophytes and trimerophtyes are often referred together as the psilophytes. Cladistically the rhyniophtyes seem likely to prove of diverse relationships and some (e.g. *Renalia*) may be more closely related to the lycophytes than to trimerophyte taxa. The trimerophytes themselves are more homogeneous, but because they seem to have no unique defining characters they apparently constitute only a relatively primitive grade of nonlycophyte vascular plant.

The diversification of the trimerophyte grade has been viewed in terms of competitive superiority over the rhyniophytes based on increasing size and complexity of branching (Knoll 1984). This scenario attributes the 'success' of trimerophytes to enhanced light-gathering efficiency and also highlights the evolution of two different modes of light interception in the two major clades of tracheophytes. In the lycophytes the photosynthetic capabilities of the originally naked axes seem to have been increased by the development of microphylls, and in this group large laminar photosynthetic surfaces never evolved. In contrast, in the trimerophytes the initial increase in photosynthetic area appears to have been via a different route, involving greater complexity of branching, and this in turn is thought to have provided the starting point for the independent origin of large, laminar and often pinnately compound leaves (megaphylls) or branching systems in progymnosperms, ferns and perhaps also in seed plants.

Whatever the causal factors underlying their Middle Devonian expansion, the 'trimerophytes' have occupied a pivotal position in traditional interpretations of land-plant evolution (see Chaloner and Sheerin 1979; Stewart 1983). The tracheophyte grade which they represent has been seen as an appropriate starting point from which to derive sphenopsids, ferns and ultimately the 'progymnosperms' and seed plants (Figure 7.3). In cladistic terms it is extremely difficult to resolve the exact pattern of relationships between trimerophytes and the three groups of pteridophytes to which they are thought to be closely related (Figure 7.3), and many of the critical plants involved in this question (e.g. certain Cladoxylales, Coenopteridales, Hyeniales) are rather poorly understood (see Stein *et al.* 1984).

The sphenopsids themselves are a clear and undoubtedly monophyletic group (Figure 7.3) and the earliest unequivocal representatives (e.g. *Archaeocalamites*) are from the latest Devonian/Early Carboniferous. Thereafter, the group is a consistent but rarely dominant component of fossil floras. The greatest diversity was apparently attained during the Late Palaeozoic but local Late Carboniferous extinctions seem to have been much less pronounced than in the stigmarian lycophytes. The post-Carboniferous decline in diversity occurs abruptly in the Early Permian when judged in terms of total diversity (Niklas *et al.* 1985) but much more gradually, and over a longer period, when judged in terms of within-flora diversity (Knoll 1986).

The sphenopsids are usually divided into two main groups, and while the extant genus *Equisetum* is clearly monophyletic, the Calamitales may be paraphyletic. Possible calamite synapomorphies, such as the presence of bracts in the cones between the whorls of sporangiophores, and the vascular cambium which produces only secondary xylem (unifacial), may actually be features that were lost in *Equisetum*. Whatever the precise relationship between *Calamites* and *Equisetum*, they are undoubtedly more closely related to each other than either is to the sphenophylls, the third group traditionally included in the sphenopsids.

The main similarity between the sphenophylls (Sphenophyllales) and the other sphenopsids is the whorled phyllotaxy, and the absence of other convincing shared derived features (synapomorphies) has led some authors to suggest that this group may be entirely misplaced within the sphenopsids. One alternative hypothesis is that they are perhaps more closely related to the lycopods, based on their exarch xylem development, pattern of cambial development (Cichan 1985) and potential similarities of sporangial arrangement with some *Leclercqia*-type lycopods.

The ferns are a much more difficult group to circumscribe than the sphenopsids, being very diverse both in the fossil record and in the Recent flora. As currently delimited they are a poorly defined group of plants perhaps of diverse relationships. The early ferns underwent their initial radiation in the latest Devonian and Early Carboniferous (Galtier and Scott, 1985), and the first modern group to be recognised out of this plexus of extinct forms is the Marattiales. During the Late Carboniferous the Marattiales, and particularly *Psaronius-Pecopteris* plants, were an important, sometimes dominant group in certain habitats (e.g. Figure 7.4).

During the Mesozoic a secondary evolutionary radiation of ferns seems to occur; several extant marattialean genera can be recognised (e.g. *Angiopteris*, *Marattia*) and are joined by representatives of other extant fern families including the Schizaeaceae, Gleicheniaceae, Matoniaceae, Dicksoniaceae, Dipteridaceae and Osmundaceae. These groups were undoubtedly important in certain ecological settings (e.g. fire-maintained habitats, marshes, mudflats, early succession) and were the dominant herbaceous taxa in most Mesozoic vegetation, perhaps fulfilling an analogous role to grasses in Recent ecosystems. In the Late Cretaceous many of the characteristic Mesozoic fern groups declined in importance as the angiosperms expanded and today most of these families are of low systematic diversity. However, along with the angiosperm diversification there seems to have been an increase in the structural complexity of many

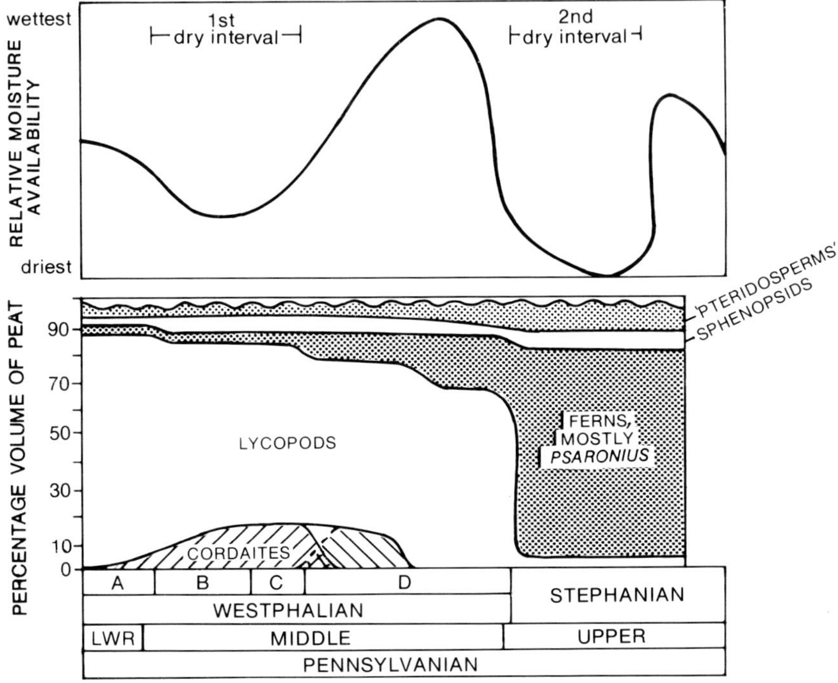

Figure 7.4. Top, inferred pattern of changing moisture availability during the Late Carboniferous (modified from Phillips and Peppers 1984). Bottom, summary of changing abundances of major plant groups in Late Carboniferous swamp vegetation based on percentage volume of permineralised peat (coal balls) (modified from Phillips and Peppers 1984).

terrestrial communities and this may have facilitated a third radiation of the 'higher' (filicalean) ferns into the damp shaded habitats of multistratal angiosperm forests.

'Progymnosperms' and primitive seed plants

The third group of tracheophytes to which fossils at the trimerophyte grade are thought to be closely related is the group sometimes called the lignophytes (Figures 7.3 and 7.5), which includes seed plants and their closest pteridophytic relatives the 'progymnosperms'. The lignophyte clade is defined by the presence of a vascular cambium that produces secondary xylem centripetally and secondary phloem centrifugally. This innovation only occurs elsewhere in the sphenophylls, although the cambia of *Sphenophyllum* and seed plants are basically different in the way that circumferential expansion occurs (Cichan 1985). The lignophyte cambium may have been one of the key features that facilitated the exploitation of an essentially modern form of arborescent habit.

The 'progymnosperms' constitute the basal grade of the lignophytes and include plants that lack seeds. Two groups have generally been recognised; the aneurophytes which resemble the trimerophytes in being homosporous and having a single solid and frequently ribbed rod of xylem tissue in the centre of the stem, and the archaeopterids (*Archaeopteris*) which show a significant modification in having the primary xylem divided into a series of longitudinal strands arranged in a ring within the stem (eustele) (Beck 1976). While it is generally accepted that the progymnosperm grade includes those pteridophytes most closely related to seed plants, relationships between aneurophytes, archaeopterids and seed plants are controversial. Initially the demonstration of heterospory in some archaeopterids indicated that they may be the closest group to spermatophytes (Pettitt 1970; Meyen 1984), but more recently the recognition of anatomical similarities between the earliest seed plants and aneurophytes has supported the opposite conclusion (Rothwell and Erwin 1987). Beck (1981) has suggested a biphyletic origin of seed plants with cordaites and conifers derived from archaeopterids, and Palaeozoic 'seed ferns' (see below) being derived from aneurophytes.

As with the rise of the trimerophytes, the 'success' of the progymnosperms has been linked to still greater size and perhaps further enhanced light-gathering efficiency (Knoll 1984). This explanation can be applied most obviously to archaeopterids which were large trees and had prominently flattened (plagiotropic) branching systems: what is less clear is the extent to which such a hypothesis might apply to aneurophytes. Aneurophytes were generally smaller, their branching systems were not extensively planated, and there is no clear webbing between the branches which would result in increased photosynthetic area. What does occur in the aneurophytes, however, is a distinct increase in the regularity of branch formation, and this may have given these plants some photosynthetic advantage over the trimerophytes. It has recently been demonstrated that the three- to four-lobed protosteles characteristic of the aneurophytes (e.g. *Rellimia*, *Tetraxylopteris*) are simply an expression of this increased regularity of branching (White 1987) rather than, as previously thought, the result of 'adaptive' processes acting directly on stelar morphology.

Despite the rise of progymnosperms and the demise of plants at the trimerophyte level neither of these groups or other Devonian plants show any more or less synchronous decrease or increase in diversity that could be interpreted as a result of mass extinction. In particular neither spore or macrofossil data show the marked Frasnian–Famennian decline in diversity that has been detected in the fossil history of marine invertebrates (Chapter 2). The marked Famennian decline seen in syntheses of spore data (Richardson and McGregor 1986) is most likely due to the absence in this study of suitable Famennian-age rocks for the preservation and isolation of spores (Richardson, personal communication). Other analyses

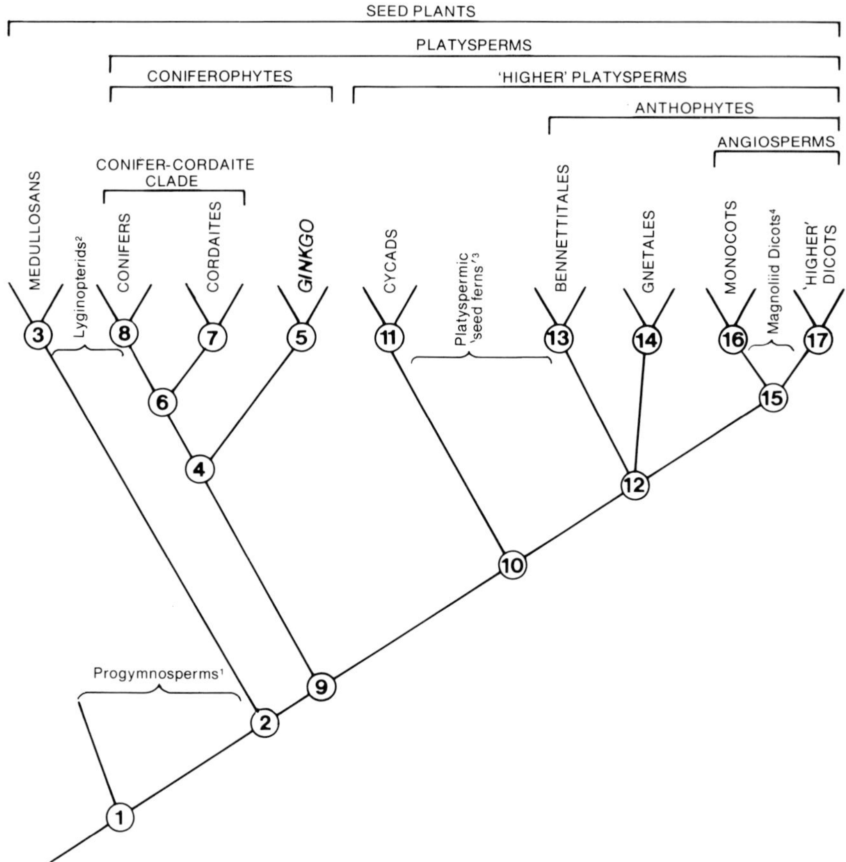

Figure 7.5. Simplified summary of the major groups of seed plants (several taxa omitted, e.g. *Pentoxylon*). Groups in upper case are possibly monophyletic (clades). Taxa given in lower case are paraphyletic (grades). 1, bifacial cambium producing secondary xylem and phloem. 2, seeds. 3, complex primary vascular structure (pseudopolystele) and *Monoletes* or similar-type pollen. 4, seed-bearing structures borne on a short shoot in the axil of a bract or leaf, simple leaves with linear or dichotomous venation. 5, seed-bearing structure a short shoot bearing two or more fused and highly modified leaves. 6, seed-bearing shoots aggregated into an inflorescence. 7, strap-shaped parallel-veined (*Cordaites*) foliage. 8, resin canals, awl-shaped leaves. 9, flattened (platyspermic) seeds, saccate pollen. 10, pollen grains with a distal aperture (also acquired independently in conifers, *Cordaites* and *Ginkgo*). 11, girdling leaf traces, cycasin. 12, non-saccate pollen, granular pollen exine structure, aggregation of reproductive parts into flower-like structures, scalariform pitting in the secondary xylem, stomata flanked by two subsidiary cells (paracytic). 13, sterile scales between the seeds (interseminal scales). 14, opposite leaves, vessel perforation plates formed by circular bordered pits. 15, double fertilisation, 4–16 nuclei in megagametophyte, vessel perforation plates formed by scalariform pits. 16, one cotyledon, parallel-veined leaves, loss of

through the Late Devonian (Scheckler 1986) suggest that the decline in diversity begins early in the Frasnian and that while this stage does not have an abnormally high extinction level it is characterised by the appearance of very few new taxa (i.e. 'speciation' rate apparently fails to 'keep pace' with background extinction).

Seed plants are defined as a monophyletic group by a series of important modifications of the heterosporous condition (Figure 7.5; Pettitt 1970). Only a single megaspore tetrad develops within each megasporangium, and three of the four megaspores abort. The single surviving megaspore is retained during gametophyte development and fertilisation. The role of dispersal in the life cycle is 'transferred' from the spore to the megasporangium which is also surrounded by an additional covering (integument). At the level of seed plants it is conventional to refer to the megasporangium wall as the nucellus and the microspores as pollen grains. The earliest seed plants (e.g. *Archaeosperma*, *Moresnetia*) are known from the Late Devonian (Stewart 1983; Fairon-Demaret, 1986). In several of these taxa the integument does not completely enclose the nucellus and pollination seems to have involved a modification of the apical part of the nucellus (the lagenostome; see Rothwell 1986). In extant and later fossil seed plants the apparent functions of the lagenostome in collecting and trapping pollen, and sealing the access to the gametophyte, are carried out by the integument.

Seed plants undergo their initial major radiation during the Early Carboniferous and several groups have traditionally been recognised in the Late Palaeozoic. The 'lyginopterids' are a heterogeneous and probably rather unnatural assemblage of primitive seed plants that includes most of the spermatophytes known from the Early Carboniferous (Stewart 1983; Crane 1985). In contrast the medullosans are a predominantly Late Carboniferous group and are more easily circumscribed by their large, distinctive pollen grains, their seeds with a double ring of vascular bundles and their unusual stem anatomy (Stewart 1983; Crane 1985). Both of these seed-plant taxa had fernlike leaves and, in the case of medullosans and some 'lyginopterids', these leaves were large, complex structures. Both of these groups and several others from the Late Palaeozoic and Mesozoic (e.g. glossopterids, peltasperms, corystosperms) are frequently referred to as 'pteridosperms' (seed ferns). Unfortunately, over the last 60 years, the gradual incorporation of newly discovered and enigmatic seed plants into this concept has made the 'seed ferns' a highly unnatural taxon, and this

---

significant secondary growth (excluding the 'anomalous secondary thickening' of some arborescent monocots). 17, tricolpate pollen. [1]Includes both aneurophytalean and archaeopteridalean forms. [2]Includes *Lyginopteris* and many other Early Carboniferous 'seed ferns'. [3]Includes *Callistophyton*, *Caytonia*, corystosperms, glossopterids, and peltasperms. [4]Approximately equivalent to the extant subclass Magnoliidae. For further details, see Crane (1985); Doyle and Donoghue (1986).

has confused phylogenetic discussions in which the pteridosperms have been treated as a natural (monophyletic) group (see Doyle 1978; Stewart 1983).

Traditionally, seed plants have been divided into coniferopsid (conifers, cordaites, *Ginkgo*, Gnetales) and cycadopsid (cycads, Bennettitales, pteridosperms) groups (see Chamberlain 1935; Bierhorst 1971), although recent cladistic analyses do not support this division (Crane 1985; Doyle and Donoghue 1986). Assuming that the seed plants are a monophyletic group, in terms of mean percentage contribution to individual fossil floras, they show a spectacular diversification beginning in the latest Devonian and continuing through the Early Carboniferous (Knoll 1986). On average, seed plants account for at least half of the species recorded in Carboniferous and later floras, and the Devonian–Carboniferous boundary most prominently marks a pronounced decline in the importance and diversity of those lignophytes with pteridophytic reproduction ('progymnosperms').

The 'success' of seed plants is usually attributed to the intrinsic biological advantages of the seed habit over pteridophytic reproduction, particularly the protection afforded by the nucellus and integument to the developing embryo, and the ability to effect fertilisation with motile male gametes in the absence of external water. Unfortunately this hypothesis has never been subject to critical examination and several lines of evidence suggest that it may not provide the full explanation. Despite the rapid rise of seed plants many pteridophyte groups (e.g. stigmarian lycopods, marattialean ferns) were able to persist and also proved highly successful in terms of both abundance and diversity in a variety of habitats during the later Palaeozoic. In addition, there are perhaps other biological consequences of the seed habit that have not been fully considered and it is possible that, rather than conferring any basic biological advantage, such a fundamental change in the reproductive system might instead have enhanced diversification rate via indirect effects on levels of speciation and extinction. In addition, the currently rudimentary state of early seed plant systematics makes it difficult to assess variations in the importance of different seed-plant clades during the early spermatophyte radiation. In the Late Carboniferous, for example, it was one particular group of seed ferns (medullosans) that were the floristically dominant seed plants in most lowland environments, and this may have more to do with the characteristics of medullosans than the biological attributes of seed plants as whole. Until considerably more is known about the structure biology and relationships of early seed plants the causal factors underlying their success will remain obscure.

Platysperms

One of the most important factors contributing to the demise of the 'pteridosperm' concept has been the recent discovery of a Late Carboniferous group of 'seed ferns', termed the callistophytes (Rothwell 1981). These are characterised by fern-like foliage, saccate pollen and distinctly flattened (platyspermic) seeds with two vascular bundles. All of these features are diagnostic of a recently recognised major clade of seed plants termed the platysperms (Figure 7.5; Crane 1985; Doyle and Donoghue 1986). The indications are that callistophytes are more closely related to conifers, *Ginkgo* and cycads than to other 'pteridosperms' such as medullosans or 'lyginopterids' with which they are often placed.

The callistophytes are not the first group of platysperms to appear in the fossil record, and the earliest representatives are members of a subclade termed the coniferophytes (Figure 7.5, traditional coniferopsids *sensu* Chamberlain 1935, minus the Gnetales). Coniferophytes comprise the conifers, cordaites and *Ginkgo*, and are distinguished from other Carboniferous seed plants by their simple (non-pinnate) foliage with a correspondingly simple, non-pinnate pattern of venation. The cordaites are the first to appear in the fossil record (Early Carboniferous, Stewart 1983; Crane 1985; Rothwell and Warner 1984), although dispersed platyspermic seeds of uncertain relationships occur in the latest Devonian of Ireland (Chaloner *et al.* 1977). The ovules of cordaites are borne on small leaf-like scales grouped together in small buds and borne in the axil of a bract. Several of these bract–bud complexes are present in each cordaite cone. Exactly the same arrangement occurs in the ovulate cones of conifers, although in highly modified form, with each bract–bud complex flattened and often fused to form the individual cone scales. This basic similarity is important evidence that the cordaites and conifers are sister taxa (Florin 1951). Ecologically cordaites are thought to have occupied predominantly drier, perhaps upland, sites in the Carboniferous landscape, although there is evidence that others periodically inhabited swamps, or were associated with marine conditions (Figure 7.4).

The cordaites are one of the plant groups affected by extinction in the Late Permian and Early Triassic, but the fate of the conifers was strikingly different. They first appeared during the Late Carboniferous (Scott and Chaloner 1983; Mapes and Rothwell 1984), gradually became important during the Permian and diversified through the Mesozoic into the seven extant families (Araucariaceae, Cupressaceae, Cephalotaxaceae, Pinaceae, Podocarpaceae, Taxodiaceae, and Taxaceae) (Miller 1977; Hart 1987). Several extinct taxa are also known from the Mesozoic, of which the most abundant in terms of diversity and abundance are the Cheirolepidiaceae. The Cheirolepidiaceae produced highly characteristic pollen grains assigned to the genus *Classopollis*, and the distribution of both macrofossils and pollen indicates that although they were significant components of the vegetation in a range of environments, they were particularly important in xeric, seasonally dry, and certain coastal habitats (Upchurch and Doyle 1981). Conifers remain an important floristic component through the Mesozoic where they typically comprise about 10–20% of the species in macrofossil floras (Lidgard and Crane 1988). In the Late Cretaceous the conifers experience a slight decline as a result of the extinction of Cheirolepidiaceae, but it is evident that the extant families continued to diversify through this interval, with many extant genera first being recorded from the early Tertiary (Miller 1977).

The genus *Ginkgo* is one of the most distinctive of extant seed plants and has traditionally been allied with the conifers (Figure 7.5) on the basis of the morphology of reproductive structures and leaves (Crane 1985). Along with the cycads, *Ginkgo* retains the primitive characteristic of motile male gametes (as at the pteridophyte grade) but gamete release occurs within the seed. The Palaeozoic history of *Ginkgo* is obscure and the supposed relationship with the poorly known Permian genus *Trichopitys* (Florin 1949) is not well supported. In the Jurassic, however, the presence of *Ginkgo* is well documented, and the genus can be traced with confidence and little morphological change in leaves and reproductive structures through the Mesozoic and Tertiary to the Recent.

Throughout the Permian, and much of the Mesozoic, other members of the platysperm clade were important components of the vegetation, and these taxa include most of the Late Palaeozoic and Mesozoic 'pteridosperms' (glossopterids, corystosperms, peltasperms and caytonias, Crane 1985; 1988). Some of these were very important in certain areas (e.g. glossopterids in the Southern Hemisphere) but their interrelationships are uncertain partly because many of their structural details are not well understood. The cycads are probably closely related to some of these platysperm taxa (Figure 7.5; Doyle and Donoghue 1986; Crane 1988a). Extant cycads comprise ten genera, with the earliest forms appearing in the Permian (Crane 1988). Subsequently the group is a significant component of many Mesozoic fossil floras but clearly declines in importance during the mid-Cretaceous (Lidgard and Crane 1988). Many Mesozoic cycads may have been very different, in terms of their habit, from the living representatives, and it is clear that the extant genera are relictual, with widely scattered distributions predominantly in tropical to subtropical regions of the world (Johnson 1959).

Anthophytes

One of the most important results to emerge from recent cladistic analyses of seed-plant relationships has been the recognition of a relatively derived group of seed plants termed the anthophytes (Figure 7.5), which comprises the angiosperms and their nearest relatives, the Bennettitales, Gnetales and *Pentoxylon* (Crane 1985; Doyle and Donoghue 1986). Increasing knowledge of fossil representatives of these groups may require revision of this pattern of relationships (Crane 1988), but with the data currently available the anthophytes are defined as monophyletic based on a variety of characters (Figure 7.5).

The diversification of the anthophyte clade, along with changes in the fern flora, is one of the features that distinguishes Palaeozoic fossil plant assemblages from those of the Mesozoic. Typically Palaeozoic floras are dominated by pinnate leaved ferns and 'seed ferns', but in the Mesozoic it is the cycads, Bennettitales, conifers, ginkgophytes and more advanced 'pteridosperms' that are the dominant floristic and vegetational elements. Recent analysis indicates that this floristic transition from what has been termed the Palaeophytic to the Mesophytic was relatively rapid in any given area, but occurred at different times in different places (Knoll 1984). The primary causal factor is thought to have been climatic, involving continuance and intensification of the drying trend first detected in the Late Carboniferous tropics and which led to the extinction of many of the stigmarian lycopods (DiMichele *et al.* 1987). Initially there seems to have been a shift toward an increasingly continental climate, with later changes resulting in increased regional variability associated with accentuated latitudinal temperature gradients and physiographic complexity (Di Michele *et al.* 1987). Diversity assessed at the family level provides a crude measure of the magnitude of extinctions associated with the Permian–Triassic transition. The number of families of vascular plants decreased gradually from the Early Permian to the Middle Triassic, whereas the decline of families of marine invertebrates seems to have occurred more rapidly in a more concentrated interval between the Middle Permian and Early Triassic (Knoll 1984; and Chapters 2 and 9).

The most conspicuous anthophyte clade to diversify during the Triassic is the Bennettitales, and this group is common in most Late Triassic–Early Cretaceous fossil floras. The Bennettitales are frequently considered with the cycads because of their similar foliage (Stewart 1983), but they are easily distinguished by their specialised, usually flower-like, reproductive structures (Figure 7.6), by the organisation of their stomata and by a variety of other features (Crane 1985; 1988). Cladistic analyses suggest that the Late Jurassic–Early Cretaceous genus *Pentoxylon* (Bose *et al.* 1985) may also be related to the Bennettitales. One of several innovations seen for the first time in the Bennettitales is the presence of a second integument around the nucellus, and this may be an important homology with angiosperms, *Pentoxylon* and possibly Gnetales (Crane 1985). The Bennettitales also show some of the most striking 'flowers' outside of the angiosperms and there can be little doubt that some of these were pollinated by insects: perhaps in a manner similar to that in extant *Magnolia*. Certain of these 'flowers' are bisexual (Figure 7.6; e.g. *Williamsoniella*, *Cycadeoidea*; Crane 1988), and in others there is circumstantial evidence of the presence of nectaries. The Bennettitales declined in importance through the Late Cretaceous (Lidgard and Crane 1988), unlike the cycads, they became extinct by the earliest Tertiary.

Dispersed pollen grains and a small number of macrofossils from the Triassic provide the earliest indications of the Gnetales (Crane and Upchurch

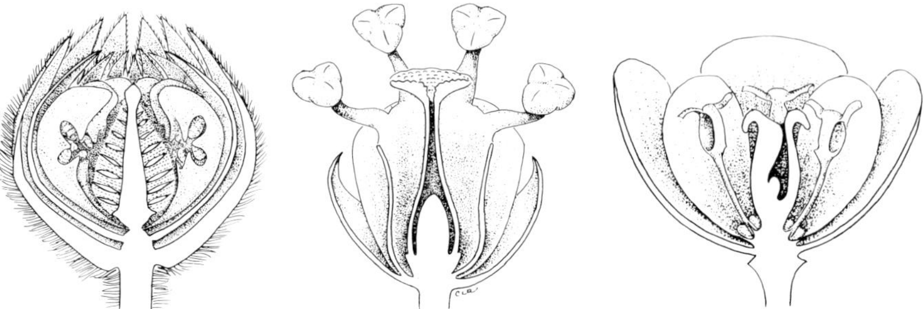

Figure 7.6. Sections through the structurally hermaphrodite (bisexual) 'flowers' produced by three different groups of plants. Each 'flower' shows the seed-producing organs in the centre surrounded both by the pollen-producing organs and structures resembling petals. Left to right: *Williamsoniella coronata*, Middle Jurassic, a bennettitalean 'flower'; *Welwitschia mirabilis*, Recent, a gnetalean 'flower' with a central ovule that does not mature into seed (note the funnel-shaped integument that produces a nectar-like pollination drop); *Berberis vulgaris*, Recent, an angiosperm flower with the seed enclosed in a carpel.

1987; Crane 1988). The Gnetales are a well-defined monophyletic group with three extant genera: *Gnetum*, a tropical climber; *Welwitschia*, a bizarre desert plant with one species endemic to southwest Africa; and *Ephedra*, an almost leafless plant of extratropical xeric habitats (Martens 1971). Like the Bennettitales, certain Gnetales also have flower-like reproductive structures (Figure 7.6), and studies of the extant taxa show that they have relatively sophisticated insect pollination systems in which 'sterile' ovules associated with the pollen organs produce a sweet pollination drop which functions as a nectary (see Bino *et al.* 1984). Based on the diversity and abundance of fossil pollen grains, the Gnetales underwent a marked radiation during the mid-Cretaceous at low palaeolatitudes (northern Gondwana—see below) before declining in importance in the middle of the Late Cretaceous (see Herngreen and Chlonova 1981).

Angiosperms

The monophyly of angiosperms is better supported than that of any other major clade of seed plants (Figure 7.5). The first unequivocal evidence of the group are dispersed pollen grains from the Early Cretaceous (Hauterivian), and these are identified as angiosperms based on their pollen wall structure which consists of a basal layer and columellae supporting a roof-like tectum (Hughes and McDougall 1987). Subsequently, during the mid-Cretaceous, angiosperms undergo a major radiation (Lidgard and Crane 1988) and rapidly become the dominant plants of the Late Cretaceous vegetation. The chronological appearance of different angiosperm pollen types in the fossil record parallels what might be predicted based on the present understanding of phylogenetic patterns among extant angiosperms. Pollen with a single distal germinal aperture (monosulcate) such as occurs in monocotyledons and relatively primitive dicotyledons appears first, followed by tricolpate, tricolporate and triporate pollen diagnostic of more derived dicotyledonous groups (Doyle 1978).

By the beginning of the Late Cretaceous angiosperms had acquired many of their characteristic structural features, including flowers with numerous or few parts, free or fused carpels, floral parts in fives (pentamerous), helical or whorled arrangements of floral parts, vessels, and broad reticulate-veined leaves (Friis *et al.* 1987; Crane 1989). Detailed studies of pollen grains, leaves and floral structures also indicate that angiosperms had already diversified into several major groups.

Monocotyledons (Figure 7.5) are represented in the mid-Cretaceous by both leaves and dispersed pollen (Doyle 1973). During the Late Cretaceous several extant groups such as palms (Arecaceae) and gingers (Zingiberaceae) can be recognised. Among the dicotyledons several families or orders of extant flowering plants are present before the end of the Cenomanian including magnolia-like plants—Magnoliaceae (Dilcher and Crane 1984), Chloranthaceae (Crane *et al.* 1989), sycamores or plane trees (Platanaceae) and probably the Trochodendrales (Crane 1989).

Extant dicotyledons comprise six subclasses, of which five (Caryophyllidae, Hamamelidae, Rosidae, Dilleniidae and Asteridae: 'higher' dicotyledons, Figure 7.5) contain about 80% of extant angiosperm species (Cronquist 1981). The sixth subclass (Magnoliidae, Figure 7.5) is a paraphyletic assemblage of relatively primitive dicotyledons that includes taxa such as the magnolia family (Magnoliaceae), cinnamon family (Lauraceae), custard-apple family (Annonaceae), as well as the water-lilies (Nymphaeaceae) which appear to be more closely related to the monocotyledons than to other dicotyledons (Donoghue and Doyle 1989). Within the 'higher' dicotyledons the monophyly and relationships of the constituent subclasses are uncertain. The Hamamelidae (Crane and Blackmore 1989) includes such families as the oaks (Fagaceae), birches and hazels (Betulaceae) and walnuts (Juglandaceae) and is certainly paraphyletic and probably polyphyletic as usually circumscribed. One of the more primitive members of the Hamamelidae, and a critical family in the phylogeny of the dicotyledons, is the Platanaceae (London plane tree, American sycamore). The early representatives of this family exhibit a pentamerous floral organisation characteristic of most 'higher' dicotyledons and provide the earliest documentation of this important feature in the fossil record (Friis *et al.* 1988). Recent work suggests that the Platanaceae may be the sister taxon to the Rosidae, Asteridae and possibly also the Dilleniidae (Crane 1989). Both the Hamamelidae and Rosidae were unequivocally present in the Early Cretaceous (Late Albian) and other subclasses such as the Caryophyllidae and Dilleniidae may also have diverged at this early stage.

Through the Late Cretaceous and Early Tertiary the number of extant taxa that can be recognised with confidence steadily increases (Muller 1984), and this systematic modernisation is reflected in the plant communities which become increasingly like those of today (Crane 1987). In addition, the range and sophistication of plant–pollinator interactions steadily increases (Crepet and Friis 1987). The earliest known angiosperm flowers from the mid-Cretaceous (Albian–Cenomanian) were generally radially symmetrical with free perianth parts, but by the end of the Cretaceous bilaterally symmetrical flowers, fused petals, epipetalous stamens and a variety of nectaries had all developed (Friis and Crepet 1987). The fossil record of insects indicates that the main pollinators of early angiosperms were probably certain beetles, flies and primitive moths with chewing mouthparts (micropterigids) perhaps co-opted from pollination interactions which were already occurring with other groups of seed plants (e.g. Bennettitales and perhaps Gnetales). However, by the latest Cretaceous and early Tertiary a great variety of more complex pollination systems had developed involving almost the full spectrum of extant pollinating insects including bees and higher Lepidoptera (Friis and Crepet 1987; Crepet and Friis 1987). In contrast, in dispersal biology, there are few indications of specialised interactions with animals before the Tertiary. The predominant Late Cretaceous herbivores were ceratopsian and hadrosaur ornithopod dinosaurs which were large, bulk plant feeders. However, in the earliest Tertiary the mammals were probably more selective herbivores and there is a marked increase in propagule size which may be linked to increasing coevolution with mammalian dispersal vectors (Tiffney 1984; Wing and Tiffney 1987).

Cretaceous–Tertiary boundary

The event(s) at the Cretaceous–Tertiary (K–T) boundary that had major effects on many animal groups (Chapters 2, 5 and 9) also appear to have exerted an important influence on the structure and composition of terrestrial vegetation, as well as the diversification of extant families of angiosperms. The exact nature of the environmental perturbations at the K–T boundary are still controversial (Chapter 9), but interpretations of evidence from pollen and spore floras, from dispersed cuticles and from macrofossils support hypotheses of mass kill of the vegetation, a brief low temperature excursion, major ecological disruption and high levels of extinction (Upchurch and Wolfe 1987). Data from fossil leaves suggest that about 75% of the Late Cretaceous species became extinct, and although estimates of extinctions based on palynofloras are less severe, there are significant changes in abundance for those pollen grains which range across the boundary. Analyses immediately above the boundary suggest that devastation of the vegetation was followed by a period of recovery that mimicked features of secondary succession exhibited by disturbed Recent vegetation, but occurred over a longer period (about 1.5 Myr). In addition, over a large area angiosperm leaf sizes markedly increased, suggesting a significant increase in precipitation (Upchurch and Wolfe 1987).

The initial post-boundary recovery phase appears to have involved recolonisation by ferns. At high and middle palaeolatitudes one of the major long-term results seems to have been the elimination of many of the common broad-leaved evergreen taxa of the Late Cretaceous, and replacement by broad-leaved deciduous taxa, such as the birch and walnut families (Betulaceae, Juglandaceae), during the Early Palaeocene. Under warmer-temperature regimes at lower latitudes the fern recolonisation seems to have been followed by the development of multilayered rainforest structurally comparable to that in certain areas of the Recent tropics, but with much lower species diversity. The overall effect of the event(s) at the K–T boundary seems to have been a further modernisation of the vegetation, at least in the Northern Hemisphere, both in terms of systematic composition and structural characteristics.

Tertiary and Quaternary extinctions

Although the details of vegetational change through the Tertiary are complex, several major features are clear (e.g. the generally deteriorating climate) and the primary driving forces were at least in part tectonic. Major mountain-building (e.g. the Andes, Rocky Mountains, Himalayas, Alps) resulted in increasingly marked rain-shadow effects, continental movements facilitated dispersal and isolation of plant and animal taxa, and the movement of land masses into high latitudes is thought to have contributed to the growth of polar ice. The combined effects resulted in a climatically more complex world with increased latitudinal temperature gradients, new barriers to migration, and enhanced local continentality and oceanicity. Against this backdrop of tectonic and climatic change angiosperm diversification, particularly of more 'advanced' groups, was proceeding rapidly and ultimately led to increasing systematic endemism and floristic provinciality. Poor knowledge of the details of the angiosperm diversification currently constrains attempts to link climatic and tectonic changes with specific pulses of diversification or extinction, but two widespread intervals of enhanced extinction, and perhaps coincident diversification in other groups, are recognised: the Late Eocene and the Late Miocene to Quaternary.

The 'terminal Eocene event' has been linked to a pronounced and rapid temperature decline after the relatively equable climates of the Eocene. During this interval the area of rainforest was evidently much reduced and in cool-temperature regions was replaced in part by large areas of broad-leaved deciduous forests. The fossil record documents that many taxa expanded at this time, including the extant hornbeams (*Carpinus*), maples (*Acer*), beeches (*Fagus*), and oaks (*Quercus*), while those groups that underwent major extinction included many of the more important taxa of the Late Palaeocene and Eocene vegetation. The exact timing of these Late Eocene effects seems to have varied in different regions. Fossil leaf floras from western North America suggest that these changes occurred close to the Eocene–Oligocene boundary, but detailed analyses of both the pollen-spore and fruit-seed floras from the Lower Tertiary of southern England indicate that these changes may have been initiated rather earlier in the late Middle Eocene (Collinson and Hooker 1987).

In the latest Tertiary and Quaternary further extinctions occurred immediately prior to and during the onset of pre- and full glacial conditions, although some herbaceous groups clearly speciated during this interval. The majority of extinctions were regional rather than global and involved the extinction in western Europe of taxa such as the tulip tree (*Liriodendron*), kiwi fruit (*Actinidia*), sweet gum (*Liquidambar*) and *Ginkgo*. Local extinction clearly accounts for the relatively depauperate flora of central and northwest Europe compared to that of eastern North America or eastern Central Asia. Some global extinctions did occur, however, and even a few of the common plants of the latest Cretaceous and Early Tertiary (e.g. *Nordenskioldia*) were only finally eradicated at this time.

A CASE HISTORY: EARLY EVOLUTION OF ANGIOSPERMS

The early diversification of angiosperms is one of the most dramatic biotic events of the Mesozoic and has been studied intensively over the last 25 years (Hughes 1976; Doyle 1978; Friis *et al.* 1987). The diversity, complexity and abundance of angiosperm clearly increases through the Cretaceous, but studies have focused predominantly on 'key' stratigraphic sequences (e.g. Potomac Group; Doyle and Hickey, 1976), making it difficult to fully separate evolutionary and migrational effects. In addition, although there are abundant mid-Cretaceous macrofossil assemblages from middle and high palaeolatitudes, there are almost no macrofossil data from the Cretaceous tropics and this makes it difficult to assess diversification and extinction in a global context. In contrast, fossil pollen and spores (palynomorphs) are preserved over a much larger geographic area and in a greater range of environments, and currently provide the best basis from which to assess global floristic and vegetational change.

Qualitative analyses of mid-Cretaceous palynofloras have documented significant latitudinal differences in the timing and pattern of the angiosperm radiation (Hickey and Doyle 1977). The earliest unequivocal angiosperm pollen grains appear at low and middle palaeolatitudes in the Early Cretaceous (Hauterivian), and are only later (Albian) recorded from higher latitude palynofloras in both hemispheres. Thus the angiosperm radiation in the mid-Cretaceous seems to have begun at low and middle latitudes, and spread quickly towards both poles. Triaperturate pollen (three equatorially disposed apertures) that are diagnostic of 'higher' (non-magnoliid) subclasses of dicotyledons show an even more pronounced latitudinal pattern, appearing at low palaeolatitudes in the Early Aptian, at middle latitudes in the earliest Albian, and at higher latitudes still later. To the extent that such latitudinally diachronous patterns are common among early angiosperms, the details of the angiosperm radiation inferred from mid-latitudes may not be representative of the overall phenomenon of diversification.

Quantitative analyses of over a thousand pollen and spore floras from the Northern Hemisphere (Crane and Lidgard, in progress) show the same patterns revealed by qualitative analyses, but also provide important insights into the magnitude of the angiosperm radiation (Figure 7.7) and contemporary changes in other elements of the Mesozoic vegetation. Palynomorph abundance can also provide a general guide to the abundance of different plants in the 'source' vegetation, although it is difficult to estimate some of the potential biasing factors (e.g. pollen production, dispersability, preservation) which may lead to the under- or overrepresentation of particular taxa.

Classopollis pollen produced by the xeromorphic conifer family Cheirolepidiaceae is the most abundant constituent of most low-palaeolatitude floras at this time, and many of these sequences also show geological evidence of aridity (e.g. evaporites). However, when examined in more detail, it is found that the greatest local abundances of angiosperm pollen are not associated with either *Classopollis* or evaporites, and occur instead with pollen of the Gnetales and spores of ferns. Indeed, at low palaeolatitudes the increasing diversity and abundance of angiosperm and gnetalean pollen through the mid-Cretaceous is remarkably similar until around the Turonian–Santonian when gnetalean pollen grains decline both in diversity and abundance.

Outside low-palaeolatitude areas the palynological patterns are slightly different. In mid-latitude palynofloras gnetalean pollen is only sparsely represented but angiosperm pollen rapidly becomes important during the Albian. Pre-angiosperm palynofloras in these areas contain abundant *Classopollis*, as well as a diverse and abundant representation of ferns, conifers, and other gymnosperm pollen. At about the same time as the angiosperm pollen increases, fern spores and the pollen of non-coniferous gymnosperms both decline significantly in abundance and diversity. Conifer pollen is less affected but the important *Classopollis* component is rapidly lost at around the end of the Cenomanian, just as it is at low latitudes.

At high palaeolatitudes Late Cretaceous vegetation extended close to the poles forming polar forests of predominantly deciduous taxa. Pre-angiosperm palynofloras in these areas are dominated by diverse and abundant fern spores and saccate conifer pollen (probable Pinaceae, and Podocarpaceae). With the appearance of angiosperm pollen it is mainly the fern component which shows the most pronounced decline. Saccate conifer

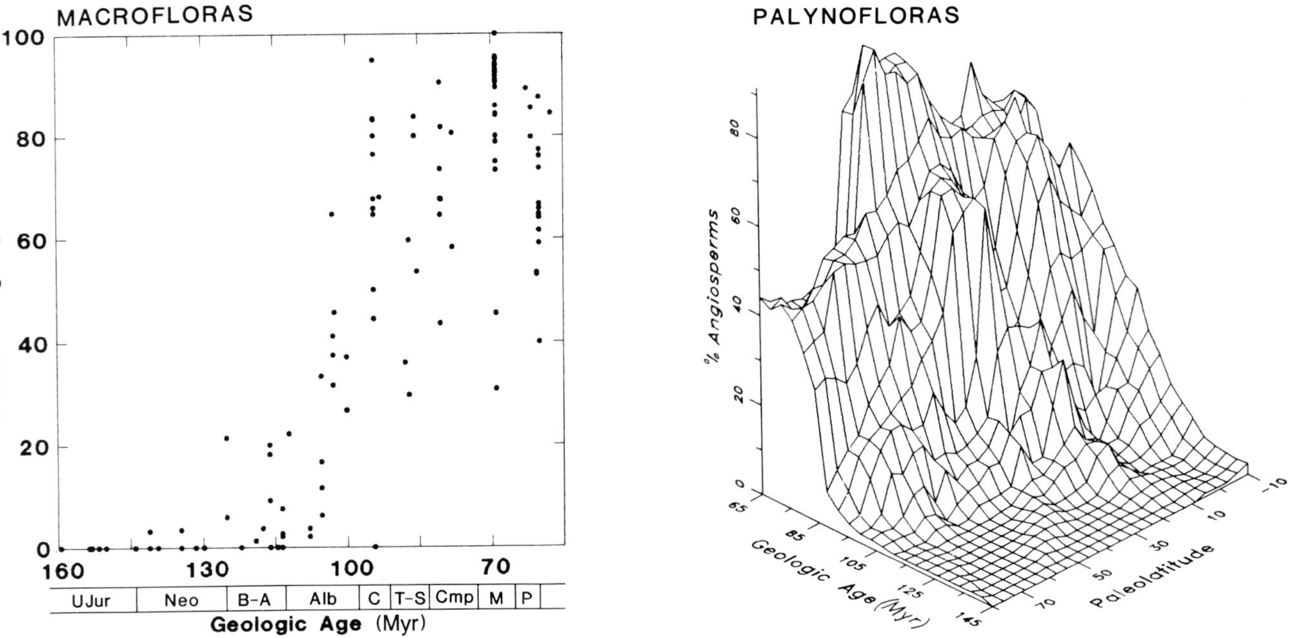

Figure 7.7. Two different measures of angiosperm diversification derived from data on within-flora diversity. Estimate of increasing angiosperm diversity from macrofloras is based on the percentage of angiosperm species in 196 macrofossil floras from mid-palaeolatitudes (Lidgard and Crane 1988). Estimate of increasing angiosperm diversity from palynofloras is based on the percentage of angiosperm pollen species in over 1000 palynofloras at various palaeolatitudes (Crane and Lidgard, in preparation).

pollen remains an important component of the palynofloras, just as it does in Recent high-latitude palynological assemblages.

Several significant points emerge from these palynological analyses concerning the dynamics and pattern of the angiosperm radiation (Figure 7.7). Most significantly, angiosperms exhibit a spectacular and remarkably rapid rise to dominance during the mid-Cretaceous, whether judged in terms of total palynomorph diversity, within-flora palynomorph diversity or palynomorph abundance. The angiosperm diversification is seen first at low palaeolatitudes but by the early Late Cretaceous (Late Cenomanian–Turonian) angiosperms appear to have been floristically and vegetationally dominant throughout most of both hemispheres. In terms of diversity and abundance it is the 'higher' (non-magnoliid) dicotyledons (as evidenced from triaperturate pollen grains) which account for most of the angiosperms encountered in Late Cretaceous palynofloras.

Although Cretaceous macrofossil floras are much more sparsely distributed than the palynofloras, quantitative analyses of macrofossil data from middle and high palaeolatitudes do provide a check on the major conclusions of the pollen-spore analyses (Figure 7.7). Results from a macrofossil data set of about 200 floras (Lidgard and Crane 1988) show similar parallel patterns of floristic change to a subset of the palynological data with similar latitudinal limits. In both data sets, through the main phase of angiosperm diversification, the absolute and within-flora diversity of ferns underwent a marked decline. Conifers show less dramatic changes, while foliage remains indicate that cycadophytes (cycads, Bennettitales) undergo a marked mid-Cretaceous decline in diversity (however it is measured). In particular, the Bennettitales were almost entirely eliminated, with only a few taxa persisting until the latest Cretaceous or earliest Tertiary.

The palaeobotanical evidence currently available from which to assess the pattern of the angiosperm radiation is much better in terms of geographical coverage than that available for most other plant diversification or extinction events, but unfortunately the patterns are not well resolved at lower systematic levels (species, genera) or in terms of local ecology. This forces any biological inferences to generalise about ecological traits of higher systematic groups and makes the causal factor(s) underlying the floristic changes still more difficult to pinpoint. Although it has been traditional in many palaeontological studies to attribute changes of this kind to competition, such hypotheses are difficult to test because direct evidence of competitive interactions is rarely preserved in the fossil record. Furthermore, simple correlation between the rise of one taxon and the decline of another is not sufficient to demonstrate a competitive effect. What needs to be demonstrated is 'ecological replacement in kind in a stable physical-climatic setting' (DiMichele *et al.* 1987, p. 173).

Ideas of competitive replacement versus changing physical conditions (extrinsic stress) lead to quite different interpretations of mid-Cretaceous vegetational change. Under a competitive replacement model, one might plausibly suggest that the cycadophytes and the ferns were affected most by the angiosperm diversification because their biology and ecology brought them into more direct conflict with the angiosperms than was the case with the conifers. With Bennettitales and ferns, evidence of biological and ecological characteristics could be adduced in support of such a hypothesis. However, under the extrinsic stress model one might equally plausibly suggest that the major vegetational changes of the mid-Cretaceous were initiated by climatic forces, themselves perhaps linked to changing tectonic conditions and sea level. This receives its strongest support from the palynological analyses which show that, in terms of abundance and diversity, angiosperms and Gnetales exhibit a co-ordinated increase, making it less likely that some advantageous innovation arose simultaneously in the two groups and resulted in dramatically enhanced competitive success. Local correlations between the abundance of pollen of the two groups in mid-Cretaceous low-palaeolatitude floras also suggest that angiosperms and Gnetales occupied similar habitats and thus probably shared certain ecological properties. In the Early Cretaceous of eastern North America, sedimentological evidence indicates that both Gnetales and the earliest angiosperms were weedy plants of disturbed ephemeral habitats (Crane and Upchurch 1987). On balance, it seems likely that ecological characteristics of both groups predisposed them to expand as environmental conditions became suitable. What remains to be explained, however, are the disparate extant diversities of the two taxa (100 as against 250 000 species). Competitive superiority may have played some role in the long-term 'success' of the angiosperms, and particularly their expansion out of the palaeotropics, but the accumulation of diversity may also have more to do with secondary effects of certain angiosperm characters (e.g. the carpel). Acting indirectly through pollination and dispersal biology, such features could have increased speciation rate or lowered extinction rate, rather than contributing directly to angiosperm competitive success (Doyle and Donoghue 1986).

CONCLUSIONS

The broad view of botanical evolution presented here shows several major evolutionary diversifications leading to the relatively rapid establishment of new 'body plans', which have then both constrained and been the focus of subsequent evolution. Like the Cambrian diversification of Metazoa, the Late Silurian–Early Devonian radiation of tracheophytes led rapidly (by the end of the Devonian) to the origin of most of the major structural features (e.g. cuticle, stomata, tracheids, leaves, cambial activity and secondary tissues, seeds, pollen-spore walls) exhibited by extant plants. However, in contrast to the palaeozoological record, the fossil history of plants does not show clearly the same critical mass extinctions which are thought to have 'reset' the evolutionary clock in many groups of animals. The most dramatic large-scale extinctions in the history of plant life, with the possible exception of that at the Cretaceous–Tertiary boundary, seem to have been rather gradual by comparison and do not always coincide with the major mass extinctions of the animal world.

The reasons for this difference must be complex but it has been suggested that basic differences in the biology of plants and animals predispose the two groups of organisms to different long-term fates (Knoll 1984; Traverse 1988). Because all plants utilise the same simple basic resources (light energy, simple nutrients, carbon dioxide and water), one argument proposes that this undemanding 'trophic uniformity' exacerbates direct competition among individuals, making it a particularly potent force in plant evolution. Further, because the biological requirements of plants are all basically similar, ecological diversification mainly consists of specialisation to particular local (largely climatic) conditions (Knoll 1984) rather than to radically different 'lifestyles'. These arguments suggest that competitive effects should be manifested in the history of plant life (Knoll 1984), but also that plants might be expected to be particularly sensitive to climatic (or edaphic) change. In the absence of possibilities for migration, such changes would lead to extinction.

Conversely, short-term environmental perturbations might be expected to have much less significance as a cause of plant extinctions than in animal groups. Perennating mechanisms and vegetative reproduction make plants especially resistant to brief environmental stress. Short-term survival of the order of one to tens of years is a basic feature of the life history of many plants, and thus geologically 'instantaneous' catastrophes on a local or global scale might have less effect on plant populations than on contemporary animals, in which brief mass mortality is more likely permanently to eradicate a population. The effectiveness of short-term extinction resistance in different plant groups and its long-term evolutionary consequences in the context of other aspects of plant biology need to be more fully explored.

REFERENCES

Banks, H.P., 1975, Reclassification of Psilophyta, *Taxon*, **24**: 401–13.

Beck, C.B., 1976, Current status of the Progymnospermopsida, *Review of Palaeobotany and Palynology*, **21**: 5–23.

Beck, C.B., 1981, *Archaeopteris* and its role in vascular plant evolution. In K.J. Niklas (ed.), *Paleobotany, Paleoecology and Evolution*, **I**., Praeger, New York. pp. 193–229.

Bierhorst, D.W., 1971, *Morphology of vascular plants*, Macmillan, New York.

Bino, R.J., Devente, N. and Meeuse, A.D.J., 1984, Entomophily in the dioecious gymnosperm *Ephedra aphylla* Forsk. (*E. alte* C.A. Mey, with some notes on *E. campylopoda* C.A. Mey). II. Pollination droplets, nectaries, and nectarial secretion in *Ephedra*, *Proceedings of the Koninklijke Nederlandse Akad. van Wetenschappen*, (C) **87** (1): 15–24.

Bose, M.N., Pal, P.K. and Harris, T.M., 1985, The *Pentoxylon* plant, *Philosophical Transactions of the Royal Society, London,* **B310**: 77–108.

Chaloner W.G. and Sheerin, A., 1979, Devonian macrofloras. In M.R. House, C.T. Scrutton and M.G. Bassett, *The Devonian System, Special Papers in Palaeontology*, **23**: 145–61.

Chaloner, W.G., Hill, A.J. and Lacey, W.S., 1977, First Devonian platyspermic seed and its implications in gymnosperm evolution, *Nature*, **265**: 233–5.

Chamberlain, C.J., 1935, *Gymnosperms, structure and evolution*, University of Chicago Press, Chicago.

Cichan, M.A., 1985, Vascular cambium and wood development in Carboniferous plants. II. *Sphenophyllum plurifoliatum* Williamson and Scott (Sphenophyllales), *Botanical Gazette*, **146**: 395–403.

Collinson, M.E. and Hooker, J.J., 1987, Vegetational and mammalian faunal changes in the early Tertiary of southern England, in E.M. Friis, W.G. Chaloner and P.R. Crane (eds), *The origins of angiosperms and their biological consequences*, Cambridge University Press, Cambridge, pp. 259–304.

Crane, P.R., 1985, Phylogenetic analysis of seed plants and the origin of angiosperms, *Annals of the Missouri Botanical Garden*, **72**: 716–93.

Crane, P.R., 1987, Vegetational consequences of angiosperm diversification. In E.M. Friis, W.G. Chaloner and P.R. Crane (eds), *The origins of angiosperms and their biological consequences*, Cambridge University Press, Cambridge, pp. 107–44.

Crane, P.R., 1988, Major clades and relationships in the 'higher' gymnosperms. In C.B. Beck (ed.), *The origin and evolution of gymnosperms*, Columbia University Press, New York, pp. 218–72.

Crane, P.R., 1989, Paleobotanical evidence on the early radiation of nonmagnoliid dicotyledons, *Plant Systematics and Evolution*.

Crane, P.R. and Blackmore, S. (eds), 1989, *Evolution, systematics and fossil history of the Hamamelidae*, Clarendon Press, Oxford.

Crane, P.R., Friis, E.M. and Pedersen, K.R., 1989, Reproductive structure and function in Cretaceous Chloranthaceae, *Plant Systematics and Evolution*.

Crane, P.R and Upchurch, G.R., 1987, *Drewria potomacensis* gen. et sp. nov., an early Cretaceous member of the Gnetales from the Potomac Group of Virginia, *American Journal of Botany*, **74**: 1722–36.

Crepet, W.L. and Friis, E.M., 1987, The evolution of insect pollination in angiosperms. In E.M. Friis, W.G. Chaloner and P.R. Crane (eds), *The origins of angiosperms and their biological consequences*, Cambridge University Press, Cambridge, pp. 181–201.

Cronquist, A., 1981, *An integrated system of classification of flowering plants*, Columbia University Press, New York.

Dilcher, D.L. and Crane, P.R., 1984, *Archaeanthus*: an early angiosperm from the Cenomanian of the western interior of North America, *Annals of the Missouri Botanical Garden*, **71**: 351–83.

DiMichele, W.A. and Phillips, T.L., 1985, Arborescent lycopod reproduction and paleoecology in a coal-swamp environment of late Middle Pennsylvanian age (Herrin Coal, Illinois, U.S.A.), *Review of Palaeobotany and Palynology*, **44**: 1–26.

DiMichele, W.A., Phillips, T.L. and Peppers, R.A., 1985, The influence of climate and depositional environment on the distribution and evolution of Pennsylvanian coal-swamp plants. In B.H. Tiffney (ed.), *Geological factors and the evolution of plants*, Yale University Press, New Haven, CT, pp. 223–56.

DiMichele, W.A., Phillips, T.L. and Olmstead, R.G., 1987, Opportunistic evolution: abiotic environmental stress and the fossil record of plants, *Review of Palaeobotany and Palynology*, **50**: 151–78.

Donoghue, M.J. and Doyle, J.A., 1989, Phylogenetic analysis of angiosperms and the relationships of Hamamelidae. In P.R. Crane and S. Blackmore (eds), *Evolution, systematics and fossil history of the Hamamelidae*, Clarendon Press, Oxford.

Doyle, J.A., 1973, Fossil evidence on early evolution of the monocotyledons, *Quarterly Review of the Biology*, **48**: 399–413.

Doyle, J.A., 1978, Origin of angiosperms, *Annual Reviews of Systematics and Ecology*, **9**: 365–92.

Doyle, J.A. and Donoghue, M.J., 1986, Seed plant phylogeny and the origin of angiosperms: an experimental cladistic approach, *Botanical Review*, **52**: 321–431.

Doyle, J.A. and Hickey, L.J., 1976, Pollen and leaves from the mid-Cretaceous Potomac Group and their bearing on early angiosperm evolution. In C.B. Beck (ed.), *Origin and early evolution of angiosperms*, Columbia University Press, New York, pp. 139–206.

Edwards, D. and Edwards, D.S., 1986, A reconsideration of the Rhyniophytina Banks. In R.A. Spicer and B.A. Thomas (eds), *Systematic and taxonomic approaches in palaeobotany*, Systematics Association, Special Volume **31**, Clarendon Press, Oxford, pp. 197–220.

Fairon-Demaret, M., 1986, Some uppermost Devonian megafloras: a stratigraphical review, *Annales de la Société Géologique de Belgique*, **109**: 43–8.

Florin, R., 1949, The morphology of *Trichopitys heteromorpha* Saporta, a seed plant of Palaeozoic age, and the evolution of the female flowers in the Ginkgoinae, *Acta Horti Bergiani*, **15**: 79–109.

Florin, R., 1951, Evolution in cordaites and conifers, *Acta Horti Bergiani*, **15**: 285–388.

Friis, E.M. and Crepet, W.L., 1987, Time of appearance of floral features. In E.M. Friis, W.G. Chaloner and P.R. Crane (eds), *The origins of angiosperms and their biological consequences*, Cambridge University Press, Cambridge, pp. 145–79.

Friis, E.M., Chaloner, W.G. and Crane, P.R. (eds), 1987, *The origins of angiosperms and their biological consequences*, Cambridge University Press, Cambridge.

Friis, E.M., Crane, P.R. and Pedersen, K.R., 1988, The reproductive structure of Cretaceous Platanaceae, *Det Kongelige Danske Videnskaberne Selskab Biologiske Skrifter*, **31**: 1–55.

Galtier, J. and Scott, A.C., 1985, Diversification of early ferns, *Proceedings of the Royal Society of Edinburgh*, **B86**: 289–301.

Grierson, J.D. and Bonamo, P.M., 1979, *Leclercqia complexa*: earliest ligulate lycopod (Middle Devonian), *American Journal of Botany*, **66**: 474–6.

Hart, J.A., 1987, A cladistic analysis of conifers: preliminary results, *Journal of the Arnold Arboretum*, **68**: 269–307.

Herngreen, G.F.W. and Chlonova, A.F., 1981, Cretaceous microfloral provinces, *Pollen et Spores*, **23**: 441–555.

Hickey, L.J. and Doyle, J.A., 1977, Early Cretaceous evidence for angiosperm evolution, *Botanical Review*, **43**: 3–104.

Hickey, R.J., 1986, The early evolutionary and morphological diversity of *Isoëtes*, with descriptions of two new neotropical species, *Systematic Botany*, **11**: 309–21.
Hughes, N.F., 1976, *Palaeobiology of angiosperm origins*, Cambridge University Press, Cambridge.
Hughes, N.F. and McDougall, A.B., 1987, Records of angiospermid entry into the English Early Cretaceous succession, *Review of Palaeobotany and Palynology*, **50**: 255–72.
Johnson, L.A.S., 1959, The families of cycads and the Zamiaceae of Australia, *Proceedings of the Linnean Society of New South Wales*, **84**: 64–117.
Kenrick, P. and Edwards, D., 1988, A new zosterophyll from a recently discovered exposure of the Lower Devonian Senni Beds in Dyfed, Wales, *Botanical Journal of the Linnean Society*, **98**: 97–115.
Knoll, A.H., 1984, Patterns of extinction in the fossil record of vascular plants. In M. Nitecki (ed.), *Extinctions*, University of Chicago Press, Chicago, pp. 21–68.
Knoll, A.H., 1986, Patterns of change in plant communities through geological time. In J. Diamond and T.J. Case (eds), *Community ecology*, Harper and Row, New York, pp. 126–41.
Knoll, A.H., Niklas, K.J., Gensel, P.G. and Tiffney, B.H., 1984, Character diversification and patterns of evolution in early vascular plants, *Paleobiology*, **10**: 34–58.
Lidgard, S.H. and Crane, P.R., 1988, Quantitative analyses of the early angiosperm radiation, *Nature*, **331**: 344–6.
Mapes, G. and Rothwell, G.W., 1984. Permineralized ovulate cones of *Lebachia* from late Paleozoic limestones of Kansas, *Palaeontology*, **27**: 69–94.
Martens, P., 1971, Les gnétophytes, *Handbuch der Pflanzenanatomie*, **12**: 1–295.
Mattox, K.R. and Stewart, K.D., 1984, Classification of the green algae: a concept based on comparative cytology. In: D.E.G. Irvine and D.M. John (eds), *Systematics of the green algae*, Academic Press, London, pp. 29–72.
Meyen, S.V., 1984, Basic features of gymnosperm systematics and phylogeny as evidenced by the fossil record, *Botanical Review*, **50** (1): 1–113.
Miller, C.N., Jr, 1977, Mesozoic conifers, *Botanical Review*, **43**: 218–80.
Mishler, B.D. and Churchill, S.P., 1984, A cladistic approach to the phylogeny of the 'bryophytes', *Brittonia*, **36**: 406–24.
Muller, J., 1984, Significance of fossil pollen for angiosperm history, *Annals of the Missouri Botanic Garden*, **71**: 419–43.
Niklas, K.J., Tiffney, B.H. and Knoll, A.H., 1985, Patterns in vascular land plant diversification: and analysis at the species level. In J.W. Valentine (ed.), *Phanerozoic diversity patterns: profiles in macroevolution*, Princeton University Press, Princeton, NJ, pp. 97–128.
Pettitt, J., 1970, Heterospory and the origin of the seed habit, *Biological Reviews*, **45**: 401–15.
Phillips, T.L. and Peppers, R.A., 1984, Changing patterns of Pennsylvanian coal-swamp vegetation and implications of climatic control on coal occurrence, *International Journal of Coal Geology*, **3**: 205–55.
Phillips, T.L., Peppers, R.A. and DiMichele, W.A., 1985, Stratigraphic and interregional changes in Pennsylvanian coal-swamp vegetation: environmental inferences, *International Journal of Coal Geology*, **5**: 43–109.
Richardson, J.B. and McGregor, D.C., 1986, Silurian and Devonian spore zones of the Old Red Sandstone continent and adjacent regions, *Geological Survey of Canada, Bulletin*, **364**: 1–79.

Rothwell, G.W., 1981, The Callistophytales (Pteridospermopsida): reproductively sophisticated Paleozoic gymnosperms, *Review of Palaeobotany and Palynology*, **32**: 103–21.

Rothwell, G.W., 1986, Classifying the earliest gymnosperms. In R.A. Spicer and B.A. Thomas (eds), *Systematic and taxonomic approaches in palaeobotany*, Clarendon Press, Oxford, pp. 137–61.

Rothwell, G.W. and Erwin, D.M., 1985, The rhizomorph apex of *Paurodendron*: implications for homologies among the rooting organs of Lycopsida, *American Journal of Botany*, **72**: 86–98.

Rothwell, G.W. and Erwin, D.M., 1987, Origin of seed plants: an aneurophyte/seed-fern link elaborated, *American Journal of Botany*, **74**: 970–3.

Rothwell, G.W. and Warner, S., 1984, *Cordaixylon dumusum* n. sp. (Cordaitales). I. Vegetative structures, *Botanical Gazette*, **145**: 275–91.

Scheckler, S.E., 1986, Floras of the Devonian–Mississippian transition. In R.A. Gastaldo (ed.), *Land plants: notes for a short course*, *Studies in Geology*, **15**, Department of Geology, University of Tennessee, pp. 81–96.

Scheihing, M.H. and Pfefferkorn, H.W., 1984, The taphonomy of land plants in the Orinoco Delta; a model for the incorporation of plant parts in clastic sediments of Upper Carboniferous age of Euroamerica, *Review of Palaeobotany and Palynology*, **41**: 205–40.

Scott, A.C. and Chaloner, W.G., 1983, The earliest fossil conifer from the Westphalian B of Yorkshire, *Proceedings of the Royal Society, London*, **B220**: 163–82.

Stein, W.E., Wight, D.C. and Beck, C.B., 1984, Possible alternatives for the origin of Sphenopsida, *Systematic Botany*, **9**: 102–18.

Stewart, W.N., 1983, *Paleobotany and the evolution of plants*, Cambridge University Press, Cambridge.

Tiffney, B.H., 1984, Seed size, dispersal syndromes, and the rise of angiosperms: evidence and hypothesis, *Annals of the Missouri Botanical Garden*, **71**, 551–76.

Traverse, A., 1988, Plant evolution dances to a different beat. Plant and animal evolutionary mechanisms compared, *Historical Biology*, **1**: 277–301.

Upchurch, G.R. and Doyle, J.A., 1981, Paleoecology of the conifers *Frenelopsis* and *Pseudofrenelopsis* (Cheirolepidiaceae) from the Cretaceous Potomac Group of Maryland and Virginia. In R.C. Romans (ed.), *Geobotany II*, Plenum, New York, pp. 167–202.

Upchurch, G.R. and Wolfe, J.A., 1987, Mid-Cretaceous–early Tertiary vegetational and climate: evidence from fossil leaves and woods. In E.M. Friis, W.G. Chaloner and P.R. Crane (eds), *The origins of angiosperms and their biological consequences*, Cambridge University Press, Cambridge, pp. 75–105.

White, D.C., 1987, Non-adaptive change in early land plant evolution, *Paleobiology*, **13**, 208–14.

Wiley, E.O., 1981, *Phylogenetics. The theory and practice of phylogenetic systematics*, Wiley, New York.

Wing, S.L. and Tiffney, B.H., 1987, The reciprocal interaction of angiosperm evolution and tetrapod herbivory, *Review of Palaeobotany and Palynology*, **50**: 179–210.

Chapter 8

VERTEBRATE FLIGHT AND THE ORIGINS OF FLYING VERTEBRATES

Jeremy M.V. Rayner

The appearance of major locomotor adaptations in vertebrates is associated not only with morphological but also with considerable behavioural and physiological specialisation. Early amphibians beginning to move on land had to develop strengthened limbs with independent movement. They also had to propel themselves efficiently against a different substrate, and acquire physiological systems capable of coping with enhanced levels of metabolism and with the different conditions of life in air rather than water. Similarly, as a more erect stance developed in archosaurs and in mammals, locomotion speeds increased, levels of metabolism became yet higher, and locomotion formed a more and more significant component of activity. In several lineages bipedalism evolved to separate the use of the limbs for locomotion and for such purposes as the gathering and handling of food, grooming, and communication. The sequence of events in the evolution of most vertebrate locomotor specialisations is understood in outline, but the fossil record provides direct evidence only of morphological adaptations, and usually gives only weak or indirect evidence of soft tissues. Physiological and behavioural correlates of locomotion can be adduced only from hypotheses about likely ecosystem and community structure, and from comparisons with modern animals.

These problems are particulary acute when studying the origins of flight in vertebrates. True flapping flight has appeared in three groups, pterosaurs, birds and bats. Each has different morphology, but all have found mechanical solutions to the physical problems of propulsion in air. Two of these groups (birds and bats) remain extant, but have limited early fossil records. By contrast, the fossil record of pterosaurs is more comprehensive, but there are few indications of their relationships with other archosaurs, and no extant representatives from which function can be directly deduced. Within each of the three groups the problems of phylogeny, of flight evolution and of the development of locomotion are closely interlinked. The identification of novel adaptations is a key factor in reconstructing their phylogenies, and ultimately in classification of the groups and their relationships. Since the dominant characters which define birds, bats and pterosaurs are all related to flight, it is natural to look to the mechanical aspects of flight to define the sequence of events in the appearance of these three groups of animals.

FLIGHT

The three groups of fliers are characterised primarily by their ability to fly. Although superficially evident, this statement raises a number of problems. Is flight the sole relevant character or character complex defining a group? How should animals which (in some sense) show intermediate flight adaptations be classified? (This question is particularly important for the avian fossil record.) And how should flight itself be defined? Unless these problems are clearly resolved discussion of the appearance of novelties with the onset of flight risk becoming circular and unhelpful (birds fly, therefore flight is defined as an attribute of birds, which they may share by convergence with other groups). Flightlessness (widespread in birds but not known in bats and pterosaurs) is not a major factor in this discussion: the group Aves contains all descendants of flying birds, even though some may secondarily have lost their powers of flight.

This difficulty may be avoided if flight *and the flying vertebrate groups* are defined in terms of the physical forces which an animal can achieve, rather than by conventional phylogeny alone. Then, the adaptive stage at which animals in a phylogenetic sequence first achieve flight can be identified in principle, and the groups Aves, Chiroptera and Pterosauria can be defined as animals at or beyond that stage. Flight then provides the complex of synapomorphies that characterise the group.

Physical forces and locomotion

Animals in locomotion must generate physical forces which transmit energy into their environment. Generally these forces are the result of reciprocating movements of the body and limbs. Owing to the large magnitude of the forces and energy rates involved, selection acts to favour systems in which locomotion is most economic and/or efficient, and which permits the animal to exploit its environment and to survive and reproduce. The forces required are twofold: a vertical *lifting* force to support the animal's weight against gravity, and a horizontal *thrusting* force to provide propulsion against resistive forces such as friction or drag. Different physical media require different solutions to the generation of these forces. Swimming animals balance most or all of their weight by buoyancy, and are concerned primarily with generating thrust forces; to minimise drag either the body is streamlined or the animals travel slowly. Walking and running animals are more influenced by gravity, which is countered by compressive forces in the limbs against the ground, while propulsion is provided by friction; in movement on land much energy is concerned with moving the limbs relative to the body.

Flight has much in common with swimming, but its major problem is that air is a far less forgiving environment than water, and provides virtually no buoyancy force. Flying animals must rely entirely on aerodynamic forces generated at the surface of their wings to provide all thrust and weight support. This imposes severe mechanical constraints. There is only one mechanism—aerofoil action—which is capable of producing forces of the required magnitude (Figure 8.1); this requires wings of large surface area, which are relatively thin in cross-section and have high aspect ratio (wing span divided by wing chord or breadth), but which are stiff enough to

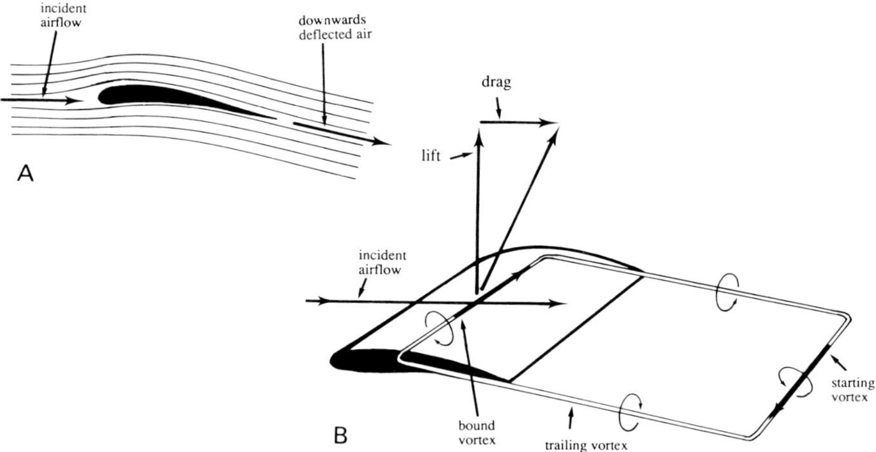

Figure 8.1. The section of the wings of flying vertebrates is an aerofoil. This profile has a rounded leading edge and a pointed trailing edge, is relatively thin, and is usually cambered so that upper and lower surfaces are asymmetrical. A: The streamlined, tapered aerofoil section deflects incident air downwards, and generates a transverse lift force by a reaction to the downward momentum. B: A steadily moving aerofoil is associated with a vortex system consisting of a bound vortex on the wing and trailing vortices behind the wingtips. A transverse starting vortex is shed whenever the magnitude of the circulation or strength of the bound vortex changes. The lift force is proportional to the product of circulation and velocity. There is also a drag force, parallel to the incident airflow, representing the energy required to generate the wake vortices and to overcome friction on the aerofoil surface. (From Rayner 1988a.)

transmit the aerodynamic forces they generate to the body. Bauplan (phylogenetic) constraints have ensured that the morphological solutions to wing design (skeletal and wing surface architecture) in the three groups are different, but the narrow limits on effective aerofoil action ensure that birds and bats (and therefore presumably also pterosaurs) are mechanically and aerodynamically very similar. These limits are responsible for the large size of the wings compared to the dimensions of limbs of other vertebrates, and for the relatively small range of size in vertebrate fliers (Norberg and Rayner 1987; Rayner 1987; 1988a).

Aerofoil action requires relative movement between the wing surface and the surrounding air. Thus, flying animals must flap their wings, unless they are travelling relatively fast. The thin, usually cambered wing surface with rounded leading and sharp trailing edges (Figure 8.1) causes air passing the upper surface to accelerate, and decelerates air passing under the lower surface. This imparts a downwards acceleration to air behind the wing, and the reaction of this downwards airflow is experienced as a pressure difference between the sides of the wing. This is the *lift* force which is responsible for weight support and thrust in flight; lift acts perpendicular to the relative air flow (equivalent to the movement of the aerofoil or the local movement of the air adjacent to the wing), and is accompanied by a *drag* force, generated by friction and pressure forces, acting parallel to the line of flight. With a well-designed streamlined aerofoil drag is usually much smaller than lift: lift to drag ratios of the order of 10–15 are typical for animal wings. There is additional drag due to the body of the animal, and to any trailing legs or tail; any lift from these other surfaces is small and is generally negligible in terms of the overall forces, although it may contribute usefully to balance and stability.

The passive aerofoil and gliding

The central limitation on the use of the aerofoil is that drag always retards the animal, and a force must be applied to overcome it. An aerofoil can never generate a force with a positive component along the local direction of motion. Many organisms exploit passive aerofoil or hydrofoil action in air and/or in water for self-stability, dispersal or transport (Vogel 1981). Various plants and marine invertebrates control orientation by aerofoil lift, and many plant and tree seeds use aerofoil action to retard fall and exploit horizontal wind currents. Moreover, many animals are adapted for *gliding*.

Gliding animals are found in each of the five vertebrate classes, and are common in the fossil record (mainly in reptiles) (Rayner 1981, 1986a; Norberg 1985a). Gliding requires little more than the development of a broad aerodynamic surface; this has been achieved by enlarging limb extremities (flying fish and frogs, some lizards), spreading a membrane between fore and hind limbs (gliding marsupials, rodents and dermopterans), and by lengthening the ribs and using them to support a membrane

(*Draco*, various fossil reptiles). To maintain speed a glider must use potential energy, and therefore must lose height relative to the air (the same is true of gliding and soaring birds and bats while they are not flapping their wings). The sole method by which a horizontal thrust force can be produced is by flapping the wings so that mean lift has a horizontal component which can act as thrust.

Flight and flapping wings

Flight as a phylogenetic character may most usefully be characterised by the ability to flap the wings to achieve a mean horizontal thrust in steady, level flight. By this definition gliders other than birds or bats are not capable of flight, but this locomotor adaptation is important in the context of flight evolution for it is the probable precursor of flight in birds and bats.

Compared with gliding, true flight makes considerable demands; the influence of these are immediately evident in the long limbs, robust pectoral and pelvic girdles, and enlarged flight musculature of birds and bats. There is no difficulty in tracing comparable adaptations in fossil animals. The purpose of flapping is to generate a horizontal thrust while maintaining a sufficient vertical force to balance weight. In birds and bats there is a mosaic of constraints which delimit the geometry of the wingbeat; many of these are associated with force and energy-flow optimisation (Rayner 1987; 1988a), and the broad similarity in wingbeat kinematics and aerodynamics between birds and bats indicates that at this level phylogenetic constraints are relatively weak.

The most significant constraint on flapping flight is that thrust can be generated only during the downstroke, when the wings move both downwards relative to the body and forwards relative to the air (Figure 8.2): the mean lift in this phase is directed upwards and forwards. High-speed photography and aerodynamic experiments based on visualisation of the vortex wakes during flapping show that in a wide range of birds and bats the wing is more or less flat and is fully extended during the downstroke to maximise lift (Rayner 1980; 1987, 1988a). The aerodynamic effect of the downstroke is remarkably similar in birds and bats, despite their widely differing wing structure.

By contrast the upstroke shows more variation, associated with both wing design (wing skeleton, wing outline size and shape) and with flight speed. At low speeds, and also in animals with short and rounded (low aspect ratio) wings at higher speeds, the upstroke is aerodynamically inactive, and no lift is generated. As the wing is raised it is brought close to the body to minimise inertial forces and drag, and the wing surface is usually considerably deformed and flexed. At higher (cruising flight) speeds in animals with larger, more pointed wings the upstroke becomes aerodynamically active; lift primarily acts vertically to support weight, but it also has a horizontal and backward component which retards the animal

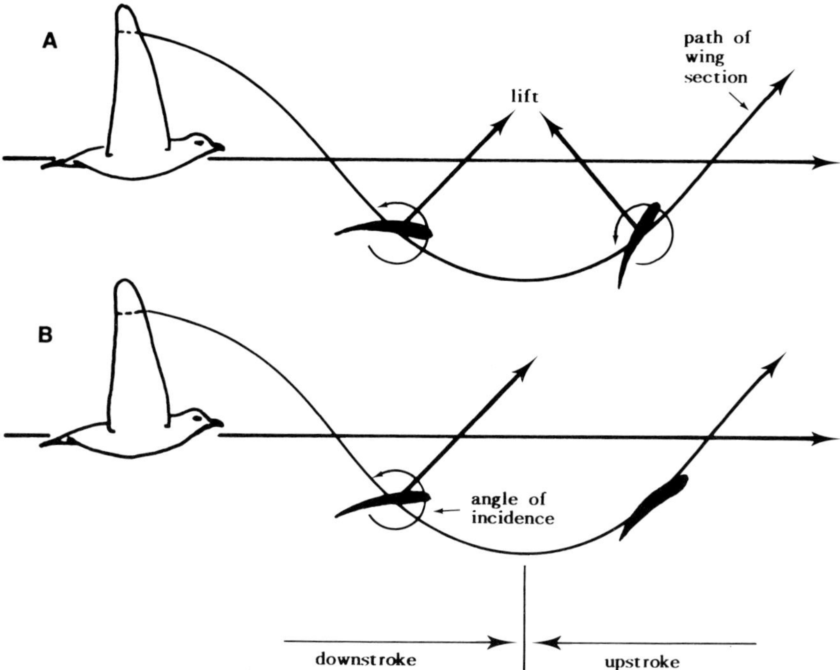

Figure 8.2. Lift generation in flapping flight. Flying vertebrates have a number of different options for use of the upstroke. A: The wing geometry, angle of incidence and circulation are assumed to remain approximately constant throughout the wingbeat; both downstroke and upstroke support the weight, but the horizontal forces from downstroke and upstroke balance, and there is no net horizontal thrust. The total upstroke lift must be reduced to provide thrust, and B: one option is for the wing to be so pitched during the upstroke that the angle of incidence is zero and there is no lift. Some animals may reduce upstroke lift by reducing incidence and circulation, but birds and bats appear to do this by flexing or sweeping the wing. Negative incidence would correspond to a horizontal thrust component of lift, but this occurs in vertebrates only in very slow flight, where the wing moves backwards relative to the air. (From Rayner 1988a.)

(Figure 8.2). The exact circumstances in which the upstroke should be used aerodynamically are not yet fully understood, but the decision is clearly based on a balance between increasing vertical force to counter weight at the expense of retardation, and minimising retardation at the expense of concentrating all weight support onto the downstroke. At low speeds drag from a lifting upstroke would be high, while small or rounded wings have relatively poor lift-to-drag ratios, and may be unable to produce sufficient useful vertical force while moving upwards relative to the air.

It is in the lifting upstroke that differences between birds and bats become most evident, and these can be traced directly to wing structure.

Wingspan must be reduced somewhat during the upstroke: if the wings remained fully extended—as they are in the downstroke—the magnitude of the horizontal retarding component of lift would equal any thrust from the downstroke, and there would be no net horizontal force. Yet the purpose of flapping is to produce a positive thrust. Owing to bauplan differences in wing structure, birds and bats shorten the wingspan in different ways (Rayner 1986b). The bat wing surface is formed of an elastic membrane which retains its aerodynamic integrity when the humerus and radius are flexed and the carpal joints are brought towards the body. The wing shortens, but the hand wing does not deform greatly and the leading edge of the wing remains approximately straight. The digits must be kept tense since, if the hand wing were allowed to flex, the membrane would billow. In birds the distal portion of the wing can readily flex, and in most species the wingtip is swept back at the carpal joint and follows a circular or elliptic path when seen in lateral view; the avian hand wing can continue to function as an efficient aerofoil as the feathers slide over one another. Little useful shortening is obtainable from flexing the upper arm.

The method by which the wing is flexed during the upstroke is dictated by the morphology of the wing, and no doubt is closely correlated with patterns of muscle motor activity in the two groups. It is most surprising that aerodynamically, despite the wide differences in the geometry of the wingtip path, the two groups are very similar. The vortex wake in both cases consists of continuous trailing vortices behind the wingtips, and the strength of the vortices on the wings remains constant throughout the wingbeat (Rayner 1987; 1988a). In terms of flight mechanics (not necessarily of physiology) birds and bats of comparable size and dimensions are very similar. This finding reinforces the suggestion that mechanical constraints on flight are very severe. There are strict limits on wing dimensions, reflecting deterioration of certain aspects of mechanical performance (e.g. speed, range, energy, manoeuvrability, agility) with sub-optimal wing design (Norberg and Rayner 1987; Rayner 1988a). These constraints would have been no less severe in the early stages of the evolution of flight, and are likely to have had a more significant effect on an animal with imperfect flight adaptations. Thus, early fliers are likely to have used wingbeats similar to those of present birds and bats, and it should be possible to reconstruct their design and performance with this in mind. Similarly, pterosaurs are likely to have been similar aerodynamically to birds and bats of comparable size.

THE PHYLOGENY OF FLIGHT AND THE FOSSIL EVIDENCE

Pterosaurs

The pterosaurs (Upper Triasic—late Cretaceous) are the oldest of the three groups of flying vertebrates, and in many ways the most enigmatic

(Wellnhofer 1978; 1980; Padian 1983; 1985). Although they are relatively well represented in the fossil record (Figure 8.3), there is at present considerable argument over various aspects of their functional morphology and biomechanics. Earlier pterosaurs—which possessed tails—belonged to the paraphyletic group Rhamphorhynchoidea, while later forms, belonging to the Pterodactyloidea, were tailless. Both sub-orders occur together in Jurassic deposits such as those of Solnhofen in Bavaria. Known rhamphorhynchoids were fairly consistent in size, most being comparable in dimension to pigeons or gulls. Pterodactyloids were more variable; some earlier representatives (mid-Jurassic) of the genus *Pterodactylus* were as small as sparrows (Figure 8.3C, D) and other later Cretaceous members of the group, such as *Pteranodon* and *Quetzalcoatlus* reached enormous sizes, up to eight times heavier than any present-day flying bird.

The pterosaur wing is supported by a leading edge spar formed of the extended metacarpals of the fourth digit (Figure 8.3). The wing surface is believed to have been formed by a membrane, which had at least certain similarities to the wings of bats. Problems with the interpretation of pterosaurs centre on reconstruction of the wing and of the animal's stance. Historically pterosaurs have been considered as bat-like, with an inverted

---

Figure 8.3. (Pages 196–7.) Representative pterosaur fossils. These specimens illustrate the main sub-orders of pterosaurs: the tailed Rhamphorhynchoidea from the Triassic and Jurassic, and the tailless Pterodactyloidea from the Jurassic and Cretaceous. All four specimens are from the Jurassic lithographic limestone of Solnhofen in Bavaria, the same location as the six specimens of *Archaeopteryx* (Figure 8.5). (All photographs JMVR.) A: *Rhamphorhynchus longicaudus* von Meyer, a typical rhamphorhynchoid, from the Bayerische Staatssammlung für Paläontologie und Historische Geologie, Munich. Impressions of the wing membrane are visible adjacent to the wing digits, but not in the region of the legs. The tail carries a sail-like terminal fin, which was probably important for stablity (× 0.64). B: Detail of left wing of *Rhamphorhynchus gemmingi* Zittel, also preserved at Munich, seen from the ventral side. The region shown stretches transversally across the wing, close to the joint between second and third wing phalanges. This famous specimen is exceptionally preserved, and shows stiffening fibres embedded within the wing membrane. The fibres are generally straight and continuous, but curve slightly as they approach the trailing edge of the wing. Note the hollow cross section of the wing digit bones. (×4.18) C: *Pterodactylus kochi* Wagner preserved in the Naturhistorisches Museum, Vienna. Few pterodactyloid fossils have impressions of the wing membranes, but this fine specimen clearly shows a membrane attached to the femur slightly proximal to the knee joint on both legs. The tibia and hind foot are free of the membrane. Fibres may be distinguished within the membrane. For description of this fossil see Wellnhofer (1987). Note that the humeri are disarticulated, although the skeleton of the trunk is not disturbed: this is common with dorso-ventrally compressed pterosaur fossils. (× 1.66). D: *Pterodactylus elegans* from Munich. The animal was preserved on its side and has been laterally compressed. The humerus is not disarticulated, but the neck has bent sharply back, probably as a result of post-mortem contraction of the *ligamentum nuchae*. (×3.01)

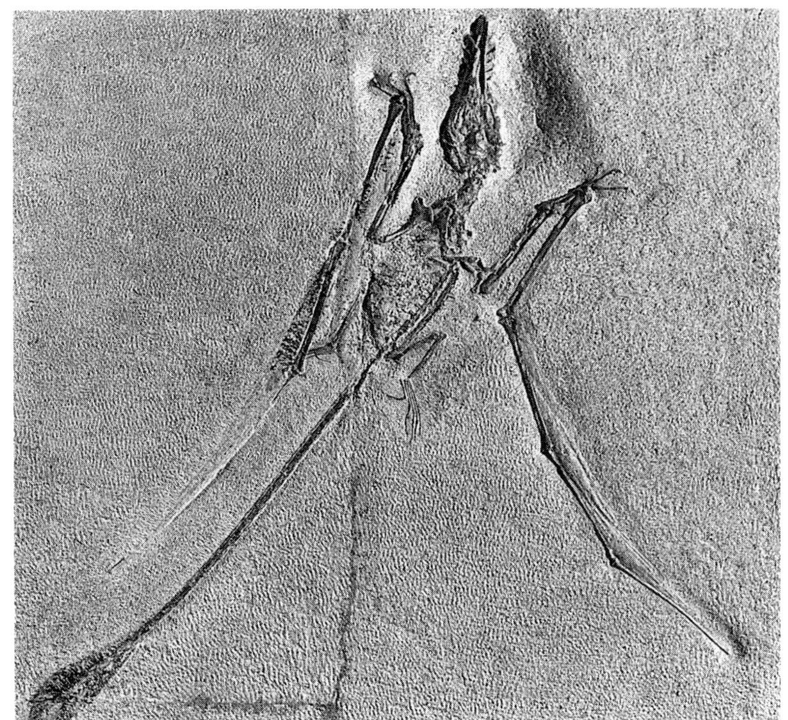

A

B

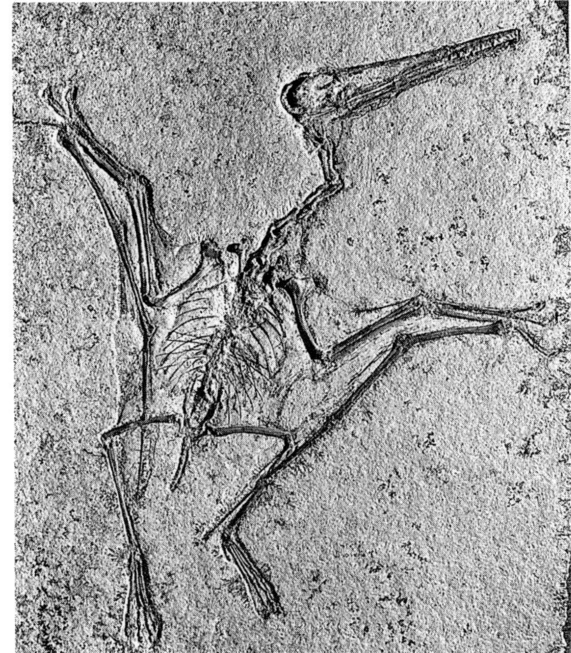

C

D

hanging posture at rest, and minimal ability to move on the ground; this view is no longer supported (Padian 1987). Current debate focuses on whether a pterosaur should be reconstructed as 'bat-like', with a membranous wing attached both to the fourth digit and to the distal portion of the hind limb, but with a sprawling stance (Figure 8.4A; Wellnhofer 1987; Unwin 1988), or as more 'bird-like', with a wing free of the hind limb, and an erect bipedal stance (Figure 8.4B; Padian 1983). The wing of a 'bird-like' pterosaur would be supported solely at the leading edge, and would have required additional structures to give aerodynamic integrity. Some exceptionally well-preserved specimens (Figure 8.3B) show fibres embedded within the surface of the wing; these probably stiffened the membrane, and functioned comparably to the shafts of avian feathers, with which they share similar patterns of orientation (Padian 1983; Rayner 1989). The radically differing stance associated with the two reconstructions implies major differences in behaviour and in the way the wings functioned, although aerodynamically the resulting force patterns as the animals flew must have been similar. Reconstruction with a fully erect bird-like stance avoids the enormous problems in envisaging a bat-like animal with suspended or sprawling posture, and the enormous size of some late Cretaceous pterodactyloids. The humerus/scapulo-coracoid articulation in both rhamphorhynchoids and pterodactyloids is remarkably bird-like, and suggests flapping movements and wing deformation similar to those of contemporary birds (Padian 1983; 1987a; Rayner 1989).

This debate has not yet been fully resolved, although the bird-like form seems favoured. The major gap in present knowledge is the complete structure of the entire pelvis and femur: many pterosaur fossils are compressed, either laterally or dorsally (Figure 8.3), and three-dimensional material is rarely complete.

Pterosaur relationships could give valuable guidance, but the fossil record is not sufficient to resolve them. Padian (1985) argued that the small ornithosuchian archosaur *Scleromochlus* from the Triassic of Scotland is the sister group of pterosaurs. Thus pterosaurs are a monophyletic group of archosaurs, but are not dinosaurs (*pace* Bakker 1986). Padian (1983), favouring a terrestrial stance in pterosaurs, proposed a cursorial origin of the group, while Wild (1984) supported an arboreal gliding ancestor analogous to gliding lizards. Until more is known of the fossil record and relationships of proto-pterosaurs the linked questions of pterosaur evolution and the origin of their flight must remain open; it would, however, be natural to model the origin of flight in a bird-like pterosaur as being similar to that in birds (see below). The mode of life of pterosaurs is hard to interpret since it is unlikely that the fossil record is complete. However, the long, thin wings of those which have been preserved (almost invariably in marine sediments) indicate pelagic or littoral habits for most species. Some of the smaller pterodactyloids may have been insectivorous, while some rhamphorhynchoids with relatively shorter wings (such as *Dimorphodon*, Figure 8.4B) may have had more in common with wading birds (Wellnhofer 1980; Rayner 1989).

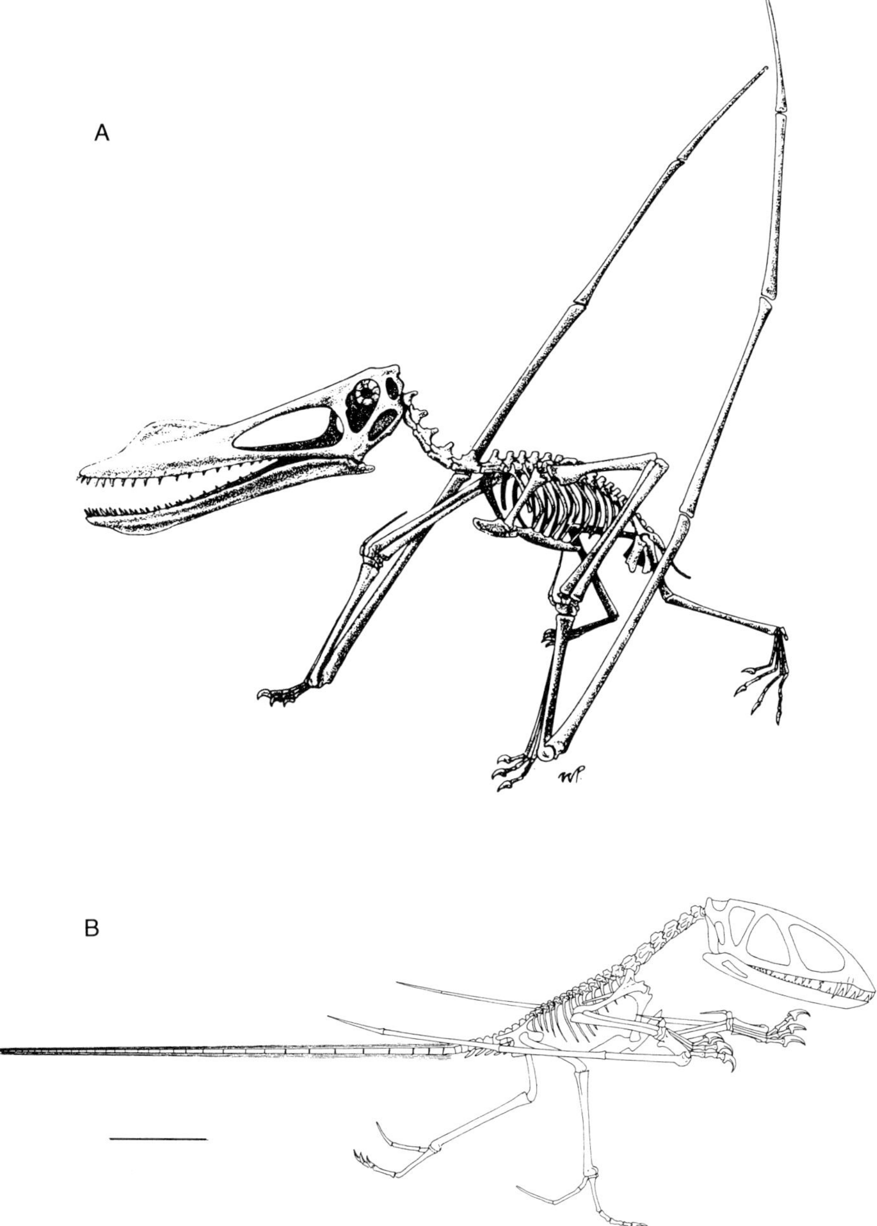

Figure 8.4. A. Skeleton of the large pterodactyloid pterosaur *Anhanguera* (wingspan 4.5 m) reconstructed by Wellnhofer (1988a) with a quadrupedal semi-erect plantigrade stance and the hindlimbs splayed out laterally. B. The rhamphorhynchoid *Dimorphodon* (wingspan 1.2 m) as envisaged by Padian (1983) as an agile running biped.

Archaeopteryx *and avian origins*

Birds are the dominant group of vertebrate fliers. Their origin and relationships and the evolution of their flight have been the subject of considerable study, and some controversy (Stephan 1987; Hecht *et al.* 1985; Padian 1986). There is overwhelming evidence that birds evolved from small bipedal theropod dinosaurs during the Middle Jurassic, probably 140–160 Myr ago (Heilmann 1927; Gauthier and Padian 1985; Padian 1987; Wellnhofer 1988b). The transformations in morphology and size from dinosaur to bird appear from the limited fossil evidence to have been fairly modest (Figure 8.5), but in the process birds acquired two features—the power of flight, and feathers—which set them apart from other dinosaurs. The sister group of birds is believed to include *Deinonychus*, a predatory cursorial theropod some 2 m in body length; other contemporary theropods such as *Compsognathus* were much smaller, and more comparable with the likely size at which flight would have evolved (Figure 8.5).

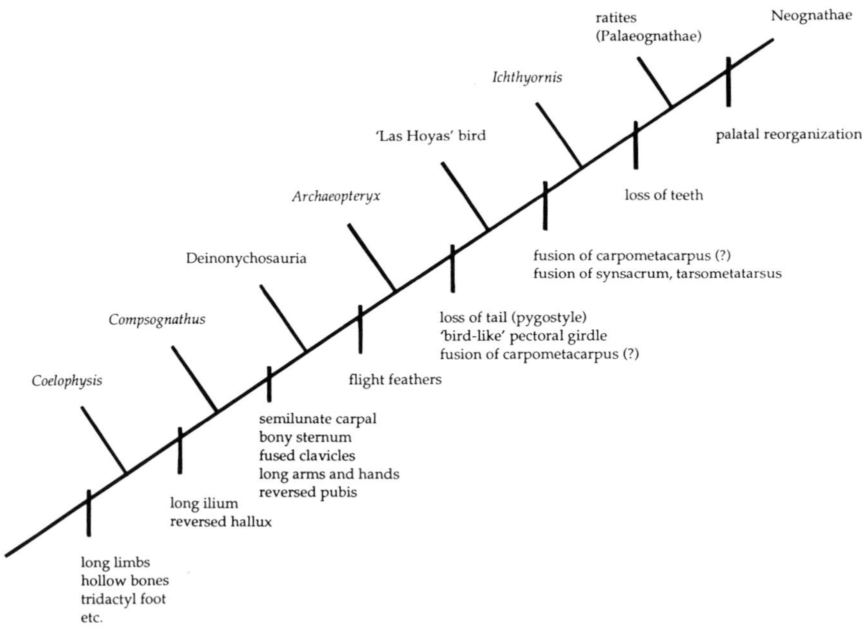

Figure 8.5. Cladogram showing the relationship of theropod dinosaurs to *Archaeopteryx* and other birds. The emphasis is on the characters, discussed in the text, which are regarded as avian or 'flight-related'. The Las Hoyas specimen is recently described from the lower Cretaceous of Spain (Sanz *et al.* 1988), and at the time of writing has not been named. Unfortunately the manus is not preserved. (After Padian 1987, based on information from Gauthier and Padian 1985 and Gauthier 1986, with further information from Sanz *et al.* 1988.)

Little is known of the first birds, and the limited fossil record may obscure their true origin. The most significant evidence is provided by the six fossils of *Archaeopteryx lithographica*, a small theropod dinosaur from the Jurassic of Solnhofen in Bavaria (Figure 8.6). *Archaeopteryx* is distinguished from other theropods by its small size (mass 0.2–0.4 kg, comparable to that of a pigeon *Columba livia*; Yalden, 1985), and it is the first theropod in which both feathers and other bird-like features which indicate an ability to fly (see below) have been identified. It is unlikely that *Archaeopteryx* is itself a direct avian ancestor, but there is little doubt that it is a sister group to the stem-group birds (Gauthier and Padian 1985), and it is therefore the best indication of the morphology of the pro-avian. Because of this intermediate dinosaur-bird status these fossils have been invested with considerable importance, but authors have been undecided as to how this evidence should be interpreted. Later in this chapter a hypothetical mechanical argument is developed to show that a gliding ancestry for flight is inescapable. This argument can be supported, for birds at least, by study of the skeletons of *Archaeopteryx*.

Archaeopteryx apparently shares the cursorial morphology of other theropods; there are no conventional arboreal features. It was a small animal, and there are strong mechanical reasons (Rayner 1985b) for expecting the first flying dinosaurs to be much smaller than its closest relative *Deinonychus*. Size is a significant correlate of phylogenetic distinctions in modern birds (for reasons probably related to flight), but it is a relatively poor character for saurischian dinosaurs, particularly because many of those which have been preserved are enormous, and only *Compsognathus* is comparable in size to *Archaeopteryx*. Moreover, small animals may be more generalised in locomotor behaviour, and the absence of arboreal features in *Archaeopteryx* does not on its own rule out climbing. Yalden (1985) has demonstated that the morphology of its fore- and hind limb claws is similar to that of arboreal birds and mammals, but the shape of these claws is also characteristic of larger, cursorial theropods.

In the context of flight evolution the most obvious questions raised by *Archaeopteryx* are whether it could flap its wings, and if so what degree of useful aerodynamic force it could generate. Four separate lines of evidence may be invoked.

1. *Pectoral girdle morphology and sternum.* The pectoral girdle of *Archaeopteryx* is similar to that of non-flying theropods; the coracoids are not the solid strut-like braces of modern birds; some specimens have a furcula, but all lack a calcified (or at least a preserved) sternum (Ostrom 1974). These features suggest relatively weak specialisation for flapping flight. In modern birds the sternum is the main origin for the supracoracoideus muscle which elevates the wing, and for some fibres of the much larger pectoralis muscle (Olson and Feduccia 1979). The majority of pectoralis fibres originate below the humeral joint and apply a positive vertical or roll moment to the humerus, depressing the wing; fibres from the sternum

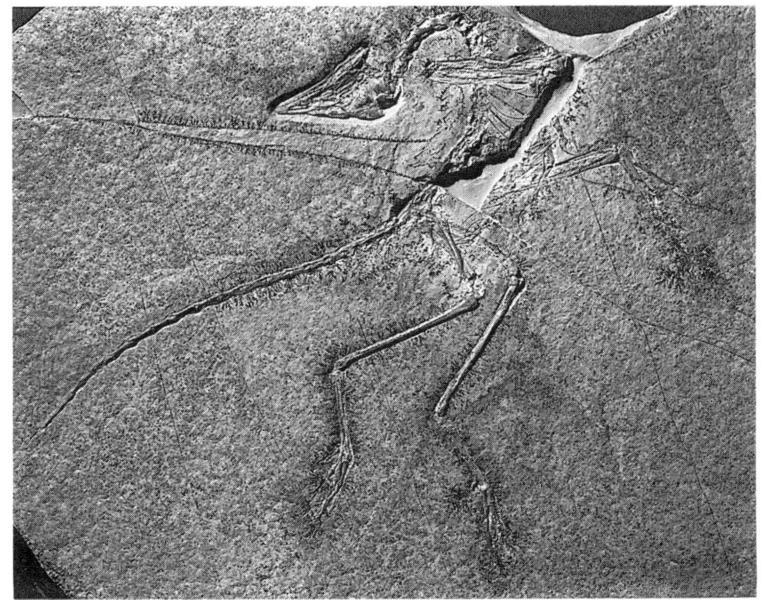

A

B

below and behind the joint attach to the forward edge of the humerus, and, in addition to depressing the wing, pronate it (i.e. rotate the ventral surface towards the ground) during the downstroke by applying a pitching moment. The sternum may be missing through poor preservation, or may have been cartilaginous, but even without it the animal may have flown well, since only the pronating fibres would have been absent. Bats do not have functional sterna, although of course their pectoral anatomy is very different from that of birds. Most theropods also lack calcified sterna (Ostrom 1976). Without a sternum *Archaeopteryx* might well have been able to depress the wing, but would have been unable to pronate it. However, extreme pronation is required only in slow flight, where large changes in the angle of incidence of the wing are needed to control the varying vortex strength (Rayner 1980; 1988a; 1988b); in fast (cruising) flight the wingbeat is primarily dorso-ventral, the wing remains near planar, and changes in incidence are more moderate. This implies that *Archaeopteryx* was unable to fly slowly, and that the keeled sternum is a derived avian character associated with the later development of slow flight and the ability to manoeuvre.

2. *Upstroke musculature.* More definite evidence is provided by the upstroke musculature. In birds and pterosaurs the wing elevator is the supracoracoideus, with origin on the sternum deep under the pectoralis, and insertion onto the dorsal side of the humerus through a cavity within the humeral joint (the *foramen triosseum*). *Archaeopteryx* did not possess this arrangement (Ostrom 1974; 1976). It may have used other muscles to elevate the wing, but if this is the case what pressures favoured the migration of a muscle to form the supracoracoideus at some later stage in bird evolution? In practice an elevator muscle is required only at low speeds (Rayner 1985a; 1986b; 1988a; 1988b), and at higher (cruising) flight speeds the wing is elevated by aerodynamic forces. It is likely that *Archaeopteryx* flew in such a way that it never needed active muscle action to elevate the wing. Again, it appears that *Archaeopteryx* could not fly slowly.

3. *Semi-lunate carpal in the wrist.* The wing skeleton also indicates poor slow-flight performance. The semi-lunate carpal forming the wrist articulation is shared with theropods such as *Deinonychus* (Ostrom 1974; Gauthier and Padian 1985; Padian 1987). This configuration apparently prevents

---

Figure 8.6. Eichstätt and Solnhofen *Archaeopteryx*. A: The Eichstätt *Archaeopteryx*, from the Jura Museum, Eichstätt. This is the smallest specimen of the genus, and is preserved in dorso-ventral compression. There are only faint impressions of the feathers. (Photograph by Dieter Kriebl.) B: The Solnhofen *Archaeopteryx*. This specimen was identified as *Archaeopteryx lithographica* in 1987, having previously been attributed to the similar-sized and contemporary dinosaur *Compsognathus*. See Wellnhofer (1988b) for a detailed description of this specimen, which is on display in the Bürgermeister-Müller-Museum, Solnhofen. Impressions of the feather shafts of the left wing are evident below the elbow joint of the right wing, running beneath the femur of the right leg. Note also the sharp and highly recurved claws of the wing digits. (×0.88; photograph JMVR.)

distortion of the wing out of its plane during the upstroke, but ideally permits sweeping of the wingtip, as during the upstroke in fast flight in most modern birds (see above). When combined with dorso-ventral movement of the humerus, this flexure pre-adapts a gliding theropod proto-bird for force generation by flapping: such a wingbeat generates thrust simply in cruising flight, without the complex twisting, folding and flexing of the wing which must be used in slow flight.

4. *Wing feathers.* Although superficially they are clear evidence of flight ability, feathers need to be interpreted with some caution. It is most likely that they evolved prior to flight for some other function (see below), and thus the presence of feathers alone does not imply flight. However, the feathers of *Archaeopteryx* are remarkably similar to those of modern birds. They have a stiffened central shaft to transmit aerodynamic forces generated over the feather vanes to the body, and this is unexpected if the feathers had no mechanical function. More significantly, the feather shaft is set asymmetrically against the vanes of the feather; this is characteristic of modern flying birds, but the feathers of most flightless birds today are symmetrical (Feduccia and Tordoff 1979). The asymmetric vanes permit the feather to distort optimally to compensate for bending in flight due to aerodynamic loads; this is important in both gliding and flapping flight.

Without elevator muscles, without longitudinal pectoralis fibres originating from a sternum, and without the ability to deform the wing during the upstroke, *Archaeopteryx* could not have flown slowly. Fast or cruising flight is less strenuous than slow flight: the forces required from the pectoralis are less extreme, no elevator muscle is needed, mechanical energy demands are less, and the wingbeat geometry is simpler. It is likely that *Archaeopteryx* would have been able to flap at reasonably high speeds, of the same order of magnitude as the flight speeds of a gliding proto-bird. This is much faster than the speeds reached by a running cursor, and at those lower speeds active muscular pronation and elevation of the wing is required. *Archaeopteryx* is unlikely to have been proficient in slow flight, or to have been very agile and manoeuvrable, and take-offs and landings were probably quite awkward (Rayner 1985a; 1985b; 1988b; 1989).

Later in bird radiation the keeled sternum and supracoracoideus would have evolved in response to pressures for slow flight and manoeuvrability (Figures 8.5). Improved flight performance would also have demanded fusing of manus, loss of the clawed wing digits and semi-lunate carpal, stiffening of the pectoral girdle, and would have permitted the disappearance of the tail. Many of these features are seen in the next oldest bird, recently discovered in the Early Cretaceous of Spain (Sanz *et al.* 1988; Cracraft 1988); the very small size (comparable to a sparrow) and the apparent primitive hindlimbs of this new find suggest that after the appearance of flight in the proto-avian ancestor of *Archaeopteryx*, selection was directed primarily towards flight performance as birds radiated into otherwise unoccupied niches.

The relationships and early radiation of bats

The situation in bats is in many ways similar to that in birds. The group has radiated widely and successfully, and has a widespread distribution; like birds, the order Chiroptera (Mammalia) is characterised by the power of flight, and virtually all synapomorphies of bats (mainly in the wing and pectoral girdle) are flight-related. The oldest known fossil bat, *Icaronycteris index* from the Eocene Green River Formation of Wyoming (Jepsen 1970) scarcely differs from modern Microchiroptera. Fossil bats, particularly from the Messel deposits of West Germany (Figure 8.7), can be exceptionally preserved, and studies of the diet and wing morphology of microchiropterans from Messel reveal a range of foraging patterns and

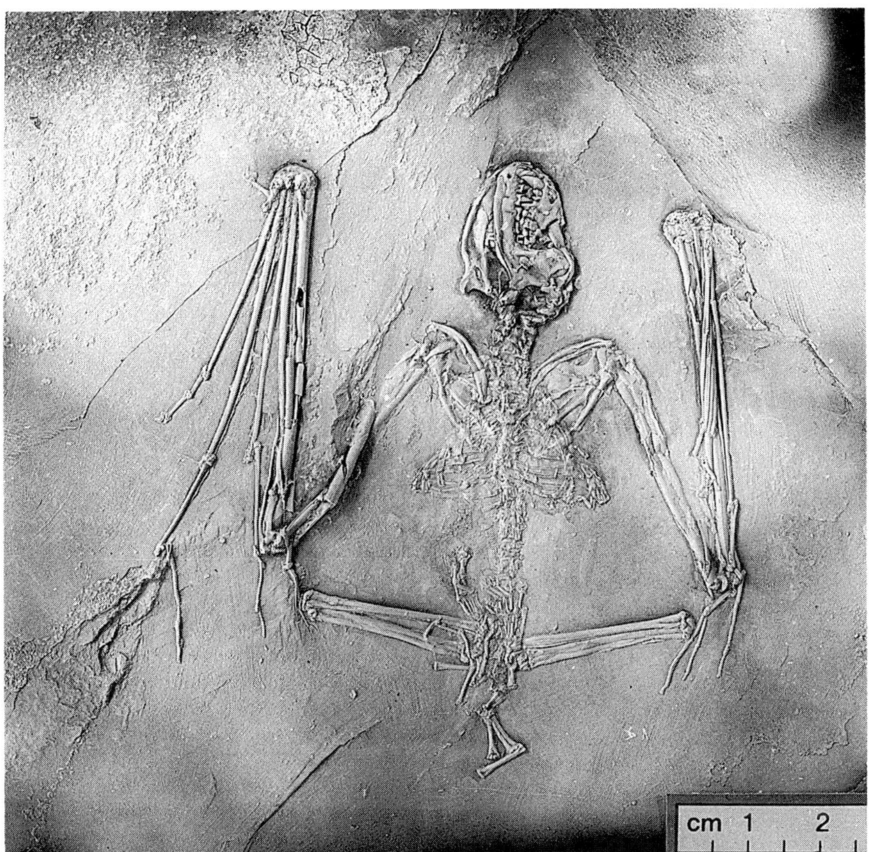

Figure 8.7. The bat *Archaeonycteris trigonodon* Revilliod from the Eocene deposits at Messel, West Germany. With the exception of the claw on the second digit (a megachiropteran character in contemporary bats) this is a fully-formed microchiropteran. (×0.63; photograph by G. Storch.)

wing designs similar to that in modern insectivorous bats. Unfortunately no intermediate fossil between bats and other mammals has been identified, and the sister group of bats is uncertain.

Bats differ from birds in many respects. They are mammals, so are hair-covered, viviparous, and have the complex skeleton and musculature of the mammalian shoulder girdle rather than the simpler form of archosaurs (Padian 1985; 1987; Rayner 1988b). The most obvious morphological feature of bats is the flight membrane, similar to the flight surface of gliding mammals, spanning both fore- and hind limbs, and supported by greatly extended hand digits. The other character which distinguishes bats from birds is the involvement of the legs and feet in the wing membrane. This has required the femur to be rotated backwards, and has been a major constraint on the radiation of bats, restricting terrestrial movement to an awkward scurrying, imposing an inverted resting posture, and preventing the adoption of certain feeding niches (Rayner 1981; 1987; Norberg and Rayner 1987). But this flight morphology carries several implications for the mode of evolution of bats.

Bats show considerable homogeneity in wing morphology, and a narrower ecological range than birds (Norberg and Rayner 1987; Rayner 1986b; 1987). Bats are classified into two sub-orders, the primarily frugivorous Megachiroptera (one family, Pteropodidae), and the primarily insectivorous Microchiroptera (about 17 families). There is a strong trend to arboreality in the group, with many species roosting and/or feeding on or close to trees, and virtually all bats are nocturnal. These characters all point to a gliding origin of flight for the group, with the hypothetical proto-bat having much in common with contemporary gliding rodents and marsupials, and with the colugo (*Cynocephalus*, Dermoptera) of South-east Asia (Rayner 1981; 1985a; 1986b). All of these gliders are nocturnal, inhabit forest or woodland, and use gliding to increase their foraging range while searching for food; most are vegetarians. *Cynocephalus* is the only gliding mammal in which the flight membrane is spread by the digits as well as the limbs. This feature, together with more specific morphological similarities, has led to claims that it is the sister group of bats, and if correct this is overwhelming evidence that bats evolved through an arboreal gliding ancestor. This is strongly supported by the subsequent ecological radiation of bats, and by the fact that a terrestrial proto-bat with a gliding membrane attached to the hindlimb and feet would have been ungainly and is unlikely to have had the ability to leave the ground.

Recently an unresolved debate has developed around the hypothesis that bats are diphyletic, with the Megachiroptera being descended from primates (via the Dermoptera), and the Microchiroptera from ancestral mammals such as insectivores. The evidence is based largely on morphology of the penis, and on characters related to the brain and visual perception (Smith 1976; Pettigrew 1986). Mega- and microchiropterans are very different in the ways they exploit their powers of flight, and laryngeal echolocation is absent from megachiropterans. It would be surprising if wings with so many similarities had evolved twice in mammals, but it is arguable that the behaviour patterns of small arboreal mammals impose a gliding origin of flight, and the mammalian bauplan allows no alternative to the fingers as support for the wing. If the Chiroptera does prove to be paraphyletic, the processes involved in the evolution of the two chiropteran lineages will have been closely similar, and Mega- and Microchiroptera will form a striking example of convergent evolution.

THEORIES OF FLIGHT EVOLUTION

The logic of evolutionary hypotheses

Not surprisingly, many hypotheses have been advanced for the appearance of flight in vertebrates. Some explain the appearance of wings and feathers by behavioural or physiological factors such as display and fighting, foraging, thermoregulation or protection from insolation (Regal 1975), or water-repellance. Few of these hypotheses can be discounted, but while they suggest behavioural correlates or precursors of flight, they do not address the major question of how an animal might have evolved the ability to flap a wing to generate useful aerodynamic forces. It is this which distinguishes birds, bats and pterosaurs from their non-flying relatives.

Three models attack this problem. Each has its adherents, and each has been the butt of valid (and at times invalid) criticisms. (The models and their advantages and disadvantages are considered in further detail in Rayner 1985a; 1985b; 1986b; 1988b; see also Norberg 1986b.)

The gliding model

The oldest, and most widely accepted, of the models proposes that birds (Marsh 1880) and bats (Darwin 1859) evolved through a gliding stage: Darwin suggested that proto-bats were not dissimilar to extant gliding squirrels. The predominant feature of the model is that proto-flappers used gravity to power their flights, probably within vegetation, but perhaps among rocks and cliffs; after gliding had become established flapping began to provide thrust to elongate the glide. The animals' use of gravity as the source of energy explains the intuitive attraction of this model, especially to the biomechanicist.

The gliding hypothesis is often referred to as the 'arboreal' model. Animals often live or move in trees, and many are able to climb; trees are a ready source of potential energy with which to start a glide. Moreover, virtually all contemporary tetrapod gliders are associated with trees. However, the term 'arboreal' has been the source of objections to the gliding model when applied to birds. (There have been few cogent objections to its validity in bats, since trees, and arboreal habit in mammals, were widespread by the Palaeocene.) The presence of trees for a Jurassic proto-bird, at least in the relevant geographical area, is unlikely; moreover, theropods show no specialised features for climbing. It is therefore preferable to use the term 'gliding' for this model, to emphasise its mechanical rather than ecological correlates; the use of trees is not ruled out, but equally is not essential. What is important is the use of descending flight powered by gravity in the intial phases of flight evolution.

The cursorial model

The most widely voiced alternative to gliding is the so-called 'cursorial' model (Nopsca 1907; 1923). This envisages the proto-flier as a running biped, making short jumps from the ground. (The difficulty of running in a terrestrial ancestor effectively excludes the application of this model to bats.) These jumps become extended, the wings are used for balance and propulsion, and the animal begins to fly. With growing understanding of flight mechanics this model fell into disfavour since gliding has intuitive advantages; the cursor must work *against* gravity, and it is apparently impossible to provide thrust with the legs while gaining any useful height jumping from the ground. Above all, the hypothesis lacks any clear selective pressure favouring the appearance of flight in a cursor.

However, realisation of the bipedal cursorial habits of theropods and of their relationship to birds forced reconsideration of the role of running in bird flight evolution. Ostrom (1979) proposed that catching flying insects while running and jumping could favour the development of wings which functioned 'like tennis rackets'. Moreover, the wingstroke needed to trap a flying insect would be similar to the arm movements of *Deinonychus* (flexure within the plane of the wing (see above)), and was similar to the flight stroke of a bird (see also Padian 1985). More recently, Caple *et al.* (1983) proposed that stability while running, perhaps at high speed while escaping a predator, is a pressure for wing enhancement. Both of these are cogent models for the appearance of wings in a cursor, and both are consistent with the apparent terrestrial habits of proto-birds. But both are implausible, in part because of the absence of contemporary analogues, but mainly because neither sensibly addresses the question of *why* the wings should be flapped, and of how the imprecise forms of wing waving needed for predation or stability should have developed into useful flapping.

Fluttering proto-flappers

The 'fluttering' model has been advanced most commonly for small microchiropteran bats (Jepsen 1970; Caple *et al.* 1983). It proposes that flight began with animals beating their wings erratically, and finding that they could generate useful aerodynamic forces: flight would thus have evolved rapidly, perhaps by saltation. The fluttering model is more elusive to mechanical formulation, and it fails because it takes no account of the extreme morphological, physiological and behavioural specialisations required for flight. The first 'flights' of a fluttering proto-flapper would have been at low speeds, where the energetic demands of flight are at their most extreme, and the wingbeat cycle is at its most complex. The model need not be considered further.

Comparison of gliding and cursorial models

One approach to a resolution of the problem of flight evolution is to compare predictions from the gliding and cursorial hypotheses. Adaptations for flight must appear in a sequence consistent with phylogeny, at least as far as it is known, and with the available fossil evidence. Further, biomechanical and physiological developments as flight improves must be gradual (in order to be tractable for the model, not because saltation evolution is *a priori* impossible) and morphologically consistent, and intermediate forms must, in some (rather undefined) sense, be viable.

Biomechanical analysis relies on estimation of quantities such as forces in the muscles and limb segments, energy rates, and transport speeds. These data can be used in two ways: first, they may form the basis for prediction of what an animal of the assumed design might have been able to achieve; and second, they allow the results of similar assumptions for the two models to be considered side by side. The first programme demands accurate knowledge of the dimensions of a hypothetical proto-flier and of the mechanical properties of its tissues; any absolute predictions may be unreliable. The second approach can be more informative; predictions from the two models should give a fair comparison of the *relative* performance of the alternatives. In this chapter only the step from the presence of a wing generating static aerodynamic forces to the flapping of that wing is considered. As argued above, it is this which distinguishes flight, and is the defining characteristic of birds, bats and pterosaurs. Assume that the wing already possesses the main features of its structure, namely the membrane of bats and the feathers of birds. The initial stage in the analysis is to consider the requirements of the two models.

Formulation of the gliding model

The first essential is to ask whether gliding in a proto-flier is consistent with its other activities. Climbing performance gives a reasonable estimate of the forces and energy rates such an animal might be able to generate, especially since the main forces in climbing are associated with the forelimbs. For instance, in order to hang in a suspended position with its limbs at an angle of 45° to the vertical, the animal would have to impose a static moment of 0.36 Mgb at the humeral joint (where M is body mass, g

acceleration due to gravity, and b the length of one forelimb). In active climbing a somewhat greater moment would be needed, perhaps approaching twice this value. The energy rates in climbing are hard to estimate without knowing climbing speed, but will certainly exceed the rate of gain of potential energy during the climb owing to the additional work needed to move the limbs.

The glides of extant gliders consist of an initial leap, followed by a drop to gain speed, before a sustained steady glide. Presumably the proto-flier would have been similar. Glide speed depends on the area of the flight surface: slowing the glide reduces the risk of accident and improves the chance of control and of landing safely, and there is pressure to increase wing area. Height loss is related to drag, which depends in part on the shape of the wing: a more elongated wing gives a shallower glide, and therefore greater range. The lift generated by the proto-wings supports the weight, and a static moment is required at the humeral joint to maintain the wing in position. This moment is estimated as $0.23\,Mgb$, which is smaller than that in hanging and vertical climbing. A climbing animal therefore already has the muscular capacity to hold its wing outstretched in gliding. The main advantages of gliding are that it is energetically economical, and relatively fast: a gliding squirrel might travel at up to 15 m s^{-1}, while its running speed on level ground is unlikely to exceed 2 m s^{-1}.

A glider loses height as potential energy is expended against drag. By flapping, the animal generates thrust, and even a small thrust will reduce the glide angle usefully. The geometry of the wake vortex in a flapping proto-flier at the normal glide speed may be used to predict the effect of incipient flapping. Wingbeats with amplitude greater than about 25° can generate useful thrust while still balancing weight. Moreover, the increment in forces at the wing root is relatively modest, and the peak positive roll moments remain comparable to those in climbing. The energy required for flight is also not excessive: a 27 g proto-bat climbing a tree at 0.7 m s^{-1} would require a minimum mechanical energy of 0.18 W; a noctule bat of the same mass requires 0.2 W in steady level flight at its cruising speed of around 6 m s^{-1}.

Formulation of the cursorial model

The cursorial model is less straightforward to analyse since existing hypotheses fail to predict the mechanical environment in which the flight stroke first appeared. The problem of getting off the ground is compounded by the need for a shift in selection forces from those favouring the hindlimbs for speed and agility in a cursor, to those favouring the forelimbs as the animal begins to fly. Did the proto-flier have separate sets of muscles for the two modes? Climbers and gliders both use the forelimb muscles in locomotion, and are not affected by this problem.

To make it comparable with the gliding model, a form of the cursorial model in which the cursor makes the best possible use of its hind limbs to fly is envisaged: while still running it *jumps* from the ground to gain height.

Its initial flaps can then take place while descending, and, as in the gliding model, are assisted by gravity. Any alternative where the initial flaps work *against* gravity would increase the mechanical and energetic demands of incipient flight, and make the model relatively less attractive.

The mechanics of running in bipedal birds, mammals and lizards are well understood. Without using the forelimbs the animals have good agility, and can move at relatively high speeds up to about $2\,\mathrm{m\,s^{-1}}$; the energy rates required at higher speeds are extreme, and can be sustained only for brief periods. If it is to jump before initiating flying, a running proto-flier must use its hindlimbs to leave the ground. At maximum exertion this must be associated with some degradation of running performance, and in a biped of mass 0.2 kg and typical theropod morphology the drop in speed is of the order of 30–40%. This implies very low gliding or flying speeds. The animal would be unlikely to extend its jumps significantly by flapping to obtain thrust since the energy demanded at such a low speed would be considerable. Moreover, acceleration to higher and more efficient flight speeds takes time, and is unlikely to be achieved during the short leaps.

A biped *could* get sufficiently off the ground to glide briefly, and it *could* save energy by so doing. It *could* benefit further by flapping. But it has insufficient energy to reach speeds at which flapping is mechanically straightforward, and the costs of flight at these low speeds are so high that the demands on the forelimb musculature become extreme. It seems that mechanically the cursorial gliding model could just work, but practically it is highly unlikely. However, this is based on the assumption that the animal travels over level ground in still air. If it were to run into a modest head wind, or if the ground were uneven, jumping and gliding could become more practicable.

The models compared

To compare the gliding and cursorial models, assume that the purpose of movement is for transport, and that the animal minimises the cost of transport (the energy to transport unit mass through unit distance). This ignores other possible benefits such as escape from predation or access to new microhabitats. Cost of transport can then be predicted based on the mechanics of flapping flight, running and climbing.

The climbing and gliding locomotion of gliding squirrels reduces cost of transport considerably compared to running (Figure 8.8). This mode is adaptively successful even if the wings are not flapped, and becomes more attractive if the wings enlarge (although long non-flapping wings may be a liability in a cluttered habitat). Adding flaps as the wings develop gives a greater advantage, with a continuous and gradual reduction in cost of transport to around one-fifth of the climbing and gliding value. Provided that the animal can take advantage of the reduced cost (by travelling further to forage, for instance) there is strong feedback encouraging development of the wings and other flight organs. If the contribution to fitness is inversely proportional to cost, feedback favouring the incipient adaptation is strongest in the early stages of wing development, and thus it would have rapidly established itself in a population.

By comparison, the costs for running and jumping and for sustained running are much higher than for climbing and gliding. The final cost of transport, assuming progression to a fully adapted flier, is the same as for the gliding model, so the total reduction in cost is considerable (Figure 8.8). But feedback from the new adaptation is weak in the initial stages, and only becomes significant in the later stages when wings and flight are fully developed and the cost reduction can be attributed to the radiation of

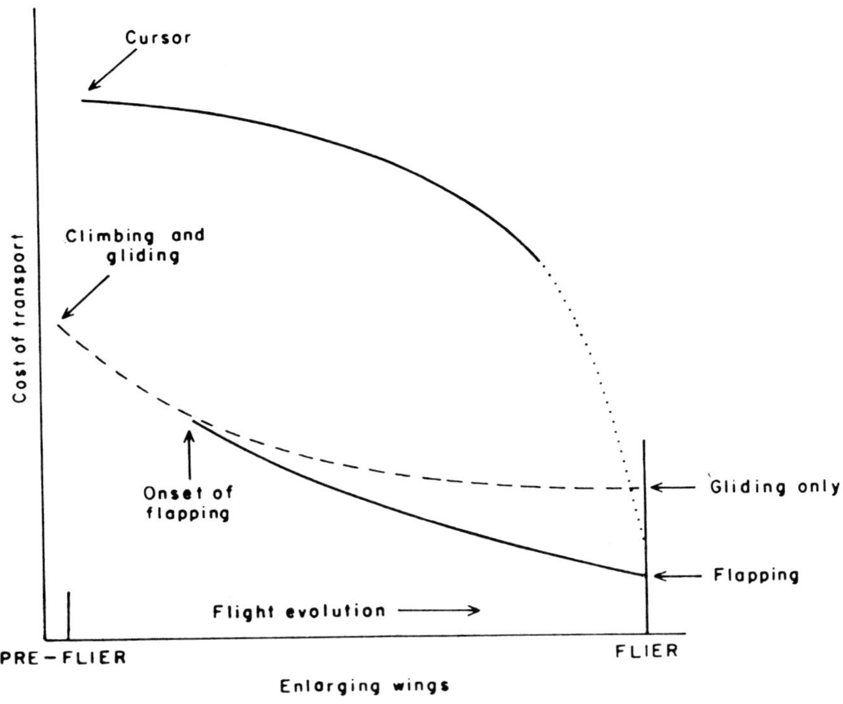

Figure 8.8. Comparison of energetics for climbing and gliding and cursorial proto-fliers. The graph shows cost of transport changes as wings enlarge and lengthen for the gliding (lower curves) and cursorial (upper curve) models. Calculation for the gliding model explained by Rayner (for bats, 1985a; 1986b; for birds 1985b); cursorial model after Rayner (1985b). The graphs are non-dimensional, but the qualitative predictions are relevant for climbers or cursors between about 20 g and 0.5 kg in mass, and there is little relative difference between the two options with size. Flapping with the short, broad wings of a passive glider is disadvantageous, but thereafter flapping brings considerable advantage, reducing total cost of transport by as much as 80%. For the cursorial model the initial energy saving is small; only as flapping flight becomes fully developed is there a significant reduction in cost; the final part of the curve (shown dotted) is notional and has not been modelled in detail. (From Rayner 1988b.)

an established adaptation rather than to the initial evolution of flight. In the early stages the cost saving is unlikely to outweigh the developmental and other costs of enhancing the forelimbs for flapping while remaining a viable running biped. Given the simultaneous problems of attaining sufficient speed and sufficiently long jumps, it is unlikely that a cursorial runner could have begun to fly. Comparison of the two models strongly implies, on mechanical grounds, that a gliding origin of flapping flight in both birds and bats is considerably more probable, unless, as explained above, uneven ground or air currents can be invoked.

Remaining problems with evolution of flight

The gliding or arboreal model is strongly indicated by estimating the biomechanical and ecological performance of hypothetical proto-fliers and by reconstructing the muscle function and locomotion of *Archaeopteryx*, but there remain real objections to it, particularly in birds. Climbing would be impossible without suitable habitat, and this conflicts with the apparent environment of the time. Proto-fliers and their sister groups would be expected to show some morphological specialisation if climbing were important to them, but *Archaeopteryx* does not. Some writers have argued that gliders are unlikely to have evolved further into flappers since gliding is an adaptive plateau in extant mammals. The logic of this is false: by the above arguments gliding is an essential pre-adaptation for flapping, and it is precisely because gliding is adaptive that there were animals in which flapping could evolve. Moreover, it is as illogical to argue that gliding proto-flappers should have become extinct through competition with flappers as it is to claim that dinosaurs became extinct because one group of them evolved into birds.

Objections to the cursorial model are more specific. The most critical is the need to provide a vertical force *against* gravity at an early stage in flight evolution. For this reason speeds in running cannot approach those at which flight is energetically or morphologically efficient. The wingbeat cycle in slow flight, the associated muscular specialisations, and the wake aerodynamics are complex and more derived than in cruising flight. Moreover, no logical selective pressure explaining the morphological and mechanical developments necessary for a cursor to fly has yet been proposed.

These conclusions are reinforced by comparison of the two models. Gliding gives more rapid returns in foraging range, locomotion efficiency and behavioural flexibility; the contribution these make to fitness provides strong positive feedback as the wings develop and begin to flap. The sequence envisaged is consistent with biomechanics and aerodynamics and with the ecology and morphology of the hypothetical proto-flier, and moreover, with the pectoral girdle and wings of *Archaeopteryx*. There can be no doubt that bats evolved through gliding, and on balance this is by far the more likely model for birds.

But these arguments are founded entirely on locomotion mechanics and transport energetics: they presume that an animal evolves to move in the most economic way compatible with its design. However, locomotion strategies may also be dictated by other factors. Gliding is used by arboreal insectivorous lizards and birds to increase manoeuvrability while foraging around a habitat. Access to otherwise unavailable food resources might select for flapping; even so, for mechanical reasons a gliding pathway to flapping still seems inevitable. Escape from predation (by speed or unpredictability) is a further possibility, but is hard to quantify in relation to selection: an incipient flight adaptation might render an animal more vulnerable, and so be selected against. Time invested in movement may be more critical than energy, so selection would favour high flight speed; this again implies a gliding origin, since gliders can travel far faster than cursors, and higher speeds are easily attained by reducing wing size.

One further possibility should be considered. The evolution of flying vertebrates has been modelled as a gradual progression through adapted and adaptive levels. Some stages pre-adapt to following stages (as gliding to flapping), and the sequence is driven by consistent, readily identified and realistic selection pressures. Intermediate stages might have been maladaptive compared to their successors, but it is important only that they were not subject to significant predation or other pressures. As an alternative, it is possible that flight evolved rapidly (by saltation). No intermediates would have existed to be found in the fossil record. (The identification of *Archaeopteryx* as a sister group to the pro-avian stem-group, and of *Cynocephalus* as the megachiropteran sister group, apparently discount this.) Or proto-flappers might have benefited from remarkable environmental conditions in which selection pressures were sufficiently strong, and mechanical constraints sufficiently weak, that comparison with aerodynamics of contemporary fliers is invalid. However, if flight did appear suddenly there is no mechanical, or evolutionary, problem to be discussed. The 'first bird' or 'first bat' would have had the same attributes as modern animals, and its success in flight would be mediated by much the same ecological pressures. Questions of its evolution become untestable and meaningless, although the phylogeny of its descendants can still be modelled by flight-related characters.

This survey has not considered all the problems of flight evolution or of the phylogeny of vertebrate fliers. Features such as the behavioural and ecological flexibility of an incipient flier, its likely predators, and the habitats that it might have used, have been ignored, as has the important role of control and stability. (Did *Archaeopteryx* have a long bony tail just because its ancestors did? Does the superficial similarity with magpies (*Pica*, Corvidae) indicate arboreal habit in proto-birds?) The biomechanical approach to functional evolution cannot unequivocally resolve these points, any more than it can prove that proto-fliers had to be gliders. Its value lies in showing that some possibilities—such as the fluttering model—are untenable, and that some are more likely than others. But as in all evolutionary debates the central difficulty lies not only in assembling the evidence, but also in assessing its worth, particularly when predictions or interpretations conflict. Mechanical arguments would be devalued if they predicted the most likely behaviour or morphology of a proto-flier to be phylogenetically or developmentally untenable. But this is not the case in either birds or bats: evidence from aerodynamics and flight mechanics is reinforced by the interpretation of *Archaeopteryx* that this approach implies, and strongly argues for a gliding origin of flight in all groups of vertebrates.

REFERENCES

Bakker, R.T., 1986, *The dinosaur heresies*, William Morrow, New York.

Caple, G., Balda, R.P. and Willis, W.R., 1983, The physics of leaping animals and the evolution of preflight, *American Naturalist*, **121**: 455–67.

Cracraft, J., 1988, Early evolution of birds, *Nature*, **331**: 389–90.

Darwin, C., 1859, *On the origin of species by means of natural selection*, John Murray, London.

Feduccia, A. and Tordoff, H.B., 1979, Feathers of *Archaeopteryx*: asymmetric vanes indicate aerodynamic function, *Science*, **203**: 1021–2.

Gauthier, J.A. and Padian, K., 1985, Phylogenetic, functional and aerodynamic analyses of the origin of birds and their flight. In M.K. Hecht, J.H. Ostrom, G. Viohl and P. Wellnhofer (eds), *The beginnings of birds*, Jura Museum, Eichstätt, pp. 185–97.

Hecht, M.K., Ostrom, J.H., Viohl, G. and Wellnhofer, P. (eds), 1985, *The beginnings of birds*, Jura Museum, Eichstätt.

Heilmann, G., 1927, *The origin of birds*, Witherby, London: reprinted Appleton, and Dover, New York, 1972.

Jepsen, G.L., 1970, Bat origins and evolution. In W.A. Wimsatt (ed.), *Biology of bats*, Vol. **1**, Academic Press, New York, pp. 1–64.

Marsh, O.C., 1880, Odontornithes: a monograph on the extinct toothed birds of North America, *Professional Papers of the Engineers Department of the U.S. Army*, **18**.

Nopsca, F., 1907, Ideas on the origin of flight, *Proceedings of the Zoological Society of London*, pp. 222–36.

Nopsca, F., 1923, On the origin of flight in birds, *Proceedings of the Zoological Society of London*, pp. 463–77.

Norberg, U.M., 1985a, Flying, gliding, soaring. In M. Hildebrand, D.M. Bramble, K.F. Liem and D.B. Wake (eds), *Functional vertebrate morphology*, Harvard University Press, Cambridge, MA, pp. 129–58.

Norberg, U.M., 1985b, Evolution of vertebrate flight: an aerodynamic model for the transition from gliding to active flight, *American Naturalist*, **126**: 303–27.

Norberg, U.M. and Rayner, J.M.V., 1987, Ecological morphology in bats (Mammalia; Chiroptera): wing adaptations, flight performance, foraging strategy and echolocation, *Philosophical Transactions of the Royal Society, London*, **B316**: 335–427.

Olson, S.L. and Feduccia, A., 1979, Flight capacity and the pectoral girdle of *Archaeopteryx*, *Nature*, **278**: 247–8.

Ostrom, J.H., 1974, *Archaeopteryx* and the origin of flight, *Quarterly Review of Biology*, **49**: 27–47.

Ostrom, J.H., 1976, Some hypothetical anatomical stages in the evolution of avian flight, *Smithsonian Contributions to Paleobiology*, **27**: 1–21.

Ostrom, J.H., 1979, Bird flight: how did it begin? *American Scientist*, **67**: 46–56.

Padian, K., 1983, A functional analysis of flying and walking in pterosaurs, *Paleobiology*, **9**: 218–39.

Padian, K., 1985, The origins and aerodynamics of flight in extinct vertebrates, *Palaeontology*, **28**: 413–33.

Padian, K. (ed.), 1986. The origin of birds and the evolution of flight, *Memoirs of the California Academy of Sciences*, **8**.

Padian, K., 1987, A comparative phylogenetic and functional approach to the origin of vertebrate flight. In M.B. Fenton, P.A. Racey and J.M.V. Rayner (eds), *Recent advances in the study of bats*, Cambridge University Press, Cambridge, pp. 3–22.

Pettigrew, J.D., 1986, Flying primates? Megabats have the advanced pathway from eye to mid-brain, *Science*, **231**; 1304–6.

Rayner, J.M.V., 1980, Vorticity and animal flight. In H.Y. Elder, and E.R. Trueman (eds), *Aspects of animal movement*, Seminar Series of the Society for Experimental Biology, **5**, Cambridge University Press, Cambridge, pp. 177–89.

Rayner, J.M.V., 1981, Flight adaptations in vertebrates, *Symposium of the Zoological Society of London*, **48**: 137–72.

Rayner, J.M.V., 1985a, Mechanical and ecological constraints on flight evolution. In M.K. Hecht, J.H. Ostrom, G. Viohl and P. Wellnhofer (eds), *The beginnings of birds*, Jura Museum, Eichstätt, pp. 279–88.

Rayner, J.M.V., 1985b, Cursorial gliding in proto-birds. In M.K. Hecht, J.H. Ostrom, G. Viohl and P. Wellnhofer (eds), *The beginnings of birds*, Jura Museum, Eichstätt, pp. 289–92.

Rayner, J.M.V., 1986a, Pleuston: animals which move in water and air, *Endeavour*, **10**: 58–64.

Rayner, J.M.V., 1986b, Vertebrate flapping flight mechanics and aerodynamics, and the evolution of flight in bats. In W. Nachtigall (ed.), *Biona Report* **5**: *Bat flight—Fledermausflug*, Gustav Fischer Verlag, Stuttgart, pp. 27–74.

Rayner, J.M.V., 1987, The mechanics of flapping flight in bats. In M.B. Fenton, P.A. Racey and J.M.V. Rayner (eds) *Recent advances in the study of bats*, Cambridge University Press, Cambridge, pp. 23–42.

Rayner, J.M.V., 1988a, Form and function in avian flight, *Current Ornithology*, **5**: 1–77.

Rayner, J.M.V., 1988b, The evolution of vertebrate flight, *Biological Journal of the Linnean Society*, **34**: 269–87.

Rayner, J.M.V., 1989, Mechanics and physiology of flight in fossil and recent vertebrates, *Transactions of the Royal Society of Edinburgh Earth Sciences*, in press.

Regal, P.J., 1975, The evolutionary origin of feathers, *Quarterly Review of Biology*, **50**: 35–66.

Sanz, J.L., Bonaparte, J.F. and Lacasa,, A., 1988, Unusual Early Cretaceous birds from Spain, *Nature*, **331**: 433–5.

Smith, J.D., 1976, Comments on flight and the evolution of bats. In M.K. Hecht, P.C. Goody and B.M. Hecht (eds), *Major patterns in vertebrate evolution*, Plenum Press, New York, pp. 427–37.

Stephan, B., 1987, *Urvögel* (3rd edn), Neue Brehm-Bücherei, **465**, A. Ziemsen Verlag, Wittenberg-Lutherstadt.

Unwin, D.M., 1988, New remains of the pterosaur *Dimorphodon* (Pterosauria: Rhamphorhynchoidea) and the terrestrial ability of early pterosaurs, *Modern Geology*, **13**: 57–68.

Vogel, S., 1981, *Life in moving fluids*, Willard Grant, Boston.

Wellnhofer, P., 1978, *Handbuch der Paläoherpetologie, Teil 19, Pterosauria*, Gustav Fischer Verlag, Stuttgart.

Wellnhofer, P., 1980, *Flugsaurier—Pterosaurier*. Neue Brehm-Bücherei, **534**, A. Ziemsen Verlag, Wittenberg-Lutherstadt.

Wellnhofer, P., 1987, Die Flughaut von Pterodactylus (Reptilia, Pterosauria) am Beispiel des Wiener Examplares von Pterodactylus kochi (Wagner), *Annalen des Naturhistorisches Museums in Wien, Serie A: Mineralogie und Petrographie, Geologie und Paläontologie, Anthropologie und Prähistorie*, **88**: 149–62.

Wellnhofer, P., 1988a, Terrestrial locomotion in pterosaurs, *Historical Biology*, **1**: 3–16.

Wellnhofer, P., 1988b, A new specimen of *Archaeopteryx*, *Science*, **240**: 1790–2.

Wild, R., 1984, Flugsaurier aus der Obertrias von Italien, *Naturwissenschaften*, **71**: 1–11.

Yalden, D.W., 1985, Forelimb function in *Archaeopteryx*. In M.K. Hecht, J.H. Ostrom, G. Viohl and P. Wellnhofer (eds), *The beginnings of birds*, Jura Museum, Eichstätt, pp. 91–7.

Chapter 9

PATTERNS OF EVOLUTION AND EXTINCTION IN VERTEBRATES

Michael J. Benton

PATTERNS OF EVOLUTION AND EXTINCTION IN VERTEBRATES

Vertebrates have not figured as strongly in analyses of patterns of diversity and extinction as have invertebrates. This is probably because of a general perception of their fossil record as less complete, both taxonomically and stratigraphically, than that of marine invertebrates in particular. This is especially the case since much of the vertebrate fossil record is terrestrial, and terrestrial environments are generally poorly represented compared to marine ones. A final problem is the overall size of the data set—there are probably only about 1400 known families of living and extinct fish and tetrapods (the land vertebrates—amphibians, reptiles, birds and mammals). This compares with over 3000 families of marine invertebrates (Sepkoski 1982). However, the vertebrate fossil record has advantages for macroevolutionary study, since the taxonomy is mature enough in many parts for the recognition of monophyletic groups, and there is scope for detailed ecological analysis.

THE NATURE OF THE VERTEBRATE FOSSIL RECORD

Vertebrates have been informally divided into 'fish' and tetrapods. The 'fish' include a variety of swimming forms in the seas and in freshwater, and the term is useful shorthand for these forms, although it is no longer generally used in classifications because it is ill-defined. Carroll (1987) provided a formalised classification of vertebrates (Table 9.1). Their broad temporal distributions and relationships are shown in Figure 9.1.

Table 9.1. *Numbers of families of fishes and tetrapods.*

Taxon	*Living and extinct*	*Total living*	*Living only*	*Singletons**
Class Agnatha	39	2	1	7
Class Placodermi	37	0	0	11
?Class Acanthodii	5	0	0	0
Class Chondrichthyes	76	40	5	5
Class Osteichthyes	618	416	207	18
Chondrostei	52	3	0	5
'Holostei'	10	2	0	1
Teleostei	531	409	206	12
Sarcopterygii	25	3	1	0
All 'fish'	775	459	213	41
Class Amphibia	102	23	1	16
Class 'Reptilia'	233	43	7	49
Class Aves	202	157	48	11
Class Mammalia	379	129	26	0
All tetrapods	916	352	82	76
All vertebrates	1691	811	295	117

*Singletons are families represented by a single species from a single geologic formation.

An outline of vertebrate evolution

The first vertebrates are known from scrappy fossils of bony plates from the Late Cambrian. These are ascribed to agnathan (literally 'jawless') fishes which flourished in freshwater and marine environments in the Silurian and Devonian as generally heavily armoured forms. They declined thereafter, and are represented today by the eel-like lampreys and hagfishes.

The only extinct clan of vertebrates, the placoderms ('platy skin'), were also heavily armoured fishes in Devonian seas and freshwaters. These were the first vertebrates with jaws—the agnathans had only suckers and rasping tooth plates. The placoderms were able to adopt fully predatory modes of life, and some became very large. *Dunkleosteus* may have reached a length of 9 m.

The Chondrichthyes ('cartilaginous fish'), including sharks, rays, and chimaeras, arose in the Devonian and went through a major radiation in the Carboniferous and Permian. These early groups largely died out in the

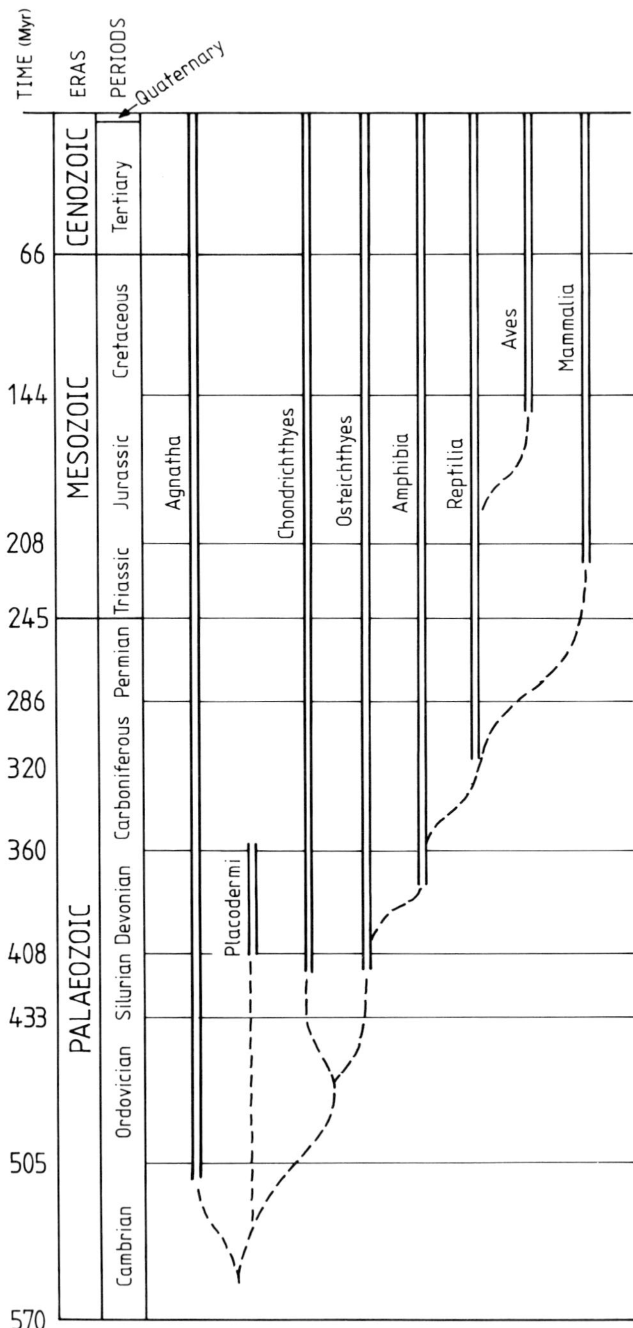

Figure 9.1. Summary of the temporal distribution of vertebrate classes and their probable relationships. (After Caroll 1987.)

Permian and Triassic, and a second radiation, when all the modern groups arose, began in Jurassic times.

The Osteichthyes ('bony fish') also arose in the Devonian, and they have become by far the most diverse and abundant group of fish—there are over 20 000 living species in 409 families. Osteichthyan evolution went through three phases of major radiation. The first, from Devonian to Late Triassic times, consisted mainly of heavily-scaled forms with simple jaws, the 'chondrosteans', of which the sturgeon and paddle fish are living representatives. This radiation also involved a variety of sarcopterygians, or lobe-finned fish, which included the ancestors of tetrapods. Living sarcopterygians include the rare coelacanth, and the lungfishes. The second osteichthyan radiation, which began in the Early Triassic and continued through the Mesozoic, was of 'holostean' forms which still had heavy scales, but whose jaw apparatus was more advanced than in 'chondrosteans'. Living forms include *Amia*. The third, and largest, osteichthyan radiation took place in the Jurassic and Cretaceous and involved the teleosts, which have light scales and very flexible jaws that together allow fast swimming and great adaptability in feeding modes. Modern teleosts include familiar forms such as cod, herring, carp, and perch, as well as more unusual fish such as flatfish, sea horses, and angler fish.

The Amphibia ('both' modes of life—water and land) include frogs and salamanders today, but these modern groups did not appear until the Mesozoic, radiating mainly in the Late Cretaceous and Tertiary. The first amphibians arose in the Late Devonian and radiated in the Carboniferous, Permian and Triassic as the heavily-built semi-aquatic temnospondyls (Carboniferous–Jurassic), the generally more terrestrial anthracosaurs (Carboniferous–Permian), and the small salamander and lizard-like lepospondyls (Carboniferous–Early Permian).

The Reptilia ('crawling' animals) arose from anthracosaurs in the Late Carboniferous. They soon split into three main lineages, the anapsids (including living turtles), the diapsids (including dinosaurs, crocodiles, pterosaurs, lizards and snakes), and the synapsids (the mammal-like reptiles essentially of the Permian and Triassic). These are all basically terrestrial groups, with some secondarily aquatic forms (turtles, crocodiles, sea snakes). The main radiations took place in the Permian (Synapsida), Late Triassic–Cretaceous (Dinosauria), and Late Cretaceous onwards (lizards, snakes).

The Aves (birds) are feathered flying diapsids which arose from smaller bipedal carnivorous dinosaurs in the Late Jurassic, and radiated modestly in the Late Cretaceous and more extensively in the Cenozoic to reach the present total of about 150 families.

The Mammalia arose in the Late Triassic from small carnivorous mammal-like reptiles and radiated only a little in the Jurassic and Cretaceous. The modern groups—Monotremata (echidna, platypus), Marsupialia (pouched forms: kangaroo, wombat, opossum) and Eutheria (placental mammals: bats, humans rabbits, rats, whales, horses, etc.) all arose around the Early Cretaceous (Late Jurassic–Early Cretaceous for the Monotremata, and mid-Cretaceous for the Marsupialia and Eutheria). The Marsupialia and Eutheria radiated modestly in the Late Cretaceous, and then diversified extensively in the Cenozoic to give the present total of about 130 families.

Relationships

It is important to establish clades, or monophyletic groups, for use in macroevolutionary studies (Cracraft 1981; Benton 1988a). Such studies generally focus on supraspecific categories, often families or orders. There is no objective way, of course, to determine the rank of a clade in the taxonomic hierarchy, e.g. whether a particular group is a family or an order. Such groups should, however, as far as can be determined, include all of the descendants of a single common ancestor (i.e. *monophyletic* is synonymous with *holophyletic* for some authors; Figure 9.2). It would clearly be pointless in most contexts to discuss the evolution of a *poly-*

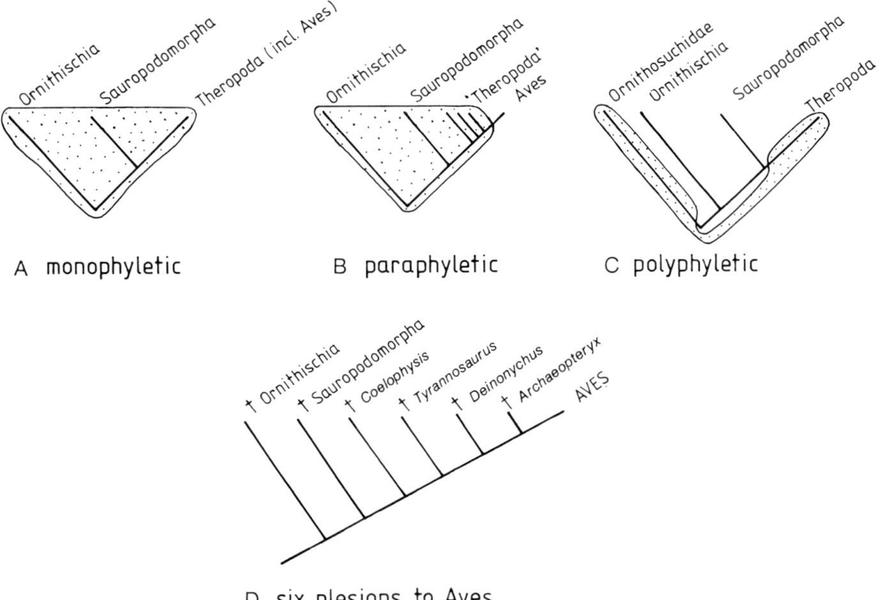

Figure 9.2. Kinds of phylogenetic group, based on a dinosaurian example. A: Dinosauria, a monophyletic group, including all descendants of one ancestor; B: Dinosauria (excluding birds), a paraphyletic group, including some, but not all, descendants of one ancestor; and, C: a polyphyletic group of carnivorous bipeds derived from more than one immediate ancestor. D: The relationships of birds to their closest extinct relatives among the dinosaurs, all of which are plesions of the crown-group Aves.

phyletic group (i.e. derived from several ancestors) since these are entirely human inventions. A *paraphyletic* group (Figure 9.2), such as Class Reptilia, is descended from one ancestor, but excludes some of the descendants (here, birds and mammals). The starting point of the clade is a real part of the phylogenetic tree, but the terminations of 'Reptilia' along the lines to mammals and to birds are artificial. Reptilia, therefore, is at least partly a human invention. Graphs of the evolutionary rate of reptiles might show, for example, that they enjoyed rapid rates of origination during Permian and Triassic times, but that these rates dropped off in the Jurassic and Cretaceous. This does not necessarily mean that reptiles were evolving in a sluggish manner, but simply that the new hairy reptiles and feathered reptiles have been arbitrarily excluded from the calculations.

The key to identifying monophyletic groups among Vertebrata is cladistic character analysis, in which patterns of relationship are established on the basis of shared derived characters (synapomorphies). Most vertebrate groups have now been tackled by one or more cladists, and attempts are also being made to analyse the links between these major groups. These latter efforts have generated most controversy (e.g. relationships of sarcopterygian fish and tetrapods, birds and reptiles, early mammals) and this has tended to obscure the fact that a great deal of agreement has become evident in smaller-scale cladograms of particular orders or subclasses. In addition, cladistic analyses of vertebrates have generally not affected the composition of family-level taxa. Even before cladistic methods were widely used, vertebrate systematists defined families on the basis of clear-cut derived characters. It has been in linking the families into orders, then the orders into classes, that character definitions have lost their sharpness, leading to the establishment of artificial taxa on the basis of primitive (plesiomorphous) characters—e.g. Chondrostei, Labyrinthodontia, Cotylosauria, Eosuchia, Thecodontia, Prototheria.

Most studies of vertebrate macroevolution have been based on families, and the new classifications, therefore, have not had as profound an effect as might have been expected. The main changes have arisen in drawing the lower boundaries of families—cladists would tend to exclude 'potential ancestors' from a family unless they display at least one synapomorphy of that family. This has pulled the dates of origin of some families forwards in time. The plesiomorphous taxa are then assigned plesion ranks, possibly equivalent to families (Figure 9.2D). This could potentially give rise to a vast proliferation of new 'singleton' families based on single ill-defined ancestral species. By convention, however, such families are excluded from calculations of origination and extinction rates until a second occurrence is discovered. For example, the Family Archaeopterygidae arose and disappeared instantaneously, being represented only by the species *Archaeopteryx lithographica* (albeit by several specimens) from rocks of a single age.

Figure 9.3 shows the broad patterns of evolution in the 'fishes' and the tetrapods. It would clearly be inappropriate to survey the relationships and

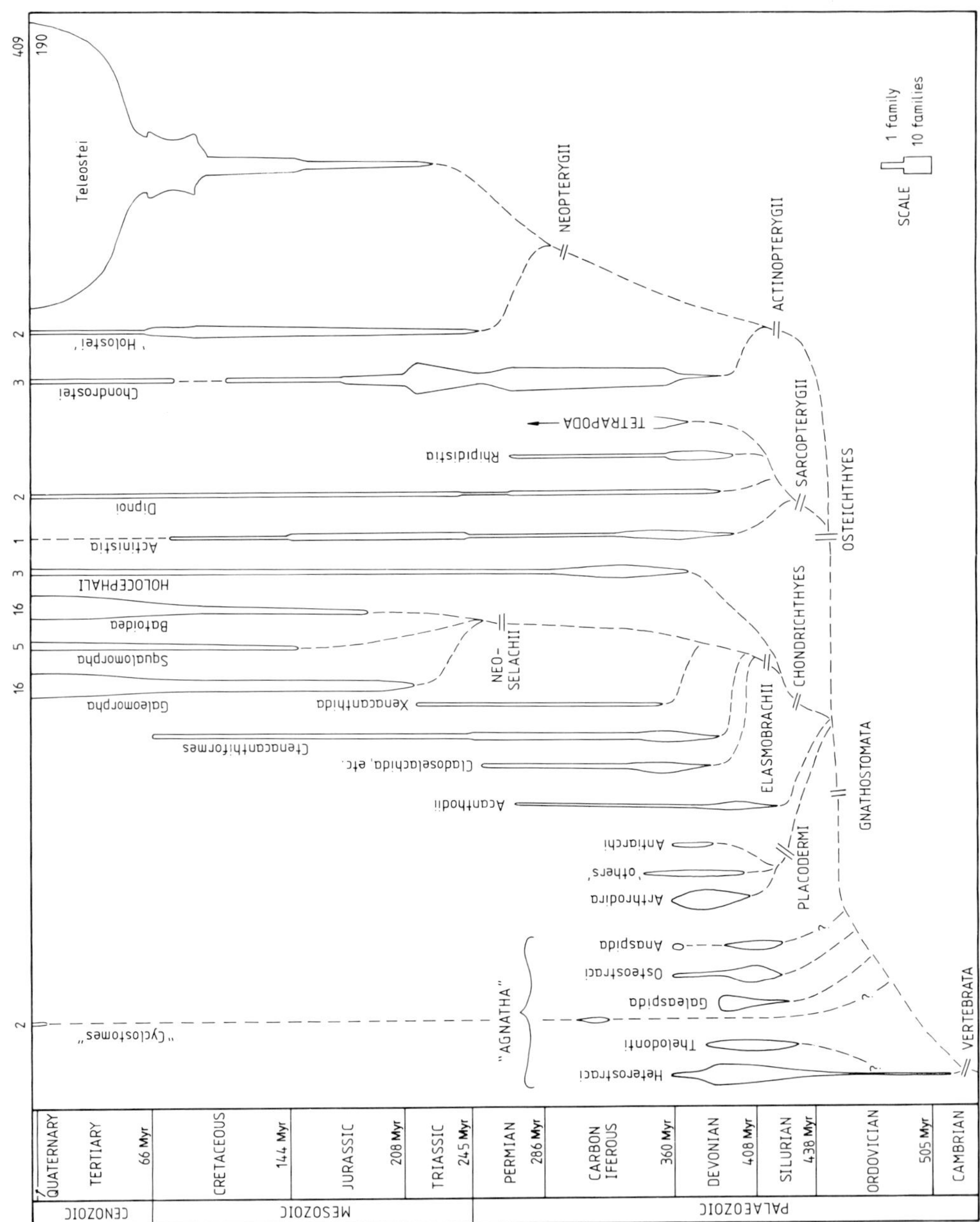

Figure 9.3. Phylogenetic trees of A: fish taxa, and B: the Tetrapoda, showing relationships, stratigraphic duration, and diversity of each group. The major groups are indicated as balloons that show the known stratigraphic range by their height, and the relative numbers of families present by their width (see scales in bottom right-hand corner). Relationships of the groups based on recent cladistic analyses

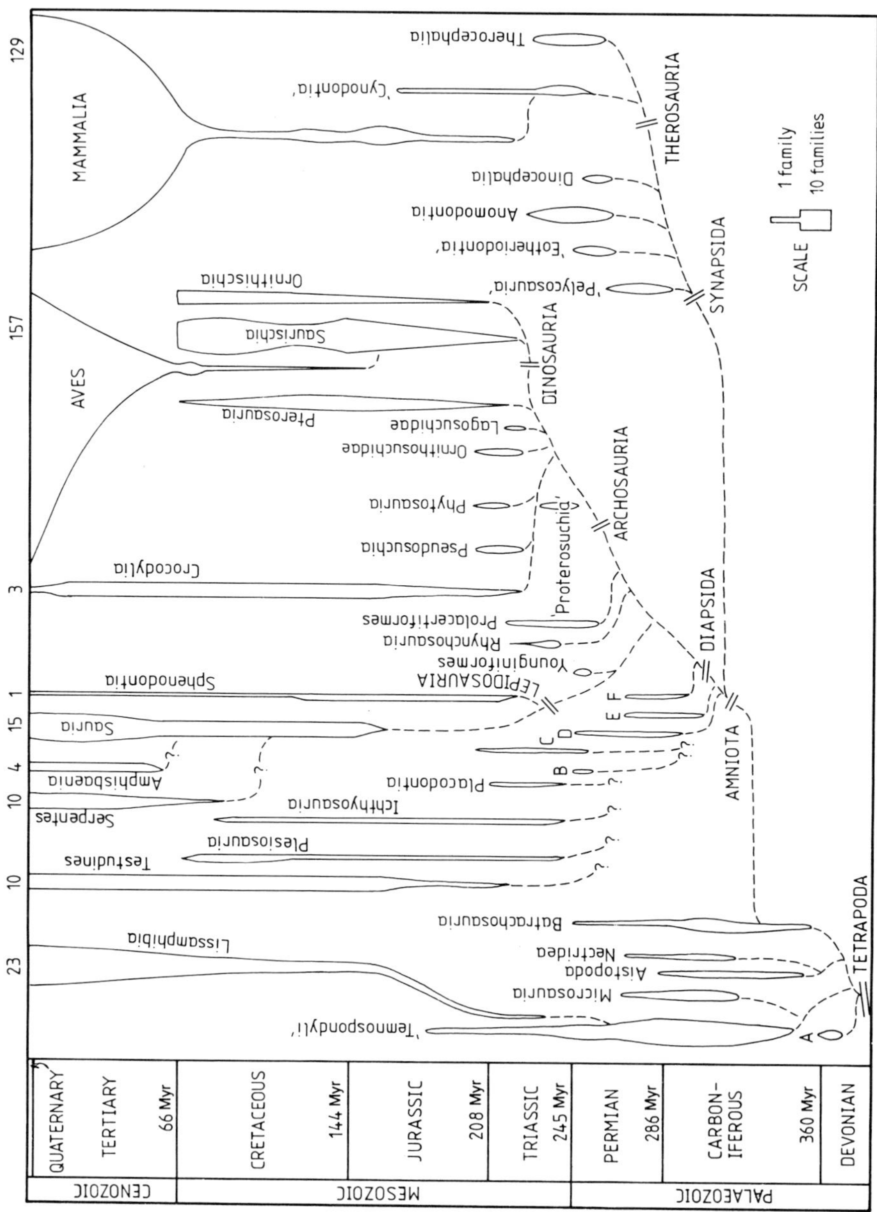

(e.g. Gaffney 1980; Kemp 1982; Benton 1984; 1985c; Gauthier 1986; Heaton and Reisz 1986; Maisey 1986; Panchen and Smithson 1987; 1988) are indicated by dashed lines. Abbreviations: A, Ichthyostegalia; B, Pareiasauria; C, Procolophonia; D, Captorhinidae; E, Protorothyrididae; F, Araeoscelidia.

evolutionary fate of all vertebrate families here. Reviews that treat major groups cladistically include: Maisey (1986) on early chordate and 'fish' relationships; Forey (1984) on agnathans; Goujet (1984) on placoderms; Maisey (1984) on chondrichthyans; Gardiner (1984) on osteichthyans; Panchen and Smithson (1987) on sacropterygians; and papers in Benton (1988b) on tetrapods.

Size of the data set

There are 459 families of living fishes (including 409 families of teleosts) and 352 families of living tetrapods (including 157 families of birds and 139 families of mammals). The total figures for living and extinct families are 775 for fishes and 916 for tetrapods, or 1691 for all vertebrates. This is smaller than the marine animal data set of 3300 families (Sepkoski 1982) used in most studies of diversification and mass extinction to date. This total of 1691 is reduced to 1396, because 295 families of vertebrates have no known fossil representatives (these are mainly teleosts, birds and mammals). A further culling of the data has been made to exclude the small number of families that have been based on a single species or a single genus found in one geological formation ('singletons', a total of 117 families). Indeed, some families have been based on single specimens, and they are best omitted until further finds are made. In effect, a singleton family has zero distribution in time—it arises and disappears in a geological instant, and cannot be sensibly included in calculations of rates of origination or extinction.

Incompleteness of the record

The relative incompleteness of the fossil record of vertebrates has been described by many authors (e.g. Pitrat 1973; Thomson 1977; Padian and Clemens 1985; Benton 1985a; 1985b; 1987; see Chapter 5). The record of fishes is generally assumed to have been better than that of non-marine tetrapods, largely since aquatic environments are assumed to be more frequently preserved than terrestrial environments. However, there is a gap between the Llanvirnian and the Llandoverian which is presently devoid of fish fossils. Particular groups of fishes are also absent from long stretches of geologic time. For example, the Myxinidae (hagfishes) appear in the Late Carboniferous, but are unknown between that time and the present day. Coelacanths (Actinistia) are unrepresented from the Late Cretaceous to the present, and there is a mid-Cretaceous gap (Barremian–Santonian) in the record of chondrostean bony fish. The record of non-marine tetrapods is demonstrably poor. Some stratigraphic stages, for example the Aalenian (Middle Jurassic), have yielded no identifiable tetrapod fossils anywhere in the world, and other stages (e.g. Gzelian

(Carboniferous); Toarcian, Bajocian, Callovian, Oxfordian (Jurassic); Berriasian–Aptian, Cenomanian–Santonian (Cretaceous)) have yielded very few remains.

It is possible to estimate the completeness of the vertebrate record in a broad way by examining the numbers of families present per stage. The Simple Completeness Metric (SCM, Paul 1982; Benton 1987; 1988a) compares the numbers of families that are known to be present compared to the numbers thought to be present. The SCM is based on the fact that vertebrate families span several stratigraphic stages. The family may be represented by fossils throughout its entire duration, or there may be gaps spanning one or more stratigraphic stages where fossils are absent. Jablonski (1986) termed this the Lazarus effect, where a taxon apparently disappears, and then reappears higher up in the sequence. The more incomplete the fossil record is for a particular stage, the more Lazarus (hidden) taxa there will be. The SCM ranges from 0% (no fossils at all, e.g. Aalenian) to 100% (all families represented by fossils), e.g. Visean. Most other stages have SCM values between 50 and 100% but values fall below 50% in the Early–Middle Jurassic (Toarcian–Bajocian), the Late Jurassic (Oxfordian), and the Late Cretaceous (Turonian–Santonian).

Advantages of using the vertebrate fossil record in evolutionary studies

High probability of identifying clades

Rates of evolution, origination, and extinction must be analysed, as noted above, on the basis of monophyletic groups. Vertebrates have proved highly amenable to cladistic analysis, in contrast to fossil invertebrate groups (with the exception of echinoderms and arthropods). The significance of this problem has been highlighted by Patterson and Smith (1987), who discovered that as much as 76% of the standard data set on fossil echinoderms and fish (Sepkoski 1982) used in most recent studies of mass extinctions, is invalid because the families are paraphyletic, polyphyletic, monogeneric or monospecific. A study of various data sets of non-marine tetrapods (Maxwell and Benton 1987) suggested that the major improvements in our knowledge of the evolutionary patterns of tetrapods has stemmed more from the rigorous identification of monophyletic groups than from the discovery of new fossils.

Scope for ecological analysis

Many detailed studies of the functional morphology and palaeoecology (autecology) of single species of fossil vertebrates have been carried out, and these often allow detailed reconstructions of their modes of life. Studies have also been made of whole faunas (synecology). This work offers great potential for detailed palaeobiological interpretations of aspects of extinction events. It may be possible, for example, to compare 'extinction-prone' and 'extinction-resistant' taxa for a broad range of potential ecological correlates including size, diet, position in food chains, locomotory adaptations, reproductive mode, growth rate, habitat preference and geographic distribution.

Refined generic- and species-level taxonomy
Because *Homo sapiens* is a non-marine tetrapod, zoologists have devoted more attention to the systematics of vertebrates than they have to the systematics of brachiopods, annelids, pogonophorans or hyolithids. Our understanding of the relationships and the bounds of living vertebrate species is probably more mature than that of any other group of organisms. This should allow more confident extrapolation of such concepts into the past, and thus better identification of fossil genera and species, better censuses of these taxa, and better phylogenetic reconstructions, thereby improving the usefulness of such data for macroevolutionary research.

DIVERSIFICATION

The data

Several authors have plotted graphs of the diversity of vertebrate families and orders through time (e.g. Charig 1973; Pitrat 1973; Bakker 1977; Thomson 1977; Olson 1982; Padian and Clemens 1985; Colbert 1986) based largely on data from the classic source works of Romer (1966) and Harland *et al.* (1967). More reliable, however, are studies based on a new compilation of data on families of tetrapods (Benton 1985a; 1985b; 1987; 1988a). The data set on fish families used here is also new, compiled from Sepkoski (1982; revisions 1986) and Carroll (1987), with modifications from other recent sources. These new compilations differ significantly from those derived from Romer (1966) and Harland *et al.* (1967) in several ways. First, new records up to the end of 1985 are included. This has affected the date of origination or extinction of as many as 50% of families. Second, the latest cladistic classifications have been incorporated, as far as possible, and attempts have been made to test that all families are clades. Third, the stratigraphic resolution of family distributions has been improved. As far as possible, the dates of origination and extinction of each family have been determined to the nearest stratigraphic stage, usually by examination of the primary literature. The stage is the smallest practical division of geologic time for this compilation (relevant stages vary from 2 Myr to 19 Myr in length, with a mean duration of 6 Myr). This allows more detailed analysis than simply relying on the Lower, Middle and Upper divisions of geologic periods in Romer (1966), Carroll (1987) and elsewhere.

Diversification of fish taxa

The diversity of fishes has increased markedly through time (Figure 9.4A) from a known level of 0–1 families in the latest Cambrian and Ordovician to levels of 1–17 in the Silurian (radiations of ostracoderm agnathans and acanthodians (late Silurian only)), and 25–46 in the Devonian (ostracoderms, placoderms, acanthodians, sarcopterygians). These groups declined markedly in the Late Devonian, but diversity levels in the Carboniferous remained at about the same level (31–46 families) because of radiations of chondrichthyans and chondrostean bony fish. Fish diversities overall fell in the Permian to Early Cretaceous interval to levels below 40 families, with mean values of about 28. The Permian decline was caused by the loss of families of acanthodians, chondrosteans, and sarcopterygians, and a major late Permian extinction of chondrichthyans (11 to 3 families). The 'holostean' bony fish never achieved great diversity in the Mesozoic, but a progressive diversification of teleosts began in the Late Jurassic, although they had been present at low diversity (2–3 families) since the Middle Triassic, and the neoselachian sharks began to radiate during the Jurassic as well. These two groups, and particularly the teleosts, radiated dramatically in the Late Cretaceous when overall fish diversity leapt to 61–85 families, and again in the Tertiary, when the major jump (from 87 to 186 families) took place during the Early and Middle Eocene. The present total of 459 families was not approached in the Pleistocene (232 families) because of the large number of teleost families with no known fossil record as yet.

Diversification of tetrapods

The diversity of tetrapods has increased through time, with a particularly rapid acceleration in the rate of increase from the Late Cretaceous (Campanian) onwards (Figure 9.4B; Benton 1985a; 1985b). Three major diversity assemblages have been identified (Benton 1985b), each of which appears to have been dominant for a time, before giving way to another: I (labyrinthodont amphibians, 'anapsids', mammal-like reptiles) dominated from Late Devonian to Early Triassic times at diversity levels of typically 20–40 families; II (early diapsids, dinosaurs, pterosaurs) dominated during the Mesozoic at diversity levels of 20–50 families; and III (the 'modern' groups—frogs, salamanders, lizards, snakes, turtles, crocodiles, birds, mammals) dominated from Late Cretaceous times to the present day, rising rapidly from overall diversities of 50 to 89 in the Maastrichtian to successive peaks of 158 in the Early Eocene, 234 in the Late Oligocene, and 279 in the Late Miocene.

MASS EXTINCTION

Methods

Extinction and origination rates were calculated stage by stage for fish and for tetrapod families based on the new data sets. Total extinction (R_e) and

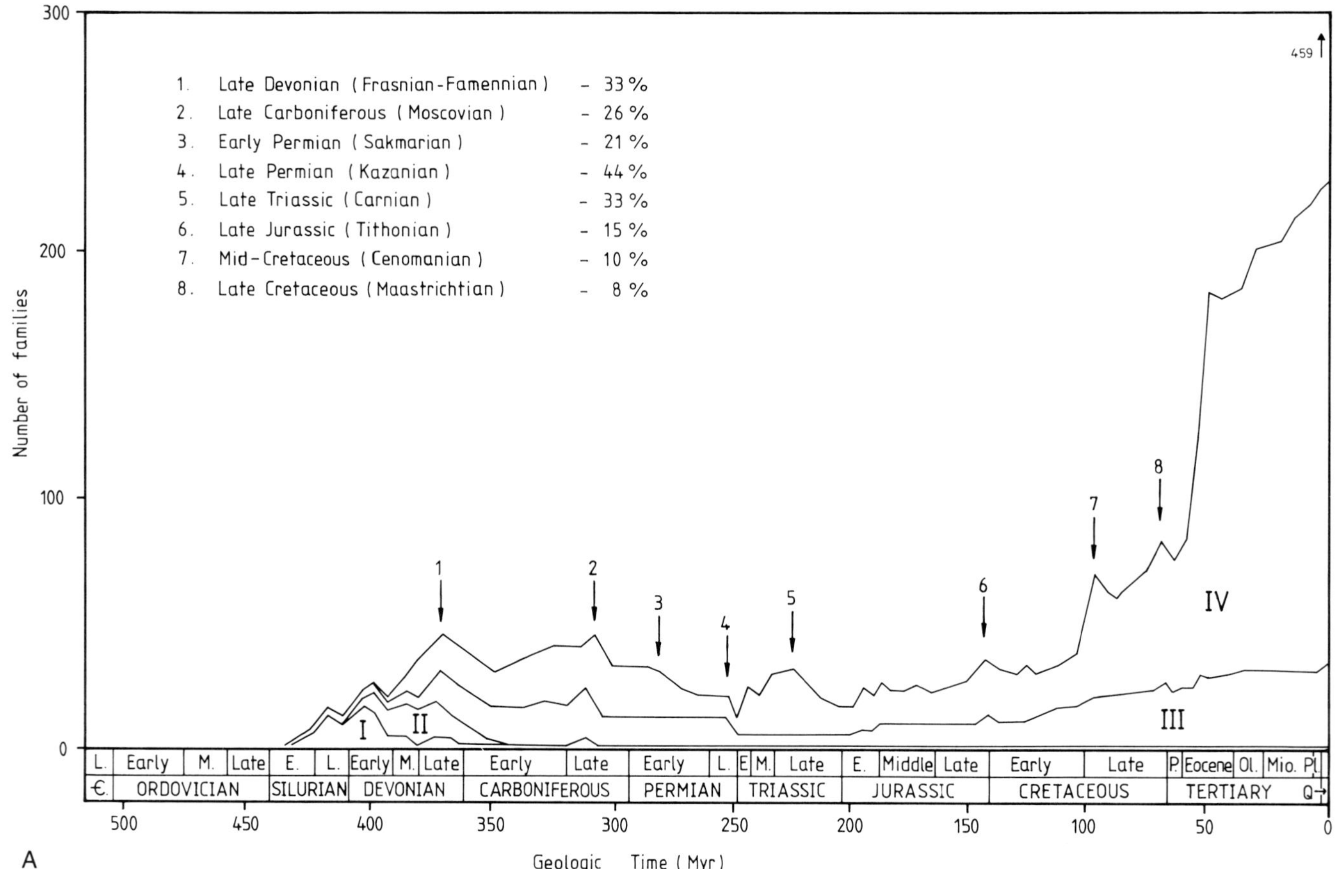
1. Late Devonian (Frasnian-Famennian) - 33 %
2. Late Carboniferous (Moscovian) - 26 %
3. Early Permian (Sakmarian) - 21 %
4. Late Permian (Kazanian) - 44 %
5. Late Triassic (Carnian) - 33 %
6. Late Jurassic (Tithonian) - 15 %
7. Mid-Cretaceous (Cenomanian) - 10 %
8. Late Cretaceous (Maastrichtian) - 8 %
Number of families
300
200
100
0
459
1
2
3
4
5
6
7
8
I
II
III
IV
L.
Early
M.
Late
E.
L.
Early
M.
Late
Early
Late
Early
L.
E
M.
Late
E.
Middle
Late
Early
Late
P.
Eocene
Ol.
Mio.
Pl.
€.
ORDOVICIAN
SILURIAN
DEVONIAN
CARBONIFEROUS
PERMIAN
TRIASSIC
JURASSIC
CRETACEOUS
TERTIARY
Q
500
450
400
350
300
250
200
150
100
50
0
Geologic Time (Myr)
A

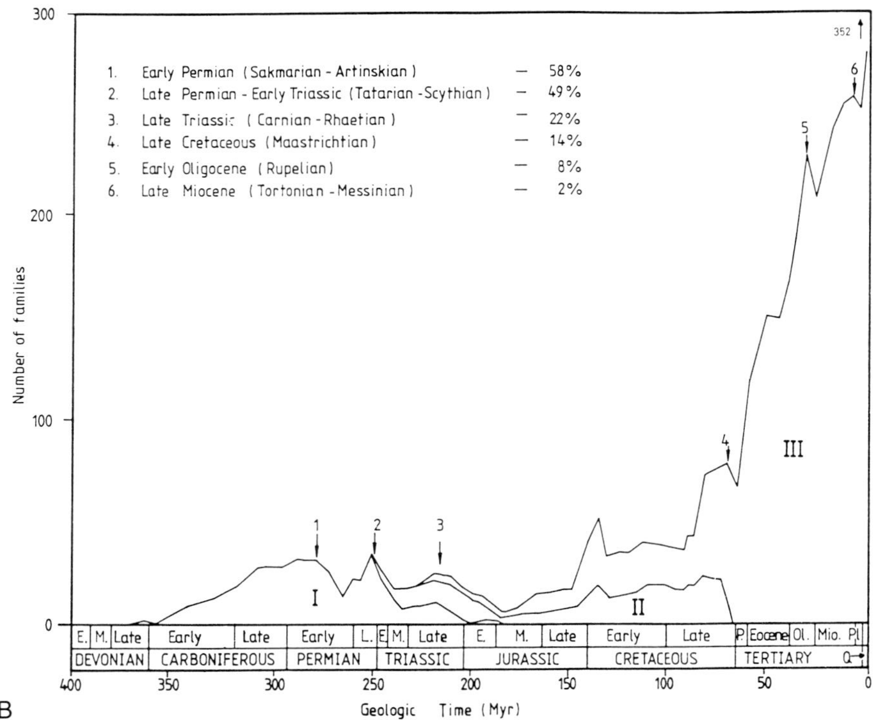

Figure 9.4. Standing diversity against time for families of A: fishes and B: tetrapods. The upper curve shows total diversity against time, and apparent mass extinctions are indicated by drops in diversity, their commencements numbered 1–8 and 1–6, respectively. The relative magnitude of each drop is given in terms of the percentage of families that disappeared. The time-scale is that of Palmer (1983). Four fish clades are indicated (Figure 9.4A): I, Agnatha; II, Placodermi; III, Chondrichthyes (+ Acanthodii); IV, Osteichthyes. Three assemblages of families of tetrapods succeeded each other through geologic time (Figure 9.4B): I, II, III (see text for details). Abbreviations: E, Early; L, Late; M, Middle; Mio, Miocene; Ol, Oligocene; P, Palaeocene; Pl, Pliocene; Q, Quaternary.

total origination (R_o) rates were calculated as the number of families that disappeared or appeared, respectively, during a stratigraphic stage, divided by the estimated duration of that stage (Δt):

$$R_e = \frac{E}{\Delta t} \quad \text{and} \quad R_o = \frac{O}{\Delta t},$$

where E is the number of extinctions and O is the number of originations. Per-taxon extinction (r_e) and origination (r_o) rates were calculated by dividing the total rates by the end-of-stage family diversity N (see Chapter 2):

$$r_e = \frac{E}{N\Delta t} \quad \text{and} \quad r_o = \frac{O}{N\Delta t}.$$

The per-taxon rates can be seen as the 'probability of origin' or the 'risk of extinction'. In these calculations, the recent summary geologic time-scale of Palmer (1983) was used for stage lengths in millions of years.

Mass extinction events (see Chapter 2) are times when large numbers of taxa of diverse taxonomic and ecological position appear to die out in a geologic instant (Jablonski 1986). No clear numerical definition of mass extinctions has yet been possible, but indications are provided by major drops in overall diversity, and times of unusually high extinction rates.

Diversity drops

The fish record (Figure 9.4A) shows numerous minor drops in overall diversity, and it is difficult to determine which were mass extinctions and which merely fluctuations of no particular significance. Clearly, not every drop can be a mass extinction. Figure 9.4A indicates those intervals (nos. 1–8) when a decline of about 10% or more in total diversity took place. These drops all correspond to mass extinction intervals that have been identified on the basis of other data sets. They are considered chronologically below.

There appear to be six declines in diversity in the tetrapod record (Figure 9.4B, nos. 1–6) that are attributable to mass extinction events. The other drops (early Jurassic, end-Jurassic, mid-Cretaceous) probably indicate mainly a change in the quality of the fossil record (Benton 1985a; 1985b), and mass extinctions cannot be assumed here. These three episodes correspond to times when the SCM described above gives particularly low values. These serious gaps in the fossil record of tetrapods do not appear to be reflected to the same extent in the fish record.

Origination and extinction rates

Figure 9.5 shows total origination and extinction rates (for these graphs for tetrapods alone, see Benton 1985a; 1985b; 1988a). The total rates were found to be dependent on two non-random sources of error. The first of these is variation in the total numbers of taxa available to give rise to new taxa, or to become extinct. Early parts of both records show very low diversity (1–10 families), whereas the Tertiary portions are two orders of magnitude higher; this must bias the rate values. The second non-random source of error is Lagerstätten effects. The total origination rates generally track the total extinction rates quite closely; peaks in both rates might have been produced in part by episodes when the fossil record is better than usual, corresponding to particular fossil Lagerstätten such as the Sakamena Group (Late Permian), the Solnhofen Limestone (Tithonian), and the Monte Bolca fish beds (Eocene). The improvement in the record boosts the apparent number of family originations and extinctions (Hoffman and Ghiold 1985). The per-taxon rates remove this bias in part. Thus when extinction and origination rates are recalculated relative to the numbers of taxa available (Figure 9.5), the rates do not track each other so closely, although 'Lagerstätten peaks' remain in the Ufimian, Tithonian, and Coniacian for tetrapods (Figure 9.5B).

There are particularly high per-taxon extinction rates at times of mass extinctions corresponding to the Famennian, Kazanian, and Carnian events for fish (nos. 1, 4, 5: Figure 9.4A) and the Artinskian, Tatarian, and 'Rhaetian' events for tetrapods (nos. 1, 2, 3: Figure 9.4B). Per-taxon extinction rates are barely elevated at the times of the Moscovian, Sakmarian, Tithonian, Cenomanian and Maastrichtian events for fish (nos. 2, 3, 6, 7, 8: Figure 9.4A), and the Maastrichtian, Rupelian, or Late Miocene mass extinctions for tetrapods (nos. 4, 5, 6: Figure 9.4b). These mass extinctions correspond to depressed per-taxon origination rates in most cases (Figure 9.5; see Benton 1985b).

Mass extinction events

The history of fishes and of tetrapods has been punctuated by at least eight and six mass extinction events respectively (nos. 1–8; 1–6: Figure 9.4A and 9.4B, respectively). Some of these overlap with each other, and with extinction events reported for other groups of organisms (see Chapters 2, 5). The fossil record of vertebrates is probably not complete enough to test the hypothesis of periodicity of mass extinctions (Raup and Sepkoski 1984; 1986), but data from the Triassic record appear to contradict the idea (Benton 1986a; 1988a). The vertebrate events are the following.

Late Devonian (Frasnian–Famennian)

Thirty-five families of fish died out during these two stages, which was a high rate of loss from a starting point of only 46 in the Frasnian. The main extinctions were among agnathans (the last heterostracans and osteostracans), placoderms (loss of 16 families, and virtual annihilation), and sarcopterygians (loss of ten families of rhipidistians and lungfish), with smaller losses among chondrichthyans and chondrosteans. This presumably

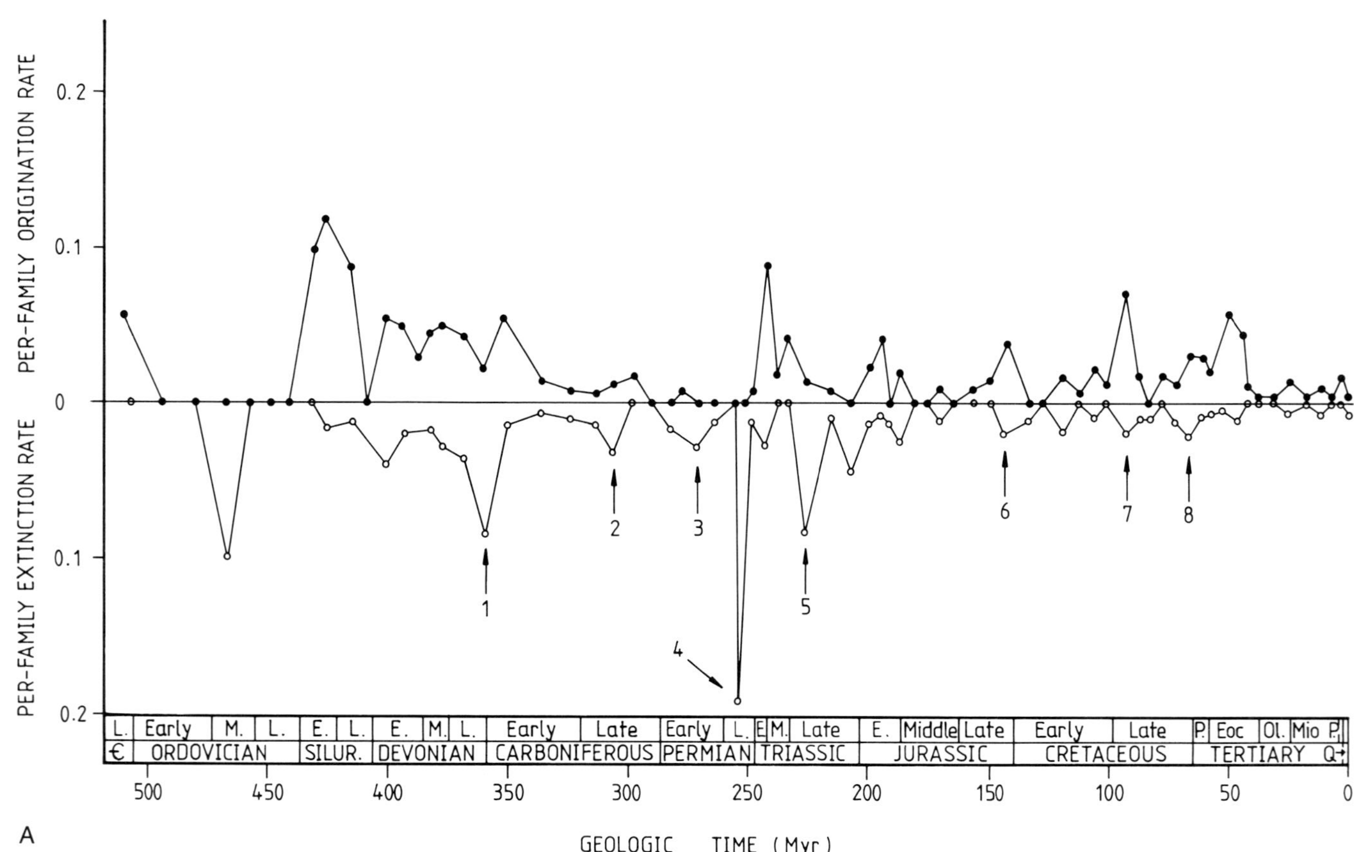
PER-FAMILY ORIGINATION RATE
PER-FAMILY EXTINCTION RATE
0.2
0.1
0
0.1
0.2
L. Early M. L. E. L. E. M. L. Early Late Early L. E M. Late E. Middle Late Early Late P. Eoc Ol. Mio P.
€ ORDOVICIAN SILUR. DEVONIAN CARBONIFEROUS PERMIAN TRIASSIC JURASSIC CRETACEOUS TERTIARY Q
500 450 400 350 300 250 200 150 100 50 0
GEOLOGIC TIME (Myr)
1
2
3
4
5
6
7
8
A

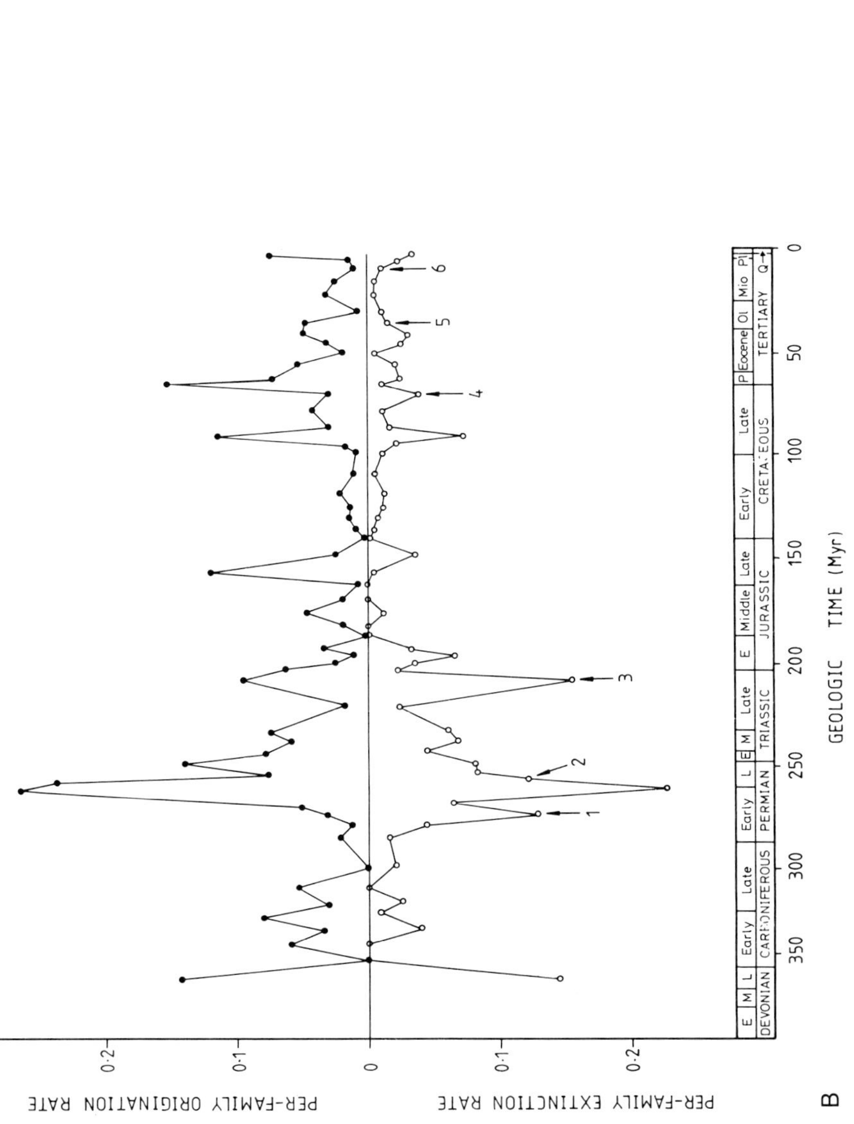

Figure 9.5. Per-taxon rates of origination and extinction for families of A: fishes, and B: tetrapods. Rates were calculated stage by stage for 73 stages between the Late Cambrian and the Pleistocene (fishes), and 56 stages from the Late Devonian (tetrapods). The Miocene was divided into Early, Middle, and Late units only, and the Pliocene was treated as a single time unit. Origination rates are plotted above the zero line (increases), and extinction rates below (declines).

corresponds to the Late Devonian extinction events among marine invertebrates (House 1985; McGhee *et al.* 1986; see Chapter 2).

Late Carboniferous (Moscovian)

Fifteen families of fish (mainly primitive chondrichthyans, as well as a few acanthodians, chondrosteans and sarcopterygians) disappeared during this stage. Six families of tetrapods (a cross-section of basal amphibians) also disappeared. These drops may correspond to the late Carboniferous events noted tentatively by Sepkoski and Raup (1986, p. 23) among marine animals. The end-Namurian event (in the preceding Bashkirian stage) noted by Saunders and Ramsbottom (1986) is not shown by either the fish or the tetrapod data. McGhee (see Chapter 2) notes two minor extinction events among marine invertebrates, during the Visean–Serpukhovian interval (early Carboniferous) and in the Stephanian (i.e. Kasimovian or Gzelian, late Carboniferous), but neither of these is indicated in the record of fossil vertebrates.

Early Permian (Sakmarian–Artinskian)

Seven families of fishes (mainly palaeonisciform chondrosteans) were lost in the Sakmarian stage, with only one further loss in the Artinskian. Among tetrapods, however, 15 families (a variety of amphibians, anapsid reptiles and synapsids) were lost at the same time. This marked a major drop in the diversity of the 'sail-backed' pelycosaur synapsids which were replaced as dominant land reptiles by the therapsids in the late Permian. This event is not one of the postulated periodic events of the Palaeozoic (Sepkoski and Raup 1986; see Chapter 2).

Late Permian (Kazanian–Tatarian)

The fish data show major losses in the Kazanian stage (loss of 12 families of chondrichthyans and chondrosteans), with only one family dying out in the terminal Permian Tatarian stage. However, tetrapod extinctions are focused mainly in the Tatarian, with the loss of 27 families (many lineages of amphibians, anapsid and diapsid reptiles, and especially synapsids). Indeed, this event virtually wiped out the dominant therapsid mammal-like reptiles (loss of 15 out of 19 families), and probably triggered the subsequent rise of the archosaurs. These events correspond to the first of Raup and Sepkoski's (1984; 1986) major periodic extinctions, although it is unclear whether this is supposed to be focused in the Kazanian and/or Tatarian stages (a total time-span of 10–11 Myr). McGhee (see Chapter 2) shows that the end-Permian 'event' in the seas extended through most of the last 10 Myr of that period.

Early Triassic (Scythian)

About 5 Myr later, another smaller extinction event seems to have taken place among tetrapods, with the loss of 13 families of amphibians and mammal-like reptiles. There was a small mass extinction event at this time also among marine invertebrates (Raup and Sepkoski 1984; 1986), but little effect on the fishes. Sepkoski and Raup (1986) argued that this extinction peak, and the one at the end of the Carnian, were caused by sampling errors, but McGhee (see Chapter 2) suggests that both are probably real extinction events during which ammonoids and other marine invertebrates died out.

Late Triassic (Carnian–'Rhaetian')
Sixteen families of fishes died out during the late Triassic, most of these (13) in the Carnian stage (mainly chondrosteans). Among tetrapods, it seems that there were two discernible events, one at the end of the Carnian, and one at the end of the 'Rhaetian' (this latter stage is often now included in the Norian: see Benton 1986a; 1986b; Olsen and Sues 1986; Olsen *et al.* 1987). Ten families of tetrapods (diapsids, therapsids, and marine forms) died out in the Carnian, and eight (mainly amphibians and thecodontians) in the 'Rhaetian'. These extinctions, though few in number, seem to have mediated major faunal changes in the sea and on land. The loss of many chondrosteans was followed by a small increase in holostean diversity, and later of teleosts. The loss of Triassic marine reptiles was followed by great radiations of ichthyosaurs and plesiosaurs in the Early Jurassic. On land, the loss of most therapsids, and all thecodontians and rhynchosaurs, was followed by a two-phase radiation of the dinosaurs in the Late Triassic and the Early Jurassic.

Late Jurassic (Tithonian)
Fishes declined marginally by six families (mainly chondrichthyans and teleosts) at this time, and tetrapods more markedly (Figure 9.4B). However, in the latter case at least, this may be the effect of a relatively poor early Cretaceous fossil record (see above). This event does not stand out as clearly as it does in Raup and Sepkoski's (1984; 1986) marine data.

Mid-Cretaceous (Cenomanian)
A slightly larger drop, of ten families of teleosts and 'holosteans', took place during the Cenomanian stage, another of the postulated periodic extinction events (Raup and Sepkoski 1984; 1986). The tetrapod record shows no clear drop at this time.

Late Cretaceous (Maastrichtian)
The Cretaceous–Tertiary boundary (K–T) event is surely the best known mass extinction (see Chapters 2, 5), and not least for its effects on the reptiles (dinosaurs, pterosaurs and plesiosaurs all died out then). However, in terms of the relative loss of families, this event was smaller than all of those that preceded it. Among fishes, 11 families, mainly of teleosts, died out (out of a total of 85 present). A larger relative drop took place among tetrapods (loss of 36 out of 89 families), but the losses affected only the key groups already noted. As with most of the extinction events, most

major vertebrate taxa were apparently virtually unaffected: chondrichthyans, bony fish (except teleosts), amphibians, turtles, lizards, crocodiles, birds, and placental mammals. For both fishes and tetrapods, the Maastrichtian stage was marked also by high rates of origination (appearance of 18 and 21 new families, respectively) which reduced the overall decline during this stage.

Early Oligocene (Rupelian)

This relatively minor event affected only the tetrapods, with the loss of 28 (out of 234) families, mainly of mammals. It has been noted also by Prothero (1985) for North American land mammals, but does not correspond to one of the periodic marine events.

Late Miocene (Tortonian–Messinian)

This event also affected only the tetrapods, with the loss of 21 families, mainly among mammals. It does not match the periodic marine events.

Periodicity

In general, the vertebrate fossil record is not adequate to test Raup and Sepkoski's (1984; 1986) theory of extinction periodicity (see Chapters 2 and 5). Most of the extinctions postulated above (the Late Devonian (?), Late Carboniferous (?), Late Permian, Late Triassic, Late Jurassic, Middle and Late Cretaceous) match marine mass extinctions identified by those authors. However, some (the Late Devonian, the Late Carboniferous, Late Permian and Late Triassic) do not match very well, and others (the Early Permian, Early Triassic, Early Oligocene and Late Miocene) do not fit the 26 Myr cycles at all. Further, many of the 26 Myr extinctions (see Chapter 2) seem to be absent from the vertebrate data (viz. Early Jurassic (Pliensbachian), Middle Jurassic (Callovian?), Early Cretaceous (Barremian–Aptian?), Late Eocene (Priabonian), Middle Miocene (Langhian–Serravallian)). Note, however, that Sepkoski and Raup (1986) found only limited evidence for the Middle Jurassic and Early Cretaceous events, which are necessary to fill gaps in the 26 Myr periodicity pattern. Overall, the vertebrate data are suggestive, but by no means conclusive, evidence against periodicity.

REFERENCES

Bakker, R.T., 1977, Tetrapod mass extinctions—a model of the regulation of speciation rates and immigration by cycles of topographic diversity. In A. Hallam (ed.), *Patterns of evolution as illustrated by the fossil record*, Elsevier, Amsterdam, pp. 439–68.

Benton, M.J., 1984, The relationships and early evolution of the Diapsida, *Symposium of the Zoological Society of London*, **52**: 575–96.

Benton, M.J., 1985a, Mass extinction among non-marine tetrapods, *Nature*, **316**: 811–14.

Benton, M.J., 1985b, Patterns in the diversification of Mesozoic non-marine tetrapods, and problems in historical diversity analysis, *Special Papers in Palaeontology*, **33**: 185–202.

Benton, M.J., 1985c, Classification and phylogeny of the diapsid reptiles, *Zoological Journal of the Linnean Society*, **84**: 97–164.

Benton, M.J., 1986a, More than one event in the late Triassic mass extinction, *Nature*, **321**: 857–61.

Benton, M.J., 1986b, The late Triassic tetrapod extinction events. In K. Padian (ed.), *The beginning of the age of dinosaurs*, Cambridge University Press, Cambridge, pp. 303–20.

Benton, M.J., 1987, Mass extinctions among families of non-marine tetrapods: the data, *Mémoires de la Société Géologique de la France*, **150**: 21–32.

Benton, M.J., 1988a, Mass extinctions and the fossil record of reptiles: paraphyly, patchiness and periodicity. In G.P. Larwood (ed.), *Extinction and survival in the fossil record*, Systematics Association Special Volume **34**, Oxford University Press, Oxford, pp. 269–94.

Benton, M.J. (ed.), 1988b, *The phylogeny and classification of the tetrapods, Vols 1, 2*, Systematics Association Special Volumes **35A** and **35B**, Oxford University Press, Oxford.

Carroll, R.L., 1987, *Vertebrate paleontology and evolution*, Freeman, New York.

Charig, A.J., 1973, Kurten's theory of ordinal variety and the number of continents. In D.H. Tarling and S.K. Runcorn (eds), *Implications of continental drift to the earth sciences*, Vol. **1**, Academic Press, London, pp. 231–45.

Colbert, E.H., 1986, Mesozoic tetrapod extinctions: a review. In D.K. Elliott (ed.), *Dynamics of extinction*, Wiley, New York, pp. 49–62.

Cracraft, J., 1981, Pattern and process in paleobiology: the role of cladistic analysis in systematic paleontology, *Paleobiology*, **7**: 456–68.

Forey, P.L., 1984, Yet more reflections on agnathan–gnathostome relationships, *Journal of Vertebrate Paleontology*, **4**: 330–43.

Gaffney, E.S., 1980, Phylogenetic relationships of the major groups of amniotes. In A.L. Panchen (ed.), *The terrestrial environment and the origin of land vertebrates*, Systematics Association Special Volume **15**, Academic Press, London, pp. 593–610.

Gardiner, B.G., 1984, The relationships of the palaeoniscid fishes, a review based on new specimens of *Mimia* and *Moythomasia* from the Upper Devonian of Western Australia, *Bulletin of the British Museum (Natural History), Geology*, **37**: 173–427.

Gauthier, J., 1986, Saurischian monophyly and the origin of birds, *Memoirs of the California Academy of Sciences*, **8**: 1–55.

Goujet, D.F., 1984, Placoderm interrelationships: a new interpretation, with a short review of placoderm classifications. *Proceedings of the Linnean Society of New South Wales*, **107**,: 211–43.

Harland, W.B. *et al.*, 1967, *The fossil record*, Geological Society, London.

Heaton, M.J. and Reisz, R.R., 1986, Phylogenetic relationships of captorhinomorph reptiles, *Canadian Journal of Earth Sciences*, **23**: 402–18.

Hoffman, A. and Ghiold, J., 1985, Randomness in the pattern of 'mass extinctions' and 'waves of origination', *Geological Magazine*, **122**: 1–4.

House, M.R., 1985, Correlation of mid-Palaeozoic evolutionary events with global sedimentary perturbations, *Nature*, **313**: 17–22.

Jablonski, D., 1986, Causes and consequences of mass extinctions; a comparative approach. In D.K. Elliott (ed.), *Dynamics of extinction*, Wiley, New York, pp. 183–229.

Kemp, T.S., 1982, *Mammal-like reptiles and the origin of mammals*. Academic Press, London.

Maisey, J.G., 1984, Higher elasmobranch phylogeny and biostratigraphy, *Zoological Journal of the Linnean Society*, **82**: 33–54.

Maisey, J.G., 1986, Heads and tails: a chordate phylogeny, *Cladistics*, **2**: 201–56.

Maxwell, W.D. and Benton, M.J., 1987, Mass extinctions and data bases: changes in the interpretation of tetrapod mass extinction in the past 20 years. In P.J. Currie and E.L. Koster (eds), *Fourth symposium on Mesozoic ecosystems, abstracts*, pp. 156–160, Tyrrell Museum, Drumheller, Alberta, pp. 156–60.

McGhee, G.R. Jr, Orth, C.L., Quintana, L.R., Gilmore, J.S. and Olsen, E.J., 1986, Late Devonian 'Kellwasser Event' mass-extinction horizon in Germany: no geochemical evidence for a large-body impact, *Geology*, **14**: 776–9.

Olsen, P.E., Shubin, N.H. and Anders, M.H., 1987, New Early Jurassic tetrapod assemblages constrain Triassic–Jurassic tetrapod extinction event, *Science*, **237**: 1025–9.

Olsen, P.E. and Sues, H.-D., 1986, Correlation of continental Late Triassic and Early Jurassic sediments, and patterns of the Triassic–Jurassic tetrapod transition. In K. Padian (ed.), *The beginning of the age of dinosaurs. Faunal change across the Triassic–Jurassic boundary*, Cambridge University Press, Cambridge, pp. 321–51.

Olson, E.C., 1982, Extinctions of Permian and Triassic nonmarine vertebrates, *Special Paper of the Geological Society of America*, **190**: 501–11.

Padian, K. and Clemens, W.A., 1985. Terrestrial vertebrate diversity: episodes and insights. In J.W. Valentine (ed.), *Phanerozoic diversity patterns*, Princeton University Press, Princeton, NJ, pp. 41–96.

Palmer, A.R., 1983, The Decade of North American Geology 1983 Time Scale, *Geology*, **11**: 503–4.

Panchen, A.L. and Smithson, T.R., 1987, Character diagnosis, fossils and the origin of tetrapods, *Biological Reviews*, **62**: 341–438.

Panchen, A.L. and Smithson, T.R., 1988, The relationships of the earliest tetrapods. In M.J. Benton (ed.), *The phylogeny and classification of the tetrapods*, Systematics Association Special Volume **35A**, pp. 1–32.

Patterson, C. and Smith, A.B., 1987: Is the periodicity of extinctions a taxonomic artefact? *Nature*, **330**: 248–52.

Paul, C.R.C., 1982, The adequacy of the fossil record. In K.A. Joysey and A.E. Friday (eds), *Problems of phylogenetic reconstruction*, Systematics Association Special Volume **21**, Academic Press, London, pp. 75–117.

Pitrat, C.W., 1973, Vertebrates and the Permo-Triassic extinctions, *Palaeogeography, Palaeoclimatology, Palaeoecology*, **14**: 249–64.

Prothero, D.R., 1985, Mid-Oligocene extinction events in North American land mammals, *Science*, **229**: 550–1.

Raup, D.M. and Sepkoski, J.J., Jr, 1984, Periodicity of extinctions in the geologic past, *Proceedings of the National Academy of Sciences, U.S.A.*, **81**: 801–5.

Raup, D.M., Sepkoski, J.J., Jr, 1986, Periodic extinctions of families and genera, *Science*, **231**: 833–6.

Romer, A.S., 1966, *Vertebrate paleontology*, 3rd edn, University of Chicago Press, Chicago.

Saunders, W.B. and Ramsbottom, W.H.C., 1986, The mid-Carboniferous eustatic event, *Geology*, **14**: 208–12.

Sepkoski, J.J., Jr, 1982, A compendium of fossil marine families, *Contributions in Biology and Geology, Milwaukee Public Museum*, **51**: 1–125.

Sepkoski, J.J., Jr and Raup, D.M., 1986, Periodicity in marine extinction events. In D.K. Elliott (ed.), *Dynamics of extinction*, Wiley, New York, pp. 3–36.

Thomson, K.S., 1977, The pattern of diversification among fishes. In A. Hallam (ed.), *Patterns of evolution as illustrated by the fossil record*, Elsevier, Amsterdam, pp. 377–404.

Chapter 10

EVOLUTION, CREATIONISM, AND SCIENCE EDUCATION

Rhondda E. Jones

From the 1930s until quite recently, it was relatively unusual for a book like this one to include a chapter about the creationist alternative to an evolutionary history of the Earth and its biota. Now, however, the recent revival of fundamentalist attacks on evolutionary theory would make such an omission impossible without appearing naïvely oblivious of current events.

A rehearsal of the reasons why the creationist alternative has been overwhelmingly rejected by the scientific community is superfluous here, as the arguments have been documented very well elsewhere (see below). But it is just as important to appreciate why the debate still continues in the public arena, and why it is worthwhile for scientists and scholars with a clear understanding of the current status of evolutionary theory and its internal controversies to participate actively in the debate. Most scientists will not find this a rewarding way to expend their intellectual effort. The arguments tend to be repetitive and tedious, and the participants almost never change their positions. It is important to realise, however, that onlookers do.

The purist reason for participation in the public debate is that a theoretical construct with the explanatory power, importance, broad scientific acceptance, and genuine elegance of evolutionary theory should be a part of the public as well as the scientific culture. The only way that it will become a part of the public culture is if it is communicated to the public by the scientific community. A more pragmatic reason for participation is to ensure that the decision whether evolutionary theory has a place in school science curricula is based on educational criteria rather than on a desire by textbook publishers to maximise their market. This need is illustrated by history. Evolution may have won the battle for public opinion in the Scopes trial, but in many of the world's school systems it then proceeded to lose the educational war. The decades after 1925 saw the steady disappearance of evolutionary material from school biology textbooks published in America (Grabiner and Miller 1974; May 1984). Several states banned the teaching of evolution in public schools in the 1920s, and these bans were not lifted until the 1960s. Consequently, textbook publishers who wanted access to a national market had little choice but to leave out evolutionary material, and almost all of them did so. Indeed, Grabiner and Miller (1974) provide a way to diagnose secondary school texts published in this period: 'Merely look up the word "evolution" in the glossary or index: you almost certainly will not find it.'

THE CREATIONIST DEMANDS

In the United States, science teaching in a wide range of disciplines received a considerable boost after the Soviet Union demonstrated its technological superiority in 1957 by launching Sputnik. Biology received its share of attention. Evolution is the central paradigm of modern biology, and many school textbooks produced after this event reversed the trend of the previous decades and accurately reflected its significance. The response from the fundamentalist lobby was immediate, and has continued in various guises ever since. However, the tactics used have become much more sophisticated than they were in the 1920s. Then, the policy was to eliminate any teaching about evolution from school science classes. Now, the more common tactic is to appeal to a sense of fairness by demanding equal time in school science classes for evolutionary theory and 'creation science'. As Montagu (1984) pointed out, for most biologists and geologists this is tantamount to demanding that wherever chemistry is taught, alchemy should get equal time.

Because the United States Constitution does not allow religious instruction in public schools, arguments justifying equal time for creation have taken two forms. The first claims that the doctrine of special creation as outlined in Genesis can be presented simply as science, without reference to God (Morris 1974a). This argument generally considers that only two models are possible: a Genesis-style creation model; and an evolutionary model. Consequently, any argument or dispute about evolutionary theory is *ipso facto* an argument for Genesis-style creation. Alternative stories from other human cultures explaining the origin of the living and non-living world are incorporated by asserting that every such story is really a variant of the evolutionary model. Darwin would have been a little surprised.

The second argument, which is less often used by American creationists because of the Constitutional barriers to school religious instruction, is to say that if creationism is intrinsically religious, then so is 'evolutionism'.

The relevant religion is said to be atheistic (or 'secular') humanism (Bird 1978 *in* Overton 1982; Bergman 1980). At the very least, this deviates strongly from what common sense would regard as a reasonable definition of religion. Members of non-fundamentalist Christian churches, most of whom find no conflict between evolutionary theory and their faith, tend to find it particularly offensive. It is important to emphasise that for most creationists, the scientific issues are secondary to social, moral and religious issues. Evolution is perceived as not merely wrong, but evil: 'Evolution is the root of atheism, of communism, nazism, behaviorism, racism, economic imperialism, militarism, libertinism, anarchism, and all manner of anti-Christian systems of belief and practice' (Morris 1972, p. 75). This has no part in the debate about the scientific merits of evolutionary theory and creationism, but as Kitcher (1983) pointed out: 'Those who have been beguiled into thinking that a high school course in evolutionary biology is the gateway to a life of violence and depravity are not likely to ponder the scientific credentials of the theory of evolution.' Indeed this provides a further reason for biologists and geologists who do not exhibit obvious signs of depravity to participate in the public debate, in that their resulting visibility might help persuade the wavering that teaching evolutionary theory is not the moral equivalent of drug peddling.

The demand for equal time has had some success with American school boards and state legislatures, but not with the federal courts. In 1981, the legislatures of Arkansas and Louisiana passed bills mandating a 'balanced treatment' for creation and evolution in high school science classes; both laws have since been overturned as unconstitutional (Overton 1982; Norman 1987). However, the response from American textbook publishers to the existence of controversy has followed a depressingly familiar pattern, with convoluted attempts to avoid saying anything that might offend the fundamentalists (Jukes 1986).

The history of the creation–evolution dispute has taken a rather different course outside North America. In Britain—which, after all, produced both Darwin and Wallace—a voluble creationist movement certainly exists, but it has never enjoyed either the visibility or the public support that its American counterpart has received. The reason is probably that fundamentalist beliefs have never been as strong in Britain as they have in the United States. A 1979 Gallup poll suggested that most Americans held creationist views (Morris 1983), and more recent surveys (cited in Skehan 1986b) estimate that the population of the United States includes about 73 million people with essentially fundamentalist beliefs. The results of a survey published in the *New York Times* in 1982 indicated that 44% of Americans accepted the statement that 'God created man pretty much in his present form at one time within the last 10 000 years'. It would be surprising if a comparable British survey were to produce similar results; and it is almost inconceivable that a British prime minister would say of evolution, as Ronald Reagan did during his 1980 presidential campaign:

Well it is a theory, it is a scientific theory only, and it has in recent years been challenged in the world of science and is not believed in the scientific community to be as infallible as it once was believed. But, if it was going to be taught in the schools, then I think that the biblical theory of creation, which is not a theory but the biblical story of creation, should also be taught.

WHAT *IS* CREATIONISM?

It is important to be clear about what has come to be regarded as the creationist position. Many people believe in a Creator without adopting the fundamentalist worldview, but the term 'creationist' has come to mean someone who espouses the most literal possible interpretation of the biblical creation stories. The most articulate—and certainly the most extensively published—'scientific creationist' arguments come from a body called the Institute for Creation Research, based in San Diego (see for example Morris 1972; 1974a; 1974b; 1975; 1976; 1982; 1983; 1984; Gish 1977), and from a related institution, the Creation Research Society. Creationists, in this fundamentalist sense, believe that the earth is less than 10 000 years old, was created in six days of 24 hours each, and has been deteriorating ever since. They believe that all the different 'kinds' of living thing were created separately. A 'kind' has no consistent one-to-one correspondence with any recognised taxonomic category. Sometimes it means a single biological species (e.g. *Homo sapiens*), and sometimes a group of related species (e.g. the different species of Galapagos finches may belong to the same 'kind', see Baker 1980). A 'kind' may also correspond roughly to a genus, a family, or even a class (Gish 1977). Thus a demonstration of microevolutionary events, including speciation, usually would not involve the transformation of 'kinds'. Most sedimentary rocks and the fossils they contain are believed to result from a single catastrophic deluge (Noah's flood). There are some behavioural and hydrological hypotheses to explain why, for example, fish appear lower in the fossil record than mammals (though I am unaware of any attempt to explain the absence of marine mammals from lower strata). The origin of human cultural diversity also receives attention; different human races and languages are considered to be a consequence of events associated with the Tower of Babel.

These basic tenets are summarised in a document which the Creation Research Society requires scientists to sign when applying for membership. This document says, in part:

> The Bible is the written word of God, and because we believe it to be inspired throughout, all of its assertions are historically and scientifically true in all the original autographs. To the students of nature, this means that the account of origins in Genesis is a factual presentation of simple historical truths.

> All basic types of living things, including man, were made by direct creative acts of God during Creation Week as described in Genesis. Whatever biological changes have occurred since creation have accomplished only changes within the original created kinds. The great Flood described in Genesis, commonly referred to as the Noachian deluge, was an historical event, worldwide in its extent and effect.

This is not the place to defend evolutionary theory against the attacks of the creationists, nor to evaluate the creationist claims to scientific respectability. One of the most positive aspects of the fundamentalist resurgence is that the scientific community is taking the threat that it represents seriously. Among the results has been the publication of a number of well-written books and articles for the non-specialist, as well as more technical literature. These publications provide discussions of evolutionary biology and geology and detailed responses to the creationist critiques of specific aspects of evolutionary theory (see for example, Bridgstock and Smith 1986; Futuyma 1983; Godfrey 1983; Gould 1983; Nelkin 1982; Wilson 1983; Young 1985); discussions of just what constitutes a 'theory' in scientific discourse and how it is applied to the creationist and evolutionary literature (Ruse 1982); analyses of the philosophical bases of both creationist thinking and evolutionary theory (Ruse 1982; Kitcher 1983); critiques of creationist theology (Frye 1983; Skehan 1986a; 1986b); and a variety of other topics. In 1984, the US National Academy of Sciences published a short pamphlet which covered most of the scientific issues in a way comprehensible to the lay reader.

The scientific and theological substance of the debate, therefore, has been dealt with elsewhere. But what are the effects on students when creationist educational demands are met?

CREATIONISM IN AUSTRALIA

Until a few years ago, creationist rhetoric and impact was as inconspicuous in Australia as it has been in Britain and Europe. But now, American-style creationism has arrived and, with assistance from visiting American evangelical creationists, is receiving a good deal of attention from the media. Much of the history and substance of the debate in Australia has been documented by Bridgstock and Smith (1986). The Creation Science Foundation (CSF) was established in Brisbane in 1980 and by 1986 had a full-time staff of 14. It publishes a glossy quarterly magazine *Creation ex Nihilo* and a newspaper *Creation Science Prayer News*. These publications present a similar thesis to comparable American material: that is, that evolutionary theory is not only incorrect, but its acceptance is responsible for most of the moral decay and social deterioration which the creationists perceive in contemporary society.

Australia, like most other countries except the United States, does not prohibit the teaching of religion in state schools. Primary and secondary education is the responsibility of state governments, and the government attitude toward creationist educational policies has varied enormously between states. In Queensland, for example, the newly resurgent demands that creationism be taught in school science classes received sympathetic and very highly publicised support from the politically and socially conservative state government of the day. The Queensland Minister for Education during the mid-1980s, Mr Lin Powell, repeatedly stated his expectation that Queensland state schoolteachers would provide a positive discussion of creationist views in biology classes. (By contrast, the Education Minister for the state of New South Wales made it clear that he would regard teaching creationism in secondary science classes as evidence of incompetence in the teacher.)

However, the Queensland Board of Secondary School Studies, which determined the secondary science curriculum for state schools, certainly did not incorporate creationism in its approved science curriculum (it did include evolution), so Mr Powell's remarks were not binding on state secondary schools. None the less, most of Australia, including many Queensland teachers, believed that they were. In 1985 and 1986, my own discussions with the undergraduate zoology students I teach at a university in northern Queensland made it clear that a proportion of them, though far from a majority, had indeed been taught creationism at some point in their secondary science education.

IS CREATIONISM SCIENCE?

There are many alternative definitions of what science is, but all of them include testablity, that is, some acknowledgement that current theories can be changed by future discoveries; and a commitment to finding out how the world works by studying the natural world itself.

Evolutionary theory meets both these requirements. It has changed a great deal since Darwin's day, and almost every field of scientific enquiry has contributed observations which have either supported or modified the theoretical structure. Creationism meets neither criterion. The words may be borrowed from science, but they are used to explain away rather than incorporate the observations of generations of scientists, and the final authority is a sacred text. As Kitcher (1983) pointed out, the writings of creationists suggest strongly that there are no possible observational findings which might cause them to reject the historical truth of the Genesis narrative. In the 1982 Arkansas trial which eventually declared the teaching of creationism in public schools to be unconstitutional, the creationists were challenged to nominate any piece of original creationist research which had even been submitted to, let alone accepted by, a recognised scientific journal. They were unable to do so (Overton 1982).

THE EFFECTS OF TEACHING CREATIONISM ON STUDENTS

Is the scientific community making unduly heavy weather of the teaching of creationism? A number of writers have predicted adverse consequences for science education if the creationist demands for sympathetic treatment in science curricula are met (see for exmaple Futuyma 1983; Kitcher 1983; Skehan 1986a). Kitcher (1983) identified four problems: that time which should be devoted to teaching basic science should not be wasted considering a doctrine that a rational and open-minded community of professional scientists would find worthless; that it is misleading to present ideas of manifestly unequal merit to students as though they were equally valid; that it is deceitful to pretend to students that the scientific community regards evolutionary theory as anything less than the best available explanation for the origin of living things, and as one of the great achievements of natural science; and that a 'balanced' (read sympathetic) treatment of creationism along with evolution requires teachers to subvert the valuable function of teaching students the methods of scientific reasoning and critical thinking. The arguments are compelling, but even if teaching creationism in science classes is a bad thing in principle, does it influence the students? To investigate this question, I devised a questionnaire which asked Queensland students about their religious background and conviction; what they believed about the origin of living things; what they thought evolutionary theory actually said; whether they had been taught evolutionary theory and/or creationism in school science classes; and whether they believed that their parents thought the same way about evolution and creation that they did themselves. (The results are described in more detail elsewhere; see Jones 1987.) The questionnaire was completed by 613 first-year university students during their first two weeks at university, and therefore relatively uncontaminated by their university experience. The fact that they achieved university entrance meant that all these students had performed well at secondary school level; they represent a cross-section of the most able students graduating from secondary schools. About half were first-year science students, and the other half first-year education students. About two-thirds had attended state schools, and the remainder private schools. About half had been taught evolutionary theory but not creationism in their school science classes; about a quarter had been taught both (the proportion was greater in private schools than in state schools), and a quarter had been taught neither.

Almost all responses fell into five basic categories of opinion about the origins of living things:

1. Evolutionary—an evolutionary scenario without any specifications concerning a role for God.

2. Evolutionary theist—an evolutionary scenario, but with God involved at some level: determining and sustaining the physical laws by which the universe operates, or starting the whole process in the beginning, or giving

human beings a spiritual nature, or (rarely) determining the course of evolution.

3. Creationist—essentially the Genesis scenario.

4. 'Old Earth' creationist—a creationist scenario, but with the events of creation set millions or thousands of millions of years in the past rather than a few thousand years ago.

5. Internally inconsistent—that is, responses which contradicted themselves. The most common form of internal contradiction was to agree with the proposition that 'although there may have been small changes in the forms of animals and plants, the basic kinds have been present on Earth as long as life has existed' (which is essentially the creationist position) and then, a few questions later, to agree that 'the origin and development of living things has occurred entirely by natural evolutionary processes'. An even more striking, but somewhat less common, inconsistency was to choose a young age for the Earth—less than 10 000 years—and then in response to a later question to say 'fossils are the remains of plants and animals which have gradually accumulated in sediments over millions of years.'

Consistent responses

'Consistent' means 'internally consistent', not necessarily consistent with any current orthodoxy. Many respondents had unsophisticated and highly unorthodox ideas about the mechanism and course of evolutionary change. More than half believed that modern evolutionary theory would predict that 'animals and plants will change in ways which best ensure the future survival of the species'. Some were Lamarckians, who evidently thought that evolution proceeds by the transmission of beneficial acquired characteristics from parent to offspring, and there was a large group who thought that natural selection guaranteed progress and increased complexity. None of these opinions is likely to be endorsed with enthusiasm by most evolutionary biologists; these 18-year-olds were an optimistic group. But those who did not contradict themselves and gave responses which, however unorthodox, allowed clear-cut classification into one of the first four groups listed above, were classified as 'consistent'.

Most consistent respondents fell into the first two groups; that is, they chose either evolutionary or evolutionary theist scenarios. About 85% of science students and 75% of education students made these choices (see Table 10.1). The differences between science and education students disappear when different religious affiliations are considered separately, and reflect the fact that education students included both more people with strong religious beliefs of any kind, and more members of fundamentalist religious groups. Only about 5% of students believed in creation according to Genesis.

This pattern of responses appeared to be totally unaffected by the

Table 10.1. *Overall distribution of opinions given by science and education students.*

	Science students	*Education students*
Evolutionary	137	55
Evolutionary theist	83	94
Genesis creationist	12	13
'Old Earth' creationist	25	38
Other	5	3
Internally inconsistent	60	85
No opinion	1	2

teaching of either creationism or evolutionary theory (see Figure 10.1). That is, similar response frequencies were found in students taught both evolutionary theory and creationism as were found in students taught only evolutionary theory and students taught neither.

What clearly did affect responses was the student's religious beliefs. Students who said that their religious beliefs were very important or moderately important to them were classified as 'devout'; those who claimed affiliation with an established church but said that their religious beliefs were not very important or of no importance to them were classified as 'lapsed'. There was also a smaller group who said they had no affiliation with any established religion. Members of the three dominant Christian churches in Australia—Catholic, Anglican, and Uniting Church (an amalgam of Methodists, Presbyterians, and Congregationalists)—showed very similar responses. The 'devout' group favoured evolutionary theism, the 'lapsed' group, evolution alone. Those with no religious affiliation resembled the lapsed group. A high frequency of creationist beliefs occurred only among the devout members of a small group of fundamentalist churches—the Assembly of God, Seventh Day Adventists, Pentecostalists, and some, but not all, Baptists. People with a non-Christian religious affiliation were too few and too diverse to establish any patterns.

About two-thirds of the students in all groups believed that their parents' opinions were similar or identical to their own. At least among socio-economic groups sending their children to a university, Australia is not a fundamentalist society—even in Queensland. It is clear that among students (as among professional scientists), fundamentalist religious beliefs are in general a prerequisite for consistent creationist beliefs, and that exposure to creationism in secondary school science classes has not altered that relationship. Teaching creationism may be a waste of science class time which could be better spent, but it does not appear to have converted anyone to a consistent set of creationist beliefs.

Has the scientific community been making a mountain out of a molehill

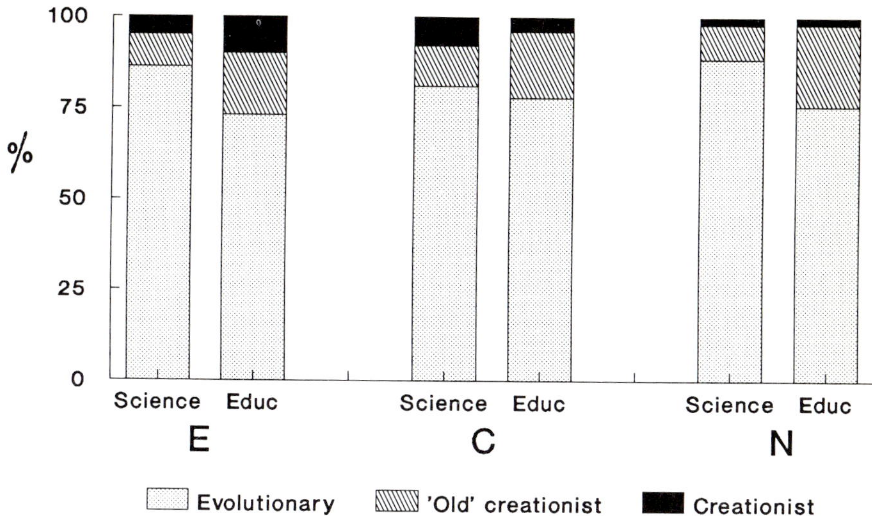

Figure 10.1. Percentages of science and education students holding evolutionary (including evolutionary theist), 'Old Earth' creationist, and creationist opinions, separated according to whether they had encountered in science classes: only evolutionary theory (E); both evolutionary theory and creationism (C); neither evolutionary theory nor creationism (N). Students giving inconsistent responses are excluded, as are the few whose opinions fell into none of the above categories.

on this issue? No, it has not! What does appear to have been affected by teaching creationism is the frequency of inconsistent responses—that is, of students who contradicted themselves.

Inconsistent responses

Preliminary analysis showed that inconsistency was spread more or less equally over all religious affiliations. But students taught creationism in secondary science classes were about twice as likely to contradict themselves as were other students (see Figure 10.2). There was also another correlation with the frequency of inconsistency: students in the 'lapsed' group were more likely to contradict themselves than either the 'devout' group or those with no religious affiliation.

Two conclusions may be drawn from this survey. First, teaching creationism leaves a substantial residue of confusion: it leaves many students able to make quite contradictory statements within the space of a few minutes,

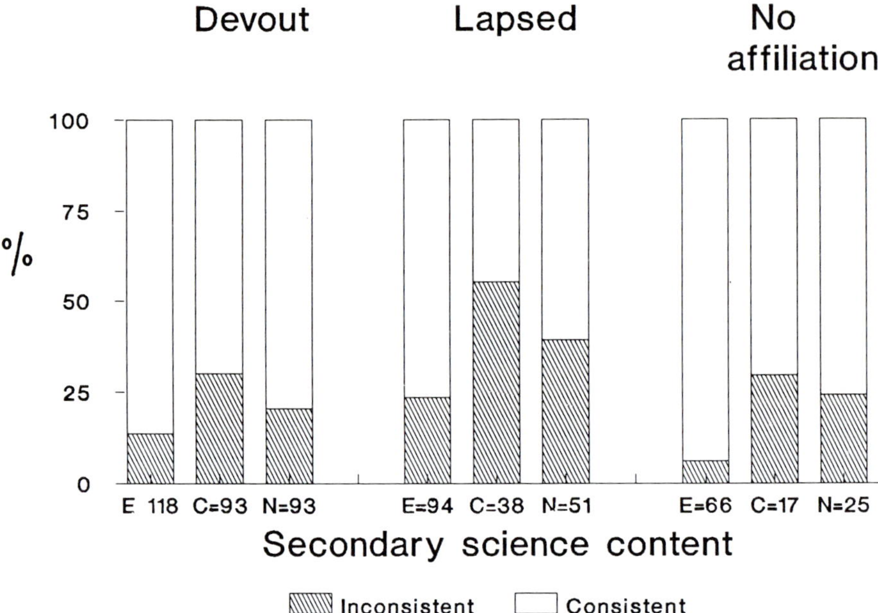

Figure 10.2. Percentages of students giving inconsistent and consistent responses, separated by secondary science content (E = evolution, C = creationism, and N = neither, in secondary science classes) and degree of religious commitment. The numbers under each column indicate the total number of respondents from which the percentage was calculated. For example, a total of 118 'devout' respondents had been taught evolutionary theory (but not creationism) in secondary science classes. Of these 118, 16 (13.6%) gave inconsistent responses. Students who did not answer questions about their religious beliefs or their secondary education are excluded.

apparently without noticing the contradiction. Second, this confusion is strongly affected by the intensity of the student's belief system. Those whose religious beliefs were important to them, and those who had given the matter enough thought to reject formal religious affiliation, were less inclined to contradict themselves than those who acknowledged a religious affiliation but said it was not important to them. It is quite remarkable that students were evidently less confused about this issue if they encountered neither creationism nor evolutionary theory in secondary science classes than they were if they had encountered both! Those in double risk categories—taught both, and with no firm religious beliefs—had a more than 50% chance of falling into this confused category.

Within the time constraints of a high-school science curriculum, it may be almost impossible for students to become informed enough for genuinely critical judgement. They would need to be able to evaluate critically evidence, for example, from comparative anatomy, biogeography, geology, palaeontology, isotope physics, astronomy, thermodynamics, embryology, and genetics. In the circumstances, the teacher may be trying to do an impossible job. However, while a lack of information may lead to poor judgement (and certainly some students had consistent but very odd ideas about evolution), it ought not to result in doublethink. That phenomenon requires further explanation.

Originally there appeared to be two possible interpretations of these results (Jones 1987). First, secondary science teaching in general does not encourage critical evaluation and examination of the evidence, but encourages instead an acceptance of whatever hypotheses students are presented with. Consequently, the presentation of two conflicting hypotheses simply results in confusion. Second, teaching creationism, with its cavalier approach to evidence, itself has this effect. An additional possibility, which is perhaps more likely than the other two, concerns the role of the teacher. The kind of teacher who considers it appropriate to teach creationism alongside evolution as theories of comparable worth is unlikely to have a strong belief in the importance of evidence or of critical thinking (or much understanding of the scientific enterprise) and hence will not transmit such understanding to students. Perhaps there are shortcomings in the training and selection of secondary science teachers.

There are two possible courses of action in dealing with creationism and science education. The first is simply to exclude creationism from any part of the science curriculum. The second is to use the opportunity it provides. Kitcher (1983) pointed out that the upsurge and new sophistication of 'creation science' is something of an asset for philosophers of science, since it provides the best-worked example we have of a 'pseudo-science', that is, an activity which purports to be modern science and may superficially sound like science, but is in its fundamentals profoundly anti-scientific. There is some attraction in using creationism as a cautionary tale for students.

REFERENCES

Baker, S., 1980, *Bone of contention: is evolution true?* Creation Science Foundation, Sunnybank, Queensland.

Bergman, J., 1980, Does academic freedom apply to both secular humanists and Christians? *Institute for Creation Research Acts and Facts*, February 1980.

Bridgstock, M., 1986, The reliability of creationist claims. In M. Bridgstock and K. Smith (eds), *Creationism: an Australian perspective*. Mark Plummer, Melbourne, pp. 60–3.

Bridgstock, M. and Smith, K. (eds), 1986, *Creationism: an Australian perspective*, Mark Plummer, Melbourne.

Frye, R.M. (ed.), 1983, *Is God a creationist? The religious case against creation-science*, Scribner, New York.

Futuyma, D.J., 1983, *Science on trial*, Pantheon Books, New York.
Gish, D.T., 1977, *Evolution: the fossils say no!*, Creation-Life Publishers, San Diego, CA.
Godfrey, L.R. (ed.), 1983, *Scientists confront creationism*, W.W. Norton, New York and London.
Gould, S.J., 1983, *Hens' teeth and horses' toes*. W.W. Norton, New York and London.
Grabiner, J.V. and Miller, P.D., 1974, Effects of the Scopes trial. *Science*, **185**: 832–7.
Jones, R.E., 1987, Evolution and creationism: the consequences of an analysis for education. *Interdisciplinary Science Reviews*, **12** (4): 324–32.
Jukes, T.H., 1986, The fight for science textbooks, *Nature*, **319**: 367–8.
Kitcher, P., 1983, *Abusing science: the case against creationism*. Open University Press, Milton Keynes.
May, R.M., 1984, Creation, evolution, and high-school texts. In A. Montagu (ed.), *Science and creationism*, Oxford University Press, New York, 306–10.
Montagu, A. (ed.), 1984, *Science and creationism*, Oxford University Press, New York.
Morris, H.M., 1972, *The remarkable birth of planet Earth*. Creation-Life Publishers, San Diego, CA.
Morris, H.M. (ed.), 1974a, *Scientific creationism*, Creation-Life Publishers, San Diego, CA.
Morris, H.M., 1974b, *The troubled waters of evolution*, Creation-Life Publishers, San Diego, CA.
Morris, H.M., 1975, *Introducing creationism in the public schools*, Creation-Life Publishers, San Diego, CA.
Morris, H.M., 1976, *The Genesis record: a scientific and devotional commentary on the book of beginnings*, Baker Book House, Grand Rapids, MI.
Morris, H.M., 1982, *Evolution in turmoil*, Creation-Life Publishers, San Diego, CA.
Morris, H.M., 1983, *Education for the real world*, Master Books, San Diego, CA.
Morris, H.M., 1984, *A history of modern creationism*. Master Books, San Diego, CA.
National Academy of Sciences Committee on Science and Creationism, 1984, *Science and creationism: a view from the National Academy of Sciences*, National Academy Press, Washington, DC.
Nelkin, D., 1977, *Science textbook controversies and the politics of equal time*, MIT Press, Cambridge, MA. and London.
Nelkin, D., 1982, *The creation controversy*, W.W. Norton, New York and London.
Norman, C., 1987, Supreme Court strikes down 'Creation Science' law as promotion of religion, *Science*, **236**: 1620.
Overton, W.R., 1982, Memorandum opinion presented in the United States District Court, Eastern District of Arkansas, Western Division, on January 5, 1982. Reprinted in A. Montagu (ed.), 1984, *Science and creationism*, Oxford University Press, New York, pp. 365–97.
Ruse, M., 1982, *Darwinism defended: a guide to the evolution controversies*, Addison-Wesley, London.
Skehan, J.W., 1986a, *Modern science and the book of Genesis*, Special Publications of the National Science Teachers Association, Washington, DC.

Skehan, J.W., 1986b, The age of the Earth, of life, and of mankind: geology and biblical theory versus creationism. In R.W. Hanson (ed.), *Science and creation*, Macmillan, New York, pp. 10–32.

Wilson, D.B. (ed.), 1983, *Did the devil make Darwin do it?* Iowa State University Press, Ames, Iowa.

Young, W., 1985, *Fallacies of creationism*, Detselig, Calgary.

Index